Media, Place and Tourism

Accessible and interdisciplinary in nature, this volume highlights the connections between media, tourism and place, bringing together the diverse perspectives, approaches and actors involved in critical issues relating to media tourism worldwide.

This book explores new avenues, adopting a global and transnational perspective and placing emphasis on the exploration, analysis and comparison of cases from around the world. Encompassing chapters from a plethora of experts, the volume discusses processes and relationships of power involved in the development and experience of media tourism. This book seeks to broaden the horizons of both the reader and existing academic research into media tourism by including research into, among other topics, Bollywood and Nollywood films, Brazilian telenovelas and South Korean K-pop culture. Illustrated with tables and figures throughout, the volume presents insights from a variety of strands of cutting-edge and empirically rich research, which are collated, compared and contrasted to demonstrate the connections between media, tourism and place around the world.

International in scope, this book is an ideal companion for academics and scholars within a wide array of disciplines, such as media studies, tourism studies, fan studies, cultural geography and sociology, as well as those with an interest in media tourism more specifically.

Stijn Reijnders is Professor of Cultural Heritage and Vice Dean of Research at the Erasmus University Rotterdam. His research focuses on the intersection of media, culture and tourism. Furthermore, he is co-founder of the Erasmus Knowledge Centre for Film, Heritage and Tourism.

Emiel Martens is Assistant Professor in Media Studies at the University of Amsterdam and Senior Researcher at the Erasmus University Rotterdam. His research focuses on the intersections between film, tourism and empire. Furthermore, he is co-founder of both the Expertise Centre Humanitarian Communication (HUCOM) and the Knowledge Centre for Film, Heritage and Tourism (FIHETO).

Deborah Castro is Assistant Professor in Media Studies at the University of Groningen. Her research interests lie in the fields of television and audience studies. She received a Marie Skłodowska-Curie Individual Fellowship to study the local community's reactions to film tourism in Spain.

Débora Póvoa is Postdoctoral Researcher at the Erasmus University Rotterdam. Her research focuses on power dynamics of media production and tourism. She is a board member of the Erasmus Knowledge Centre for Film, Heritage and Tourism (FIHETO) and film review editor of the *European Review of Latin American and Caribbean Studies*.

Apoorva Nanjangud is Postdoctoral Researcher at Radboud Institute of Culture and History, Radboud University Nijmegen. She is a cultural sociologist, with an expertise at the intersection of the media, tourism and creative sectors, particularly in the context of India.

Rosa Schiavone is PhD Candidate at Erasmus University Rotterdam. Her PhD project examines the production of place in film tourism in Scotland and explores the interconnections between popular culture, place and local and national heritage.

Contemporary Geographies of Leisure, Tourism and Mobility

Series Editor: C. Michael Hall, *Professor at the Department of Management, College of Business and Economics, University of Canterbury, Christchurch, New Zealand*

The aim of this series is to explore and communicate the intersections and relationships between leisure, tourism and human mobility within the social sciences.

It will incorporate both traditional and new perspectives on leisure and tourism from contemporary geography, e.g. notions of identity, representation and culture, while also providing for perspectives from cognate areas such as anthropology, cultural studies, gastronomy and food studies, marketing, policy studies and political economy, regional and urban planning, and sociology, within the development of an integrated field of leisure and tourism studies.

Also, increasingly, tourism and leisure are regarded as steps in a continuum of human mobility. Inclusion of mobility in the series offers the prospect to examine the relationship between tourism and migration, the sojourner, educational travel, and second home and retirement travel phenomena.

The series comprises two strands:

Contemporary Geographies of Leisure, Tourism and Mobility aims to address the needs of students and academics, and the titles will be published simultaneously in hardback and paperback.

Routledge Studies in Contemporary Geographies of Leisure, Tourism and Mobility is a forum for innovative new research intended for research students and academics, and the titles will initially be available in hardback only. Titles include:

Media, Place and Tourism
Worlds of Imagination
Edited by Stijn Reijnders, Emiel Martens, Deborah Castro, Débora Póvoa, Apoorva Nanjangud and Rosa Schiavone

For more information about this series, please visit: www.routledge.com/Contemporary-Geographies-of-Leisure-Tourism-and-Mobility/book-series/SE0522

Media, Place and Tourism

Worlds of Imagination

Edited by
Stijn Reijnders, Emiel Martens,
Deborah Castro, Débora Póvoa,
Apoorva Nanjangud and Rosa Schiavone

Routledge
Taylor & Francis Group
LONDON AND NEW YORK

First published 2024
by Routledge
4 Park Square, Milton Park, Abingdon, Oxon OX14 4RN

and by Routledge
605 Third Avenue, New York, NY 10158

Routledge is an imprint of the Taylor & Francis Group, an informa business

This project has received funding from the European Research Council (ERC) under the European Union's 2020 research and innovation program (grant agreement No CoG-2015_681663). The Editors also received funds from the Erasmus University Library and the Erasmus Trustfonds for the Open Access publication of this book.

British Library Cataloguing-in-Publication Data
A catalogue record for this book is available from the British Library

ISBN: 978-1-032-34104-0 (hbk)
ISBN: 978-1-032-34105-7 (pbk)
ISBN: 978-1-003-32058-6 (ebk)

DOI: 10.4324/9781003320586

Typeset in Times New Roman
by Newgen Publishing UK

Contents

Figures

Tables

Contributors

Deborah Castro (University of Groningen, the Netherlands)

Xin Cui (Liverpool University, United Kingdom)

Matt Hills (University of Bristol, United Kingdom)

Kyungjae Jang (Hiroshima University, Japan)

Qian Jin (Hokkaido University, Japan)

Sean Kim (Edith Cowan University, Australia)

Alexandra Kolesnik (Higher School of Economics University, Russia)

Alfio Leotta (Victoria University of Wellington, New Zealand)

Kristina N. Lindström (University of Gothenburg, Sweden)

Christine Lundberg (University of Stavanger, Norway)

Emiel Martens (Erasmus University Rotterdam & University of Amsterdam, the Netherlands)

Alisa Maksimova (Center for Advanced Internet Studies, Germany)

Wallis Motta (Liverpool University)

Apoorva Nanjangud (Radboud University, the Netherlands)

Rebecca Nedregotten Strand (University of Bergen & Volda University College, Norway)

Edmund Onwuliri (University of Abuja, Nigeria)

Débora Póvoa (Erasmus University Rotterdam, the Netherlands)

Nancy Reagin (Pace University, United States of America)

Stijn Reijnders (Erasmus University Rotterdam, the Netherlands)

Les Roberts (Liverpool University, United Kingdom)

Rosa Schiavone (Erasmus University Rotterdam, the Netherlands)

Krzysztof Stachowiak (Adam Mickiewicz University, Poland)

Timo Thelen (Kanazawa University, Japan)

Rodanthi Tzanelli (University of Leeds, United Kingdom)

Vassilios Ziakas (Leisure Insights Consultancy, United Kingdom)

Introduction

Worlds of Imagination

Stijn Reijnders, Emiel Martens, Apoorva Nanjangud, Deborah Castro, Rosa Schiavone, and Débora Póvoa

Jede Tür, egal ob Stall-, Küchen–oder auch Schranktür, kann in einem bestimmten Augenblick zur Eingangspforte in den Tausend-Türen-Tempel werden.
Die unendliche Geschichte (1979)

[Every door [...] even the most ordinary stable, kitchen, or cupboard door, can become the entrance to the Temple of a Thousand Doors at the right moment.]
(Quote taken from Michael Ende, *The Never Ending Story* (1983),
translated by Ralph Manheim)

This book focuses on media tourism: the phenomenon of people travelling to a place because of its association with a book, film, television series, video game, or other form of popular media culture. Often the site concerned is an ordinary, everyday place. For example, several cafés, mansions, and streets in Istanbul attract tourists from around the world because of their association with popular soap operas such as *Binbir Gece* (2006–2009), *Ezel* (2009–2011) and *Elif* (2014–2019) (Anaz & Ozcan 2016).

At first sight, these locations might come across as ordinary, but they contain the power to open up worlds of imagination. For fans, these "places of the imagination" can form a magical portal to an imaginary world (Zimmerman & Reijnders 2024). For example, there is a door on the South Bank of the Thames leading up to Westminster bridge. For most people, this is a simple door that provides access to a cleaning shed. For James Bond fans, however, the door means something completely different: by opening this door, secret agent 007 in *Die Another Day* (2002) gained access to an underground meeting place of MI6, where a new assignment to save the world awaited him (Reijnders 2010: 371). Likewise, the Mexico Tourism Board together with Disney Pixar has developed an eight-day 'Coco Tour'. By visiting the real-life locations that inspired the producers of *Coco* (2017), tourists are offered a glimpse into the colourful and enchanted *Land of the Dead* featured in this computer-animated fantasy film.

A more recent example concerns the HBO series *The White Lotus*. The global popularity of this series is said to have boosted tourism to both Hawaii (season

DOI: 10.4324/9781003320586-1

1) and Sicily (season 2).[1] *The White Lotus* is not only interesting as a contemporary example of media tourism but also as a reflection on it. The series offers a satirical look at the more luxurious strands of tourism, including the phenomenon of media tourism. For example, in the episode 'Bull Elephants' a son, father and grandfather together go in search of the shooting locations of the film trilogy *The Godfather*. They visit, among other places, the farm in Sicily where Apollonia (Michael Corleone's wife) is killed in the film by a car bomb. The three characters marvel at the paraphernalia in an adjacent *The Godfather* gift shop and the replication of the car from the movie, complete with an Apollonia dummy. When a local company from Sicily recently decided to organize a The White Locus Tour, this farm was of course also included.[2] By joining The White Locus Tour, tourists can now follow in the footsteps of fictional characters who in turn follow in the footsteps of other fictional characters. The fact that this fiction-in-fiction is ultimately based on a long history of very real mafia-related intimidation and violence provides the palace of mirrors in The White Lotus Tour with a macabre undercurrent. However, looking at the high visitor numbers, this does not seem to detract from its tourist appeal.

The aim of this book is to delve deeper into this palace of mirrors, this 'Temple of a Thousand Doors' that famously offers to bring you everywhere as long as it based on a genuine wish. By using media tourism as a starting point, we aim to reflect on how the connections between people's worlds of imagination and the real worlds they inhabit are made tangible through place.

In recent years, the number of studies about media tourism has significantly increased, and the range of disciplines interested in it has widened. Yet, and despite the global attention media tourism has sparked, prior empirical studies have been mostly focused on isolated examples from the Global North, in particular the United Kingdom and North America. As we will argue below, this Western focus tends to overlook the fact that the face of the media industry as well as the tourism industry has been changing rapidly on a global scale. For example, Bollywood overtook Hollywood in terms of film production and viewership in the past decade and has been leading since. At the same time, the global tourist flows can no longer be only characterized as a neo-colonial phenomenon in which the white, Western tourist consumes the 'exotic Other' in the non-West. Instead, people from the Global South are increasingly present in the global tourist flows, partly driven by images from their own media industries. These developments call for a broader, more global approach to media tourism, considering possible commonalities and differences in the development and experience of media tourism in different cultural settings, while at the same time being attentive to related themes such as migration and diaspora.

In order to take the next step and move this field of research to a higher level, a more comparative and cross-cultural approach is essential. This book aims to do so, to go beyond the limited scope of high-profile examples from the Global North and to explore more generic processes and relationships of power involved in the development and experience of media tourism worldwide. Before going into further detail about this explicit focus of our book and how each chapter contributes to this mission, it is first important to provide a broader context: what exactly is media tourism, how has it developed over time and how is it studied in academic circles?

Media tourism: Past and present

In its essence, media tourism is nothing new. For centuries, people have been interested in visiting locations associated with the lives and works of famous writers. An early example of this is the fourteenth-century diplomat and poet Francesco Petrarch. His poems were read and acclaimed throughout Europe, already during his own lifetime. As a humanist, he not only wrote about religious matters, but also about more earthly subjects such as love and the beauty of nature. Because of his predilection for nature walks, he is even called the "first tourist" (Hendrix 2007). A few decades after his death, Petrarch's house in the north of Italy (a region later rechristened as 'Arqua Pertrarca') became a secular place of pilgrimage for lovers of his poems. The locations from his poems also gained some fame, in particular those poems in which he recounted the walks he took with his muse Laura. In this sense, Petrarch was not only the "first tourist", but also one of the first literati in European history whose work led to media tourism. Learning more about similar forerunners of media tourism in other cultural traditions outside the West would contribute to deepening our understanding of the historical evolution of the nexus between media and tourism. But at present this terrain is – at least within the confines of our research field – less well documented and thus largely a *terra incognita*.

In the context of European history, media tourism gained momentum in the nineteenth century. This had partly to do with the emergence of a widely supported reading culture. Reading novels became an increasingly popular form of leisure among broad layers of the population, a trend that was to some degree facilitated by the industrialization of the printing process. These novels stimulated a desire to see the locations described, and literary authors like William Scott played a key role in the creation of an emerging tourism culture (Watson 2006). In these early decades of the fledgling tourism industry, even before the rise of mass media such as film and television, tourism and fiction were already inextricably intertwined (Beeton 2015). In this sense, from its inception, media tourism was more than just a specialized 'niche' within the emerging tourism sector. It was – and still is – close to the core of what drives the whole tourism phenomenon: the imagination of meaningful locations and the resulting desire to be there yourself (cf. Chen, Hall & Prayag 2021: 33–49; Lovell 2019:3). As Graburn, Gravari-Barbas and Staszak (2019: 1) state, it is not possible to understand tourism "without understanding how tourism simultaneously both results from *and* produces fictions". According to Alexander (2017), fiction and tourism are both based on a fundamental desire for ritual transformation. In media tourism, this desire returns in a highly concentrated form; transformation is offered on a textual and material-spatial level at the same time.

Despite this rich past, it seems justified to say that media tourism has gained considerable popularity in the past decades. Media tourism is no longer a small-scale phenomenon limited to a few hundred fans per location but is widely recognized as one of the fastest growing niches within the tourism sector. Numerical substantiation of this statement is difficult to make at a global level since such large-scale research is difficult to carry out and the boundary between media tourism and other forms of tourism is not always easy to draw. But studies at a national level do give a good indication. For example, research among tourists in the UK shows that approximately 10% of tourists have been inspired by films or TV series for their holiday destination (Connell 2012). A study by VisitBritain – the UK's tourism authority – found that film tourists now make up a quarter of all tourists from overseas. Likewise, Turkish soap operas are identified as one of the main causes for the steep rise of Latin American tourists in Turkey in recent years[3], while the Bollywood movie *Zindagi Na Milegi Dobara* (2010) – shot in collaboration with the Spanish ministry of tourism – is said to have singly caused a 32% hike of Indian tourists travelling to Spain in the year of the film's release.[4] At a local level, there is also much evidence for the considerable impact of film and television recordings on the tourist appeal of the locations concerned, a phenomenon that is captured both in the media and in academia with terms such as the *Braveheart* effect and the *Downton* effect. Nowadays, hardly any large-size city can do without a film commission and an active media tourism policy. Media tourism has become a well-known instrument of nation branding and city marketing.

Consequently, media tourism has also become a significant factor in adjacent sectors, such as the heritage sector (Tzanelli 2013; Ziakas et.al. 2024). A striking example of this is the fate of many medieval castles in Europe. Traditionally, these castles were built to keep people *out*. In 2024, many castles have become largely dependent on attracting tourists *inside* their walls, with ticket sale being the main source of income for their extremely costly maintenance. Starring in popular films or TV series works wonders: Alnwick Castle owes its popularity as a setting for Hogwarts from Harry Potter, while Highclere Castle managed to avoid bankruptcy by providing a location for the filming of Downton Abbey – the so-called Downton effect (Liu & Pratt 2019). The association with an exciting, fantastic or spooky story is not only an additional tourist attraction, but in the perception of many tourists it is also an essential part of what makes these historical locations authentic in the first place (Inglis & Holmes 2003; Lovell 2019). A modern policy plan for the management of castles and similar heritage sites is therefore not possible today without a section on media policy (Schiavone, Reijnders & Brandellero 2022).

Of course, the COVID-19 crisis from 2020 to 2021 caused a temporary dip in the rise of media tourism. While many sectors were hit by the pandemic, for the tourism industry the COVID years were simply disastrous. Tourism was restricted in almost all countries worldwide. Several countries completely closed their borders to foreign tourists, or required a quarantine period of one or more weeks, a measure that kept even the most travel-hungry globetrotters at home. At the same time, the film and television sector also suffered from the harsh measures that were taken to contain the virus. Outdoor shots were often not possible, causing productions to be delayed for months. Studio recordings were not always possible either, simply because some scenes did not relate well to the 1.5-metre rule that applied in many countries. Recordings were moved or scripts were modified.

The combination of a paralysed tourism industry and a hindered media industry brought media tourism to a temporary halt. But the first signs – as in *The White Lotus* example mentioned earlier – point to a full resurgence in media tourism. This must of course also be seen against the backdrop of a more general revival of the tourism industry – a phenomenon also referred to as 'revenge tourism', with the total number of international tourist arrivals in 2024 expected to reach unparalleled heights. Notwithstanding the opinions of those who saw the COVID-19 crisis as a great opportunity to fundamentally change the tourism industry and, above all, to make it more sustainable, the 'old' tourism seems to be making a bold return. It is to be hoped that national governments and international associations such as the European Union and the World Tourism Organization will not completely ignore this opportunity to reflect on and innovate the highly polluting tourism sector (and within it media tourism).[5]

New developments

The resurgence of media tourism also raises the question of the extent to which contemporary forms of media tourism are the same as they were before the COVID-19 crisis, or further back in time, for example, their literary predecessors from the nineteenth century. Research into the motives, behaviour, and effects of media tourists is still too early to make a proper historical comparison. At first glance, media tourism seems to be a timeless phenomenon, but that does not mean that its appearance has always remained the same. For example, media tourism is currently undergoing a number of interesting developments, which seem to stem directly from technological developments and recent trends in contemporary (digital) media culture.

Firstly, it is no longer just books, films, and TV series that attract media tourists but also increasingly new cultural products such as video games. The medium of video games has played an increasing role in popular culture since the 1990s; today the gaming industry is even bigger economically than the movie industry. Games are improving graphically and provide players with an advanced experience. The latest generation of games is praised for the level of 'immersion' and 'presence' they offer. Games like *Grand Theft Auto* create large-scale, fictional worlds in which players can move freely and shape the story themselves. Although many videogames still largely rely on imaginary game spaces that lack a real-life counterpart (Lamerichs 2019), more and more these stories are situated within recognizable geographical settings (Dubois et al. 2021). For example, the very popular game *Call of Duty: Modern Warfare II* is partly set in Amsterdam and this game offers the player a hyper-realistic experience of several streets in the Amsterdam city centre. A game like *Assassins Creed: Revolution* is set in eighteenth-century Paris, with fabulous historical and architectural precision. In that sense, it was only a matter of time before examples of media tourism based on games appeared, in which players compare images of their beloved games with the 'real' streets of Paris and Amsterdam.

Video games do not yet generate the same numbers of tourists as films and TV series, but research has shown that they might affect destination image and marketing (Dubois et al. 2021) and that, in the absence of official video game tours, fans have initiated excursions of their own based on their favourite games (Lamerichs 2019). While this medium has long been put away as a trivial pastime for teenagers – as a form of culture not worthy of celebrating outside the living room – gaming is slowly but surely rising in cultural status and asserting its grip on other dimensions of popular culture including tourism.

Secondly, digital media make it more easier to be a virtual media tourist. We are not referring so much to forms of 'armchair travel' (after all, that was already possible with the so-called old media), but to the online exploration of digital geographical representations with the aim of tracing "places of imagination" (Reijnders 2011). A good example of this is the Google Maps and Google Street View apps, which offer potential media tourists ideal opportunities to track the precise locations of movies or TV series and then share the corresponding global positioning system (GPS) locations online with other fans (Chow & Reijnders 2024). And when Google Street View is not sufficient, because it is limited to locations around motorways, online flight simulators can be used. For example, photographer, guide, and media tourist Ian Brodie describes how he used the Microsoft Flight Simulator to find out the shooting locations of the opening scene from *Paris, Texas* in a Texan desert (Brodie 2023: 23–24). It has been known for some time that many media tourists show a certain obsession with determining and pinning down the exact shooting locations of films and series, but this form of cartographic fetishism has reached new heights thanks to online tools. Of course, these aids are not only used in the living room, in preparation for a future holiday or otherwise, but also during the journey itself; smart mobiles and other portable media have fundamentally changed the travel experience of media tourists (as well as all other types of tourists) and made it more 'mediatized' (Champion et.al. 2023).

Finally, media tourism is increasingly a 'transmedia' phenomenon, which is not linked to one medium but to a complex of media expressions (cf. Jansson, 2020). According to Henry Jenkins (2006), popular culture of the twenty-first century is characterized by a convergence culture: media no longer work alongside each other, but with each other, and old and new media have become increasingly intertwined, both in terms of content and organization. A characteristic part of this is the phenomenon of 'transmedia storytelling': popular stories unfolding in different media channels. Take, for example, Harry Potter. Originally published as a book series, the world of Potter was soon rolled out in films, but there are now also Harry Potter board games, Harry Potter video games, Harry Potter slot machines, Harry Potter theatrical performances, Harry Potter musicals, and Harry Potter theme parks in Los Angeles, Orlando, and Osaka. While the same storylines and characters return in these different channels, they in turn also make a unique contribution to the entire Harry Potter universe. In addition to these commercial expressions, it is important also to mention the non-commercial, fan-driven contributions, such as

Harry Potter fan fiction in online fan forums. This 'spreading' of popular stories across different media has resulted in media tourism becoming a lot more versatile. Media tourists no longer have one book or film that drives them but pursue an entire universe. Sometimes the seed of that universe lies in a book, but sometimes also in animations or a video game.[6]

In line with the observation above, the term 'media tourism' is also deliberately chosen in this book, though sometimes in combination with closely aligned terms like film tourism, film-induced tourism, screen tourism or TV tourism. In our opinion, it is no longer tenable to reduce a complex phenomenon such as media tourism to a single medium. The visitor to the 'Wizarding World of Harry Potter' in Osaka, Japan, is not a literary tourist, nor a film tourist, nor a television tourist, but a transtextual media tourist. In the year 2024, fictional worlds disseminate across different media, with the underlying technologies apparently setting no boundaries. With the same flexibility, the tourist goes in pursuit of them and contributes by sharing experiences on Instagram or via travel blogs. In theory, it could even be argued that place is nothing more than one of many 'stepping stones' within a transmedia convergence culture (Alexander, 2017; cf., Graburn et al. 2019: 10–12). But such a postmodern reading ignores the distinction between media representations and physical reality. This distinction can be challenged on a theoretical level, but on an 'emic' level – in the perception of the media tourist – the representation–reality dichotomy continues to play a key role. As Tom van Nuenen argues in his analysis of 'hypermediated tourism' (2021), digital media are ubiquitous, but today's tourist is still driven by a 'magical' search for a direct, unmediated experience of authentic locations. Paradoxically, it is precisely new media technologies and platforms, such as Airbnb, that are seen as a means to achieving that 'authentic' experience (Nuenen 2021; cf. Lovell & Hitchmough 2020).

Understanding media tourism

The emergence of media tourism has – albeit with some delay – also had an impact on academia. The study of media tourism began in the 1990s with a few isolated studies, mostly focusing on the perceived effects of media tourism on the local community (Riley, Baker & Van Doren 1998; Riley & Van Doren 1992; Tooke & Baker 1996). But the research really took shape at the beginning of this century when Sue Beeton published the first version of her review *Film-Induced Tourism* (2005). Since then, Beeton has produced several publications over two decades, establishing media tourism as a serious subject of study (e.g. Beeton 2008, 2010, 2015, 2016). In the 2010s, research into media tourism gained momentum, when not only tourism researchers, but also geographers, literary scholars, historians, media scholars, and cultural scholars started to focus on this subject. Currently, it seems realistic to talk about a thriving, multidisciplinary field, with dozens of papers and several books on this subject being published every year.

What are the central research themes within this field? A literature study from 2012 identified, in seemingly random order, four central themes: "the film tourist" (particularly their characteristics and motivations), "the impacts of film tourism activity", "marketing, destination image and the business of film tourism", and "the appropriation and commodification of place and culture" (Connell, 2012). Although these four themes reveal a fairly rich palette, they do make it clear how, in this first phase of the research field – in the 1990s and 2000s – the tone was set through the perspectives and topics from tourism studies, in particular issues of destination management and the perceived motivations and effects of tourism. With the broadening of the research field in the 2010s, new themes were also explored. In line with the perspective of the humanities, one of the most important themes in recent years has been the experience of the media tourist and how this experience should be situated within broader cultural, cultural-sociological, and historical contexts – with Roesch (2009) and Reijnders (2011) as two early forerunners.

Notwithstanding these important developments within the field, it must be recognized that our current knowledge in 2024 still has several important gaps. In this book, we identify two major shortcomings. Firstly, it is striking that the vast majority of existing studies are based on empirical research in the Global North. Almost all our knowledge and theories about media tourism are derived from examples from Western Europe, the United States, New Zealand, and Australia, with a few important exceptions focusing on countries such as South Korea, Japan, and Brazil (cf. Kim, Long & Robinson 2009; Kim 2012; Kim & Wang 2012; Dung & Reijnders 2013; Seaton et.al. 2017, Kim & Reijnders 2018, Mason & Rohe 2019; Nanjangud 2022, Póvoa 2023). To what extent can such a one-sided focus on the Global North, and in particular the Anglo-Saxon language area, be sustained? Can a research field that is limited to specific media cultures within the Global North ever come to meaningful and well-founded statements about media tourism as a cultural, social, and economic phenomenon? The Western-centric focus of this research field has not only limited our knowledge on the particular topic of media tourism but has also refrained us from having a wider impact on important and topical debates outside the confines of our research field. We should try to de-Westernize and de-colonize media tourism research, and through that critically address wider issues of power in the domains of media and tourism that go beyond the horizon of a specific tourism niche.

An eye for diversity

It is important to note that the Anglo-Saxon media industry is not the only media industry and – depending on how you look at it – not even the largest. For example, an examination of the film industry seems to justify the proposition that Hollywood is no longer a leader but is now just one of many players in a global film market. Bollywood, the film industry based in Mumbai, now produces more feature films per year and serves a larger audience than Hollywood. Bollywood movies are watched not only in India (an immense audience, given that more than one in five people in the world live in India) but also among the Indian diaspora worldwide and countless fans abroad. The film and television industry in South Korea has also grown into a major player. The triumphant march of the 'Korean wave' started in Asia in the wake of the new streaming technologies. But K-drama has now become popular worldwide, and series such as *The Squid Game* (2021–) and films such as *Parasite* (2019) and *Decision to Leave* (2022) are also widely viewed and appreciated by global audiences. In short, the film industry went through a globalization process in the first two decades of the twenty-first century, with the result that in 2024, there exists a complex network of different media cultures that partly operate side by side, but partly also operate jointly and in which popular productions regularly transcend cultural and linguistic boundaries. There is not one media world, but a hybrid multiplicity of several 'worlds of the imagination', of which each has its own centre and influences the others.

At the same time, the tourism industry has also undergone a process of globalization. Until recently, international tourist flows largely consisted of tourists from the Global North. In the annual lists of 'international tourist arrivals', subdivided by nationality, Americans were always at the top, followed by Germans and British. However, a change has been visible since the 2010s. The proportion of Asian tourists has grown significantly. China's economic advance in particular has given tourism in and outside Asia a huge boost. More Chinese than Americans now cross borders as tourists every year.[7] The proportion of Indian tourists has also grown strongly. The development of prosperity in India since the 1990s has created a large middle class, who is claiming its place in the world economy. International tourism is a form of spending and displaying this new wealth. Bollywood films play an important role in this: they not only literally show the way to interesting countries and cultures outside the Indian subcontinent but also promote the role of cosmopolitan tourists among their Indian audiences (Nanjangud 2022). Notwithstanding this trend of globalization, many audiences worldwide still lack the financial means to visit film spots – reminding us about the relations of power and the privilege inherent in tourism.

When these two developments are combined – the globalization of the media industry and the rise of international tourism in the Global South – the time seems ripe for a fundamental step in media tourism research. It is necessary to let go of the Anglo-Saxon bias and to perform more research into forms of media tourism outside the Global North. At the same time, more research needs to be conducted into transcultural forms of media tourism, in which media tourists look beyond the boundaries of their own media culture. Media tourism research can mean more in these moments, when cultural boundaries are crossed, and new knowledge is tapped or nuances are 'lost in translation'. Only then can media tourism research go beyond the analysis of its own niche and offer a unique perspective on broader patterns and developments in and between contemporary societies, including questions of power. We hope that with this book, *Worlds of Imagination*, we can contribute to this broadening of our field of research.

Towards a holistic, multi-actor perspective

In addition, this book aims to address another research gap. Although the field of research into media tourism is now more than two decades old, our existing knowledge is still relatively limited and highly fragmented. Most of the research is based on individual case studies: high-profile, mediagenic examples of media tourism, such as *The Lord of the Rings* in New Zealand and *Harry Potter* in the United Kingdom. These are certainly 'high potentials', which have provided a wealth of data, but it is difficult to arrive at more general theories based on a handful of individual, exceptional case studies. There are still many steps to be taken in the field of theory development (Connell 2012: 1012–1013, 1025).

What also plays a role here is that most case studies focus on one specific facet of media tourism, such as the motivation of the tourist, the impact on residents or the role of the film production team. Such a focus is sometimes necessary to arrive at an in-depth analysis, but it also entails a risk. Media tourism is a complex and multifaceted phenomenon in which various social actors play a significant role. This not only involves tourists and residents, but also, for example, local policymakers, film production teams, tourism boards, local companies, and managers from the heritage sector. They all play a unique role in the creation and development of media tourism. These roles are usually not equal; there are often skewed power relationships. For example, various examples of media tourism have been documented, in which production companies had a decisive say in the creation of media productions and the resulting tourism flows, often at the expense of the needs and wishes of a local community (e.g. Póvoa, 2023). Studies that focus explicitly on one actor run the risk that this socio-political context of media tourism is left out of the picture.

More in general, many examples of media tourism are based on external, stereotypical representations of local cultures and histories, produced by foreign media companies. For example, it is well documented how Dracula tourism in Romania is based on a stereotypical, orientalist representation of Romanian history, one that is decried *culturally* by many Romanians but still facilitated because of its *economic* revenues for local tourist entrepreneurs (Reijnders, 2011). Likewise, the American series *Narcos* has painted a rather negative image of Colombia as a corrupt state, ridden by drugs and crime. This stereotype is eagerly repeated and performed in the popular *Narcos* tours. Although popular among international tourists, these tours are generally seen as 'unwanted tourism' by the local community (Van Broeck, 2018).

In order to understand the complexity of the phenomenon of media tourism, a general, holistic perspective is needed, in which media tourism is examined in its full breadth and in which attention is also paid to forms of cooperation or conflict between the various parties involved. Such a holistic perspective may be difficult to achieve within the limited space of a research paper – the usual form of reporting within this research field – but we hope that with this book we can offer the scope to rise above the level of individual cases and achieve a more comprehensive picture of media tourism, among others by addressing critical issues of power and politics in media tourism.

With that goal in mind, the book *Worlds of Imagination* is divided into four parts, each of which focuses on a certain phase in the development of media tourism and its associated actors. The first part concerns the imagination of place in popular fiction. Film and later television are considered to be the storytellers of the twentieth and twenty-first centuries, joined more recently by the genre of video games. At the same time, literature has remained a powerful source for imaginative practices ever since the nineteenth century. As mentioned above, recent years have also seen an increasing collaboration and convergence between these different media platforms, creating powerful transmedia franchises built around successful global 'brands' such as *Pokémon, Star Wars, Yu-Gi-Oh!*, or *Harry Potter*, whose worlds are rolled out in novels, movies, cartoons, and theme parks. All these stories, circulating in the media and beyond, together create a rich associative imagination of the world. Next to media producers, fans also contribute to the development of these imaginary worlds, for example by writing fanfiction or by creating geofiction maps and sharing them online (Wolf, 2017).

The five chapters from Part 1 offer important insight into the historical and contemporary dynamics involved in these processes of image-making. Setting the scene, Chapter 1 by Emiel Martens delves into the early interwoven history of film, tourism, and empire, exploring the location production of *A Daughter of the Gods* (1916), often considered the first Hollywood blockbuster with a budget over $1 million, in British colonial Jamaica. By adopting a twofold comparative perspective, connecting different histories and countries, the chapter shows the potential of a transnational perspective within film (tourism) historiography. Alfio Leotta's contribution (Chapter 2) shifts the focus to New Zealand, tracing the evolution of film tourism in the country, before, during, and after the impact of *The Lord of the Rings* trilogy on New Zealand's global popularity. Highlighting the relationships between New Zealand, film, and tourism, the chapter explores both the economic and cultural consequences of the country's popular association with fantasy lands. Notably, both Martens and Leotta highlight the connections between film tourism and colonial culture, arguing that the 'exotic' landscapes of both Jamaica and New Zealand have come to represent 'transposable Otherness'. Then, Chapter 3 by Krzysztof Stachowiak explores the imaginative geographies in Indian films set in Eastern Europe, shedding light on the simplistic and, again, exoticized portrayals of the region in these films. In addition, the chapter reveals how local film commissions and destination marketing organizations contribute to the shaping of these, often problematic, perceptions. In her chapter (Chapter 4), Rebecca Nedregotten Strand centres on Norway and proposes a potentially new form of media tourism through the utilization of radio archives to share cultural heritage while exploring landscapes. The chapter details the creation of Pastfinder, a prototype web application that uses historic audio from the Norwegian Broadcasting Corporation to enrich a landscape with cultural history and imagine a place's heritage in situ. Finally, Rodanthi Tzanelli's chapter (Chapter 5) introduces a novel perspective by examining 'environmental imaginaria' in the current Age of Extinction, exploring how themed locations and natural environments are increasingly mediated on digital platforms. This chapter maps competing arguments regarding the role of these imaginaria in the context of global crises induced by climate change, presenting them as opportunities for activism, money-making enterprises, and sites for turning visitors into pilgrims, collectively forming a 'critical zone' that challenges traditional notions of travel, tourism, and agency in the Anthropocene.

The second part of the book focuses on the spatial appropriation of these imaginary worlds, often in a tourist context. This initially involves fans, who mark certain locations in the public sphere as 'the' location from a popular novel, film, book, or game (e.g., Milazzo & Santos 2022) but also increasingly commercial organizations that offer a TV or film tour, for example, through which fictional and historical stories are linked together at a location (Schiavone & Reijnders 2022; Lovell 2019: 5). The main focus is on the evolving concept of the media tourist and the ways authors are re-evaluating the influence of media on travel experiences and affects in different contexts.

To get things started, Christine Lundberg, Vassilios Ziakas, and Kristina N. Lindström's chapter (Chapter 6) introduces the 'everyday tourist' as a new kind of traveller who goes beyond traditional tourism boundaries, incorporating everyday activities and fandoms. This approach allows for increased freedom, pluralism, and the reconstruction of social conditions through the blending of alternative travel modes and elements of popular culture. Matt Hills (Chapter 7), in his exploration of comic-cons within the framework of the experience economy, suggests a shift in perspective, viewing comic-cons not just as experiences orchestrated by professionals but as co-created by attending and blogging fans. This challenges the conventional categorization of fans and emphasizes a co-existential fandom where various elements (text, brand, place) are interconnected. Qian Jin's (Chapter 8) examination of travel practices inspired by the Japanese transmedia work *Laid-Back Camp* further delves into how audiences in China navigate the boundaries between fiction and reality, representation and practice, and different geographical locations through media consumption. The study highlights the intricate intertextuality in content tourism and provides insights into transnational, transcultural, and transmedia practices in media-induced tourism. Lastly, Nanjangud and Reijnders' chapter (Chapter 9) on Dutch Hindustanis as a Bollywood audience explores how Bollywood cinema profoundly influences the perceptions and travel decisions of diasporic communities, introducing the concept of a 'cinematic itinerary' to describe the film-induced tourism practices among this audience.

The third part of the book deals with the perspective of local residents, organizations, and action groups in relation to media tourism flows. Media tourism can be welcomed and contribute to a form of local pride, but we are increasingly seeing examples of media tourism that lead to resentment and protest, for example because of over-tourism or because people feel that local narratives about the genesis and identity of a place are being overwritten by new place narratives derived externally from popular culture (Castro, Kim & Assaker 2023). For example, in the small town of Bermeo (Euskadi, Spain), a filming location for the HBO series *Game of Thrones* (2011–2019), residents posted messages in public spaces such as "This is not Invernalia. You are now in Euskal Herria" in an attempt to manifest a regional identity.[8]

The contributors to Part 3 have used different methodologies to delve into these intricate power dialectics between popular culture and place. The chapter by Nancy Reagin (Chapter 10) examines fan tourism to homesites by taking the *Little House on the Prairie* series as a case study. In particular, Reagin reflects upon how the homesite Laura Ingalls Wilder's, understood as a place of collective memory and national history, can be drawn into debates of public history. Alexandra Kolesnik and Alisa Maximova (Chapter 11) analyse place-making processes through memorialization of Western music (e.g., The Beatles) and heritagization of late Soviet rock music in the city of Ekaterinburg, Russia. Kolesnik and Maximova also reflect upon the local enactment of global phenomena in media heritage. Drawing on conversations with different local stakeholders, Timo Thelen (Chapter 12) explores how witch-themed tourism is locally contested and negotiated in Harz, Germany. Thelen also argues how pop-cultural shapes local witch heritage to make the latter fit into visitor's expectations. Finally, the account by Deborah Castro (Chapter 13) investigates the demographic profile of local film tourism supporters in Seville, Spain. She also explores local residents' suggestions for (future) film tourism initiatives, such as the creation of film tourism related events also targeted at the local community.

The fourth and final part of this book focuses on the (potential) development of media tourism. This primarily relates to local, regional, and national governments and the committees set up by them, such as VisitMéxico, Incredible India, or Visit Sweden. Their policies partly affect the kind of imaginative practices as discussed in Part 1. Next to these governmental organizations, there are multiple other stakeholders involved, including, for example, local entrepreneurs who are setting up and trying to promote their own (media) tourism businesses.

The chapters in this section reflect this holistic perspective. Evoking discussions from Part 3, Débora Póvoa's chapter (Chapter 14) traces a chronology of film tourism development in vulnerable locations in Brazil from the perspective of different local stakeholders. By proposing an understanding of film tourism as a phenomenon of production, she identifies the various contextual factors that determine a project's success or failure and argues for a developmental approach that is respectful of local communities' wishes and needs. Xin Cui, Les Roberts, and Wallis Motta (Chapter 15), on the other hand, introduce the case of the Hengdian World Studios in China to explore the relationship between film-related tourism sites and governmental agendas regarding the promotion of national culture. Through participant observation and interviews with tourists, the authors discuss the potential of film-related tourism to showcase local traditions and cultural heritage, and critically address related issues of loss of authenticity and cultural commodification. Kyungjae Jang and Sean Kim (Chapter 16) delve deeper into policy analysis by systematically comparing the 'Cool Japan' and 'Korean wave' strategies. Through the examination of these two national policies, the authors raise critical questions as to the effectiveness of government intervention in cultural policy and the importance of setting clear scopes and scales in national campaigns regarding popular culture and tourism. Finally, Emiel Martens and Edmund Onwuliri (Chapter 17) take a discursive approach to the study of Nollywood tourism development in Nigeria. Using newspaper articles and other publications as their source material, the authors analyse how the connections between Nollywood and tourism are discussed in the public domain and evaluate the extent to which certain expectations surrounding the development of these creative industries are met in the Nigerian context.

Reality cannot be captured in clear-cut models and the phases identified in the four separate parts of this book will sometimes take place in a completely different order. The development of media tourism is infamously capricious and depends to a large degree on the local situation (Póvoa 2023). Nevertheless, we believe that the phases identified and analysed in this book offer an appropriate way to look at the development and dynamics of media tourism as a socio-cultural phenomenon in a more holistic, coherent way. Many local examples will deviate from this model, but it is precisely in the way in which they differ from the standard that the unique characteristics are revealed and a basis for comparison is created. The trick is to do justice to the uniqueness of each individual example of media tourism, delineated in time and place, but at the same time to observe the way in which these examples relate to the bigger picture. In the end, the aim of our volume is not about opening all the thousand doors, one by one, but about imagining the larger world beyond.

Notes

1 S.E. Gracia, "Is It Time for a 'White Lotus' Vacation?" *New York Times*, 25 December 2022. Downloaded on 14 February 2023 from: www.nytimes.com/2022/12/25/style/white-lotus-vacation.html

2 K. Nath, "Quiiky Tours launches 'White Lotus' guided tour in Sicily". Downloaded on 17 March 2023 from: www.traveldailymedia.com/quiiky-tours-launches-white-lotus-guided-tour-in-sicily/.

3 "Popularity of Turkish soap operas leads Latin American tourists to flock to Turkey", downloaded on 22 March 2023 from: www.hurriyetdailynews.com/popularity-of-turkish-soap-operas-leads-latin-american-tourists-to-flock-to-turkey-association-138141

4 P. Harjani, "India's tourists flock to Spain". Downloaded on 22 March 2023 from: www.cnngo.com/mumbai/life/indian-movie-boosts-spanish-tourism-694426

5 For a recent WTO initiative in this direction, see: www.unwto.org/news/unwto-and-netflix-partner-to-rethink-screen-tourism. See Tzanelli (2019) for a more in-depth, critical perspective on the relation between 'cinematic tourism', capitalism and environmental activism.

6 Examples of this are the HBO series *Last of Us* (2023–) and the film *Uncharted* (2022), which both originated from video games, and the animation *SpongeBob Square Pants*, which spun off comic books, films, theme park rides, video games, and even a Broadway musical.

7 For an overview of international tourist arrivals, see the UNWO tourism data dashboard, downloaded on 17 March 2023 from: www.unwto.org/tourism-data/un-tourism-tourism-dashboard.

8 Gara (2017). "This is not Invernalia, you are now in Euskal Herria". *Gara*, 29 November 2017. Downloaded on 1 March 2023 from: www.naiz.eus/eu/hemeroteca/gara/editions/2017-08-29/hemeroteca_articles/this-is-not-invernalia-you-are-now-in-euskal-herria

References

Alexander, L. (2017). Fictional world-building as ritual, drama, and medium. In: M.J.P. Wolf (Ed.), *Revisiting imaginary worlds. A subcreation studies anthology*, pp. 14–45. Routledge.

Anaz, N. and C.C. Ozgan (2016). Geography of Turkish soap operas: tourism, soft power, and alternative narratives. In: I. Egrezi (Ed.), *Alternative tourism in Turkey*, pp. 247–258. Springer.

Beeton, S. (2005). *Film-induced tourism*. Channel View Publications.

Beeton, S. (2008). From the screen to the field: the influence of film on tourism and recreation. *Tourism Recreation Research* 33(1): 39–47.

Beeton, S. (2010). The advance of film tourism. *Tourism and Hospitality Planning & Development* 7(1): 1–6.

Beeton, S. (2015). *Travel, tourism and the moving image*. Channel View Publications.

Beeton, S. (2016). *Film-induced tourism* (second edition). Channel View Publications.

Brodie, I. (2023). Screen tourism: marketing the moods and myths of magical places. In: E. Champion, C. Lee, J. Stadler, and R. Peaslee (Eds.), *Screen tourism and affective landscapes*, pp. 9–26. Routledge.

Castro, D., S. Kim, and G. Assaker (2023). An empirical examination of the antecedents of residents' support for future film tourism development. *Tourism Management Perspectives* 45: 101067.

Champion, E., C. Lee, J. Stadler, and R. Peaslee (2023). *Screen tourism and affective landscapes. The real, the virtual and the cinematic*. Routledge.

Chen, N.C., C.M. Hall, and G. Prayag (2021). *Sense of place and place attachment in tourism*. Routledge.

Chow, H. and S. Reijnders (exp. 2024). Tracing television scenes online: fan professionalism in K-drama location blogging. *Tourism Geographies* (accepted for publication).

Connell, J. (2012). Film tourism. Evolution, progress and prospects. *Tourism Management* 33: 1007–1029.

Dubois, L.E., T. Griffin, C. Gibbs, and D. Guttentag (2021). The impact of video games on destination image. *Current Issues in Tourism* 24(4): 554–566.

Dung, Y.A. and S. Reijnders (2013). Paris Offscreen: Chinese tourists in cinematic Paris. *Tourist Studies* 13(3): 287–303.

Gara (2017). This is not Invernalia, you are now in Euskal Herria. *Gara*, 29 November 2017. Retrieved on 1 March 2023 from www.naiz.eus/eu/hemeroteca/gara/editions/2017-08-29/hemeroteca_articles/this-is-not-invernalia-you-are-now-in-euskal-herria

Graburn, N., M. Gravari-Barbas, and J. Staszak (2019). Tourism fictions, simulacra and virtualities. In: M. Gravari-Barbas, N. Graburn, and J. Staszak (Eds.), *Tourism fictions, simulacra and virtualities*, pp. 1–15. Routledge.

Hendrix, H. (2007). The early-modern invention of literary tourism: Petrarch's houses in France and Italy. In: H. Hendrix (Ed.), *Writer's houses and the making of memory*, pp. 15–30. Routledge.

Inglis, D. and M. Holmes (2003). Highland and other haunts. Ghosts in Scottish tourism. *Annals of Tourism Research* 30(1): 50–63.

Jansson, A. (2020). The transmedia tourist: a theory of how digitalization reinforces the de-differentiation of tourism and social life. *Tourist Studies* 20(4): 391–408.

Jenkins, H. (2006). *Convergence culture: Where old and new media collide*. New York University Press.

Kim, S. (2012). The impact of TV drama attributes on touristic experiences at film tourism destinations. *Tourism Analysis* 17(5): 573–585.

Kim, S., P. Long, and M. Robinson (2009). Small screen, big tourism: the role of popular Korean television dramas in South Korean tourism. *Tourism Geographies: An International Journal of Tourism Space, Place and Environment* 11(3): 308–333.

Kim, S. and S. Reijnders (Eds.) (2018). *Film tourism in Asia. Evolution, transformation and trajectory.* Springer.

Kim, S. and H. Wang (2012). From television to the film set: Korean drama Daejanggeum drives Chinese, Taiwanese, Japanese and Thai audiences to screen-tourism. *International Communication Gazette* 74(5): 423–442.

Lamerichs, N. (2019). Hunters, climbers, flâneurs: how video games create and design tourism. In: C. Lundberg, and V. Ziakas (Eds.), *The Routledge handbook of popular culture and tourism*, pp. 161–169. Routledge.

Liu, X. and S. Pratt (2019). The Downton Abbey effect in film-induced tourism: an empirical examination of TV drama-induced tourism motivation at heritage attractions. *Tourism Analysis* 24(4): 497–515.

Lovell, J. (2019). Fairytale authenticity: historic city tourism, Harry Potter, medievalism and the magical gaze. *Journal of Heritage Tourism* 14(5–6): 448–465.

Lovell, J. and S. Hitchmough (Eds.) (2020). *Authenticity in North America. Place, tourism, heritage, culture and the popular imagination.* Routledge.

Mason, P. and G.L. Rohe (2019). Playing at home. Popular culture tourism and place-making in Japan. In: C. Lundberg and V. Ziakas (Eds.), *The Routledge handbook of popular culture and tourism*, pp. 353–364. Routledge.

Milazzo, L. and C.A. Santos (2022). Fanship and imagination. The transformation of everyday spaces into *Lieux d'Imagination. Annals of Tourism Research* 94: 1–12.

Nanjangud, A. (2022). Finding Bollywood: a comparative analysis of Bollywood tourism in a globalized world. PhD manuscript, Erasmus University Rotterdam.

Nuenen, T. van (2021). *Scripted journeys: Authenticity in hypermediated tourism.* De Gruyter.

Póvoa, D. (2023). Film tourism in Brazil: local perspectives on media, power and place. PhD manuscript, Erasmus University Rotterdam.

Reijnders, S. (2010). On the trail of 007. Media pilgrimages into the world of James Bond. *Area* 42: 369–377.

Reijnders, S. (2011). *Places of the imagination: Media, tourism, culture.* Ashgate.

Riley, R., D. Baker, and C. Van Doren (1998). Movie induced tourism. *Annals of Tourism Research* 25(4): 919–935.

Riley, R. and C. Van Doren (1992). Movies as tourism promotion: A 'pull' factor in a 'push' location. *Tourism Management* 13(3), 267–274.

Roesch, S. (2009). *The experiences of film location tourists.* Channel View Publications.

Schiavone, R. and S. Reijnders (2022). Fusing fact and fiction: Placemaking through film tours in Edinburgh. *European Journal of Cultural Studies* 25(2): 723–739.

Schiavone, R., S. Reijnders, and A. Brandellero (2022). 'Beneath the storyline': Analysing the role and importance of film in the preservation and development of Scottish heritage sites. *International Journal of Heritage Studies* 28–10: 1107–1120.

Seaton, P., T. Yamamura, A. Sugawa-Shimada, and K. Jang (2017). *Contents tourism in Japan: Pilgrimages to 'sacred sites' of popular culture.* Cambria Press.

Tooke, N. and M. Baker (1996). Seeing is believing: The effect of film on visitor numbers to screened locations. *Tourism Management* 17(2): 87–94.

Tzanelli, R. (2013). *Heritage in the digital era. Cinematic tourism and the activist cause.* Routledge.

Tzanelli, R. (2019). *Cinematic tourist mobilities and the plight of development. On atmospheres, affects, and environments.* Routledge.

Van Broeck, A.M. (2018). 'Pablo Escobar Tourism' - Unwanted tourism: attitudes of tourism stakeholders in Medellín, Colombia. In: R. Philip Stone, R. Hartmann, T. Seaton, R.

Sharpley, and L. White (Eds.), *The Palgrave handbook of dark tourism studies*, pp. 291–318. Palgrave Macmillan.

Watson, N.J. (2006). *The literary tourist. Readers and places in Romantic and Victorian Britain*. Palgrave Macmillan.

Wolf, M.J.P. (Ed.) (2017). *Revisiting imaginary worlds. A subcreation studies anthology*. Routledge.

Ziakas, V., C. Lundberg, and M. Lexhagen (2024 exp.). *Popular culture destinations: Contexts, perspectives and insights*. Routledge.

Zimmermann, M. and S. Reijnders (2024 exp.). Down the rabbit hole and back again. Dynamics of enchantment in media tourism. In: J. Lovell and N. Sharma (Eds.), *Magical tourism and enchanted geographies*. Routledge (accepted for publication).

Part 1

Imagining Place in Popular Culture

Representation, Travel and Media

1 The Runaway Production of *A Daughter of the Gods*

Film, Tourism and Empire in Early Twentieth-Century Jamaica[1]

Emiel Martens

Introduction

In 1915, the American promoter, James Sullivan, wrote a series of articles for the *Jamaica Gleaner (JG)*,[2] Jamaica's foremost newspaper, about the location shooting of *A Daughter of the Gods* (1916), at the time the biggest Hollywood production ever undertaken. The Fox film starring Annette Kellerman was entirely shot in Jamaica and Sullivan was on location to accompany the famous silent-era actress, both as her husband and manager. Sullivan emphasised the impact the moving picture would have on the island's tourism appeal. According to the promoter, audiences worldwide would get acquainted with Jamaica and inspired to "visit the beauty spots of the island shown in this picture" (*JG*, 28 October 1915). He anticipated that *A Daughter of the Gods*, which he called "the greatest film play of modern times" (*JG*, 28 October 1915), would serve as "a tremendous advertisement for the island" (*JG*, 19 October 1915). According to Sullivan, "the Kellerman Fox picture" would "acquaint the world with Jamaica", and he argued that "it now rests with Jamaicans how to present to the tourists the wonders of this island" (*JG*, 28 October 1915).

At the same time, Sullivan stressed the immediate advantages to Jamaica resulting from the location shooting of *A Daughter of the Gods*. He mentioned that the Fox Film Company was spending "a great deal of money on this island" (*JG*, 19 October 1915). According to Sullivan, the production created temporary employment to "thousands of labourers" (*JG*, 28 February 1919), including a "vast amount of supernumeraries" (*JG*, 19 October 1915), among who hundreds of "little native children" who all received "a little ready money" (*JG*, 20 December 1915). In addition, he pointed to the use of properties as well as the building of "huge laboratories" and an "outdoor moving picture studio" for the filming of *A Daughter of the Gods* and other moving pictures (*JG*, 19 October 1915). In doing so, Sullivan referred to the potential of Jamaica as a regular film location. When the location shooting was completed, director Herbert Brenon expressed his gratitude for the cooperation he received from "the public officials, business men and natives of Jamaica" (*JG*, 16 February 1916) and particularly from Governor William Manning and General Leonard Blackden, who had provided "every possible assistance" (*JG*, 11 April 1916). Both Brenon and Sullivan also specifically thanked John Pringle,

DOI: 10.4324/9781003320586-3

a "wealthy colonial planter" whose property was used for shooting "most of the important scenes" (*JG*, 19 October 1915). In turn, Pringle said that he provided permission "for the people of Jamaica", as he believed *A Daughter of the Gods* would become "the greatest advertising boom in the history of the island" (*JG*, 11 April 1916).

The many *Gleaner* articles on the location production of *A Daughter of the Gods* offer an insightful view of the close ties between film and tourism in the early twentieth century. The making and marketing of one of the first, if not the first, Hollywood "'blockbuster' spectacles" (Greene, 2018, p. 34) reveal a synergistic relationship between the two international industries as early as the 1910s. In the past 20 years, scholars have increasingly explored the connections between film and tourism. While most studies in the field of film tourism have focused on the tourist activities generated *after* the making and release of a film, Ward and O'Regan (2009), among others, have proposed to approach film tourism as a type of business tourism *during* the location production as well. In other words, Ward and O'Regan (2009) point to the ways in which governments respond to "the film

Making of a Great Moving Picture Drama by Fox Film Corporation.

Figure 1.1 A photograph in the *Jamaica Gleaner* depicting a meeting between Herbert Brenon and the then Governor of Jamaica, William Manning.

Source: JG, 22 November 1915.

producer as a long-stay business tourist, and film production itself as potentially another event to be managed and catered for" (p. 218). Both types of film tourism are often referred to as recent phenomena. However, although their size and scope indeed became unprecedented in the past two decades, both types, or at least their envisioned potential, originated almost as soon as cinema emerged.

Apart from its connections with tourism, the location production of *A Daughter of the Gods* should be related to empire, that is, the colonial enterprise that still dominated the island's (and the world's) political order throughout the first half of the twentieth century. In this late colonial period, Jamaica was a British Crown Colony with "a semi-representative government" (Black, 1991, p. 150), which it remained until the island's independence in 1962. The colonial project was extended into the realms of film and tourism as well. In the early twentieth century, and very much until the present day, both realms were forms of *leisure imperialism* (Crick, 1989) that not only adhered to empire but actively facilitated the colonial project and its power relations. The early period of both film and tourism in Jamaica overlapped with the peak of classical imperialism when the European empires tried to consolidate their stranglehold over the globe and the United States expanded into one of the world's major colonial powers. Following Shohat and Stam (1994),

> It is most significant (…) that the beginnings of cinema coincided with the giddy heights of the imperial project (…). The most prolific film-producing countries of the silent period – Britain, France, the US, Germany – also 'happened' to be among the leading imperialist countries, in whose clear interest it was to laud the colonial enterprise. The excitement generated by the camera's capacity to register the formal qualities of movement reverberated with the full-steam-ahead expansionism of imperialism itself.
>
> (pp. 100–104)

In a similar vein, Bruner (1989), among others, argues that tourism and colonialism "were born together and are relatives" (p. 439). More specifically, Tucker and Akama (2020) contend that the development of tourism in colonial states reflected "the economic structures, cultural representations and exploitative relationships" (p. 4) of empire. Zooming in on the Caribbean, Perez (1974) claims that "travel from metropolitan centres to the West Indies has served historically to underwrite colonialism" (p. 473) in the region. Since the early 2000s, several studies have appeared that examine the history of tourism in the Caribbean as a colonial practice and discourse (e.g., Strachan, 2002; Sheller, 2003; Thompson, 2006). These studies not only joined the body of research exploring the negative impacts of tourism in the Caribbean, but also complemented the literature on "the vexed history of visual culture" (Quilley & Kriz, 2003, pp. 1–14) in the region during the colonial period and beyond.

The aim of this chapter, then, is to evaluate the location production of *A Daughter of the Gods* in Jamaica on the basis of the *Gleaner* – "Jamaica's leading newspaper throughout the twentieth-century colonial period" (Rush, 2011, p. 33) – and other relevant historical newspapers and magazines, mainly from the United States, and to demonstrate the close ties between film, tourism and empire on the island in the early twentieth century. In so doing, the chapter reflects a twofold comparative perspective: between different histories and between different countries. Following Musser (2004), I find it crucial to "imagine cinema as an element (…) of other histories" (p. 105). My ongoing research into Jamaica's film history has forced me to consider the island's tourism history as well, since the sources I found made clear that productions shot on the island were almost always translated in touristic terms, that is, how can film production and exhibition help promote Jamaica as a holiday destination? At the same time, with the acknowledgement that "the dominant European/ American form of cinema" (Shohat & Stam, 1994, p. 103) was an *empire cinema* (Chapman, 2006; Chowdhry, 2000; Richards, 1973), I situate my work in the field of postcolonial cinema historiography (Ponzanesi & Waller, 2012).[3] Despite the rise of such historiography, the early cinema histories of the Caribbean have remained largely unexposed. As Hambuch (2015) argues, "scholarship in Caribbean film studies has been scarce" (p. 4), and this applies even more to the study of Caribbean film histories. More specifically, Jamaica's early film history has hardly been dealt with, particularly in relation to the island's tourism and colonial histories.

This chapter focuses on the "global conditions of production" and the "interconnected organizational cultures that characterize the film production industry" (Maltby, 2011, p. 9). Biltereyst and Meers (2006) indicate that, "given the international dimension of the film industry in terms of production, trade and consumption, the comparative mode has always been present in some form or another in film criticism and film studies" (p. 14). However, they also argue that discussions of "crossnational flows" of films and filmmakers have mainly addressed the relationships between Hollywood and European cinema or within Europe (Biltereyst and Meers, 2006, p. 14). The early relationship between Hollywood and the Caribbean has not often been explored. One of the reasons for this could be that the position of the region within the world of film was, at least until the 1970s, that of "a receiver/consumer of and a resource for Euro-American productions (in terms of its use as location)" (Cham, 1992, p. 2). The early twentieth-century utilisation of the Caribbean as a location for "foreign productions which exploit(ed) the natural/physical endowment of the tropical islands" (Cham, 1992, p. 2) has not been a common subject of investigation. However, given the transnational and interconnected nature of the film industry from the very onset, early Hollywood film history is also Caribbean colonial film history and vice versa. All in all, this chapter seeks to contribute to the discussion of the connections between film, tourism and empire, and between Hollywood, the British Empire and Jamaica, by revealing the colonialist film tourism practices and discourses of *A Daughter of the Gods*, one of the most important American moving pictures of the silent era and one of the most significant global imperial tourist films of the early twentieth century.[4]

Setting the Scene: Colonial Travelogues and the Cinema of Exotic Attractions

At the close of the nineteenth century, film succeeded photography as the most advanced visual medium of the time. The initial period of cinema has often been characterised as a *cinema of attractions*, a term introduced by Gunning (1996) to designate early cinema's fascination with novelty and curiosity about the new technology of visual display. The attractions of early cinema were largely based on its ability to present objects and events "as real as life" (*The Davenport Daily Leader*, 14 July 1896). According to Popple and Kember (2004), cinema was widely regarded as "an objectified recorder of contemporary life, an adjunct of the 'scientifically rational' art of photography, with the added dimension of movement" (p. 2). As such, cinema came to the public as the ultimate "reality capture" technology (*Winnipeg Free Press*, 1 June 1891). Taken in by its "startling realism" (*Alton Evening Telegraph*, 20 May 1899), spectators of early cinema were most fascinated by moving images of real events. Such actuality films, or *actualities*, typically depicted current affairs, official events and everyday scenes. While domestic actualities enjoyed significant popularity, early filmmakers also extensively travelled the world to record moving pictures of distant places and peoples. The shorts they brought back for viewing on screens in Europe and North America were promoted as scenic views and would later become known as *travelogues*.

Travelogues became among the most prominent pictures of early cinema and were "a regular part of the moviegoing experience from cinema's inception through the middle 1910s" (Peterson, 2005, p. 639). According to Bruno (2002), early travel films provided millions of people "a set of travelling pleasures" that turned them into "enthusiastic voyagers" (p. 77). Ruoff (2006) even proposes to view early cinema as a "machine for travel" (p. 1), arguing that watching travel pictures offered experiences similar to those produced by modern means of transportation. Kirby (1997) even asserts that travelogues came to participate in an emerging *touristic consciousness*, "a fascination exerted by foreign images" (p. 59). According to Bruno (2002), early cinema was highly shaped by this consciousness, which stemmed from "new means of transportation, architectures of transit, world expositions, (…) aesthetic panoramic practices, (…) travel photography, the postcard industry, and the creation of the Cook tours"; indeed, "film was affected by a real travel bug" (p. 76). Cinema became immediately identified as a powerful means of promoting tourism. From the onset, railway and steamship companies were interested in the use of film for the marketing of their travel packages. They entered into arrangements with film production companies to facilitate them in the making of travel films in exchange for promotion. According to Gunning (1996), "the connection between early American travel films and the transportation industry is proudly displayed in early film catalogues. (…) In all cases the transportation companies sponsored these films with the specific intention of encouraging tourism along their routes" (p. 30). Even if the majority of the cinemagoers were not yet involved in international tourist activities, travelogues still conferred "a tourist point of view on their spectators" (Peterson 2013, pp. 8–9).

From the onset, cinema was not only closely aligned with tourism but also with empire. According to Griffiths (2002, p. 232), early cinema "followed the geographical itineraries and ideological rationales of colonial expansion". The reaches of empire set the film camera in motion, both literally and ideologically. The journeys of early filmmakers, and of "Euro-American image factories" (Cham, 1992, p. 2), were dependent on, and hence complicit with, the routes of the colonial project:

> The acquisition of new territories, coupled with the explosion of tourism, meant that itinerant cameramen and production companies could set up the base in colonial expatriate communities and shoot films of native societies under the protection of the governing authority.
>
> (Griffiths, 2002, p. 234)

The popularity of war and other imperial actualities evidences early cinema's preoccupation and alignment with colonial expansion. However, travel films constituted an integral part of empire cinema as well. As Bruno (2002) argues, "in touring cities, exploring landscapes, and mapping world sites, early film also 'discovered' otherness, made it exotic, and often acted as agent of an imperialist obsession" (p. 77). Similarly, Chapman and Cull (2009) state that "images of imperial splendour" and "pictures of exotic lands and customs" were "a natural for the travelogues" (p. 1).

Around the mid-1900s, with the rise of the nickelodeons in the United States and the establishment of commercial film distribution and exhibition worldwide, cinema entered a next phase, with fiction films starting to dominate the screens. The great majority of these dramas were shot in a studio or on location close to the studio. By 1912, most of the major companies had bought land "for studios, standing sets, and back lots" in California, and by 1915 most American films were already made there (Fellow, 2013, p. 219). Concurrently, in Britain, several big film studios emerged as well. Still, while most film companies heavily relied on studio filmmaking, production companies also kept sending out crews to different parts of the world. The first film teams coming to Jamaica to shoot fiction films were the Vitagraph Company in 1910, the British and Colonial Kinematograph Company in 1913, and the Terris Feature Film Company in 1915. These foreign productions, and particularly the two short dramas directed by British actor-turned-filmmaker Tom Terriss, *Pearl of the Antilles* (1915) and *Flame of Passion* (1915), extended the links between cinema and tourism in the realm of fictional filmmaking (Martens, 2018).

Besides the hospitality he received, Terris considered the "wonderful variety of the scenery" as the "island's greatest charm" (*JG*, 14 May 1915). In addition, the filmmaker identified Jamaica's Black population as offering great picturesque possibilities, as they, "with their quaint ways, are an abundant source of good material" (*The Moving Picture World*, 16 June 1915). Terriss provided the example of "the strange baptismal ceremonies" performed by a "black Messiah", which he had caught on camera "from ambush behind a screen of cactus" (*The Moving Picture World*, 16 June 1915). The qualities highlighted by Terriss echoed the main lines of promotion that were used to attract tourists to Jamaica since the advent of tourism on the island in the 1890s: modern hospitality, tropical fecundity and exotic people. The latter, exemplified by the "quaint" and "strange" fashions of the Black population, was almost considered as picturesque as the island's tropical landscape. In fact, Terriss here joined the early "tourism image makers and travelers" who "framed the island's black population as parts of the tropical scenery" (Thompson, 2006, p. 103). At the same time, the startling manner in which the filmmaker had obtained the images, secretly without notice and consent, demonstrates the lack of humanity Black people in the colonies possessed in the "conquering eye of the motion picture camera" (Peterson, 2013, p. xvi). In doing so, Terriss followed in the tradition of early colonialist filmmakers who, "just like freebooting imperialists in their quest for plunder, (...) scurried all over the globe, frenetically gathering images – exotic, arcane, bizarre, sensational, revelatory – which became 'the reality' about the world for millions of people" (Davis, 1996, p. 2).

A few months after Terriss' departure, in August 1915, a team of about 30 "moving picture artists" (*JG*, 11 September 1915) from Fox Film Corporation, which had just been formed by American theatre chain pioneer William Fox, arrived in Kingston to produce a series of films on the island. The delegation was reportedly "the first batch of a large number of the leading moving picture actors and actresses that will come to these shores" (*JG*, 31 August 1915). Their film-making trip was intended to result in five films: *A Wife's Sacrifice* (1916), *The Spider and the Fly* (1916), *The Marble Heart* (1916), *The Ruling Passion* (1916) and *A Daughter of the Gods* (1916). The last one, an aquatic fantasy adventure set in "The Land of the Orient" (*JG*, 1 November 1915) and starring "Australia's Diving Venus, Annette Kellerman" (*The North Western Advocate and the Emu Bay Times*, 20 February 1917), was by far the most ambitious production. In fact, the four other films were primarily "by-products of the great drama" (*JG*, 3 February 1916), shot in Jamaica to "offset the expense of the big one" (*JG*, 19 October 1915). With a record budget of over one million dollars, *A Daughter of the Gods* became the most expensive film ever attempted worldwide.[5]

The Production of *A Daughter of the Gods*: Colonial Hospitality and Exotic Tropicality

After the successful runaway production of *Neptune's Daughter* (1915)[6], which was filmed on the British colonial island of Bermuda, for *A Daughter of the Gods* director Herbert Brenon decided to travel to Jamaica because of his and the island's position within the British Empire:

> Because I am a British subject and also because I made my greatest picture in Bermuda (...) I thought (...) I should make the next picture [again] in a British colony, and so I have selected Jamaica, which I think will provide us with ample scenery and every hospitality.
>
> (*JG*, 31 August 1915)

Notably, the filmmaker mentioned the same reasons for visiting Jamaica as his predecessor, Terriss. In addition, both thought of themselves as loyal British subjects who, besides considering colonial rule as self-evident, expected to receive the most hospitable reception for the advantages they supposedly brought to the island colony. In reality, the practice of hospitality at play could be seen as *colonial hospitality*. With this form of hospitality, "the *arrivant* turns into a colonizer, invader, or occupier", whereas the original hosts become "powerless guests in their own land" (Haswell et al., 2009, p. 715). While the Terriss crew already showed signs of invading guest-turned-host filmmakers, the production of *A Daughter of the Gods* pushed the "transformation of guests into hosts" (Ahmed, 2000, p. 190) to new limits. As such, Brenon's "film extravaganza" (*The Billboard*, 28 October 1916) helped establish the colonial practice of temporarily transforming landscapes and exploiting resources by Hollywood runaway productions. As Gibson (2005) notes,

> *A Daughter of the Gods* set the standard for the tradition of the impermanent imperialism of big-budget Hollywood film shoots. The film crew would arrive, taking over not only the location but the entire landscape and the local economy for the term of the film's production. Just as suddenly, like a colonial power, they would strategically withdraw, leaving the country bereft of patronage. Because most of the world had been discovered and colonised by 1915, directors, behaving like true imperialists, went to exotic locations and made their own worlds in which, for the duration of the shoot, they were absolute rulers.
>
> (p. 134)

During the location production of *A Daughter of the Gods*, Fox Film almost treated Jamaica as a *tabula rasa*, open to the transformation and domination of the environment. In doing so, they perpetuated the colonialist practices of shaping, using and controlling 'empty' spaces instigated by plantation slavery and paradise tourism (Strachan, 2002). During filming, the island's colonial administration did much to facilitate the operations of the company (*JG*, 4 March 1916). While the servicing came together on an ad hoc basis, the filming was approached as a highly profitable venture that should be accommodated in any way. Brenon allegedly even received special permission of the British government to shoot on the island (*Pinnacle News*, 20 March 1917).

Newspapers reporting on the filming of *A Daughter of the Gods* suggest the freedom the crew had to rearrange the Jamaican landscape according to the needs of the "fantastic fairy tale" (*The Billboard*, 26 October 1916). Brenon integrated major film sets into the natural environments. For example, the production team transformed the entire base of the Roaring River Falls into a miniature gnome village (*JG*, 15 October 1915). In addition, the team "diverted a river from its course" and "razed a range of hills" (*Manitoba Free Press*, 16 February 1918). In another instance, the Fox crew built an Arabic city at Fort Augusta along the Kingston Harbour shoreline. At the time, Fort Augusta was by and large "a ruin surrounding a swamp" (*JG*, 20 January 1916). Brenon obtained approval from the British Foreign Office to restore the fortress and use the surrounding wastelands for the purpose of his film. He hired local workmen to rebuild the fortress and drain the swamp. In order to clear the area from flies and mosquitoes, he ordered tons of disinfectants from New York, with the result that, according to one British reporter, "the plague-spot was turned into a pleasure resort" (*London Daily Mail*, 11 February 1916). In addition, for the film's final scenes at Fort Augusta, Brenon acquired a fleet of historic sailing vessels and brought in dozens of lions, tigers, elephants, donkeys, lizards, camels and other animals from the New York Zoo (*JG*,

THE MAMMOTH WORK OF THE FOX FILM COMPANY IN JAMAICA

MR. BRENON WORKING AT FORT AUGUSTA UNDER DIFFICULTIES.   THE SULTAN'S PALACE AT FORT AUGUSTA.

Figure 1.2 Two photographs in the *Jamaica Gleaner* depicting the work of the Fox Film Company at Fort Augusta. On the left, "Mr. Brenon working at Fort Augusta under difficulties" and on the right, "the sultan's palace at Fort Augusta."

Source: JG, 29 January 1916.

20 December 1915). Eventually, the huge sea-front set consisted of "a palace, a castle, a mosque, an Arabian slave market as well as a regular market place, and practically everything one could hope to see in an Arabian city, based to a great extent on the story of the 'Arabian Nights'" (*JG*, 31 August 1915). The city was built in three months at a cost of $350,000 (Strachan, 2002, p. 142) – only to be demolished again a few months later for the final scenes in which the huge set was consumed by flames (*JG*, 31 March 1916). Finally, the fortress was "smashed to pieces by the West Indian squadron of the British navy" (*JG*, 14 February 1916), to be left as "a waste once more" (*JG*, 29 January 1916).

The sets of *A Daughter of the Gods* were not only intended to be temporary, but also to make Jamaica stand in for an imaginary fantasyland. Brenon's picture portrayed an exotic-erotic fairy tale set in a mythical Arabian world. The film chronicled "the tale of Sultan Omar, who promises to help the Witch of Evil to destroy the mysterious Anitia if only she will revive his drowned son Omar" (Altman, 2004, p. 300). Reportedly, *A Daughter of the Gods* consisted of "a collage of fantasy, fairy tale, melodrama, and sexual display" (Erdman, 2004, pp. 97–98), featuring "lavish scenic displays" (Koszarski, 1994, p. 186) typical of the costumes pictures of the early feature scene. Evidently, the tropical environment of Jamaica provided the exotic wonderland Brenon was looking for. The island's tropicality played an essential role in the creation of the "oriental fantasy" (*The Pinnacle News*, 20 March 1917). The director explained that the "tropical scenery" of Jamaica "fit in very nicely" with the film's "Arabian fairy story" (*JG*, 31 August 1915).

All in all, the Jamaican landscape in *A Daughter of the Gods* became, like many colonial landscapes in Hollywood cinema before and after, "the stuff of dreamy adventure" in tropical paradise (Shohat & Stam, 1994, p. 124). For the purpose of Brenon's film, Jamaica was transformed into what has been called a "space of radical *dépaysement*" (Jutel, 2004, p. 60). The notion of *dépaysement*, literally translated "out-of-nation-ness", represents the simultaneous "attraction of geographical defamiliarisation" and "separation of the lost homeland" (Jutel, 2004, p. 60). In order to make Jamaica look like another world, the landscape had to be removed from its identity and history, and to be supplied with settings and meanings imported from somewhere else. The island was, like so many colonised states in the margins of empire, designed to produce "raw visual material; exotic views for the centre of the empire" (Leotta, 2011, p 18). As such, Jamaica was made part of the early imperial tourist spectacle and constructed through an orientalist tourist gaze for the enjoyment of metropolitan armchair travellers. At the same time, *A Daughter of the Gods* actualised the practice by Hollywood runaway productions of using Jamaica as *transposable otherness*, that is, the exploitation of the island as a readymade exotic tropicality (Jutel, 2004, p. 55). In so doing, the film prepared the terrain for the future branding of Jamaica as a transposable location fit to evoke a variety of tropical settings.

Undoubtedly, *A Daughter of the Gods* instigated the cinematic practice of sexualising the Jamaican landscape. From the onset, it was clear that the film

Figure 1.3 A publicity photo of *A Daughter of the Gods*, depicting Annette Kellerman unclad in a Jamaican waterfall landscape, with only her long hair covering her breasts and pubic area.

Source: Fox Film Corporation (1916, public domain).

would inscribe erotic spectacle into the island's exotic tropicality through the physical performance of Kellerman. Brenon stated that his film was "made to exploit" his star actress (*JG*, 31 August 1915), whose career was largely based on bodily display. In her role as water nymph Anitia, Kellerman was often seen unclad while strolling through a waterfall landscape with only her hair covering her breasts and pubic area. According to *Variety*, Brenon had filmed "his aquatic star in the nude on every possible occasion" (*Variety*, 20 January 1917). The "personal scenes" with the actress at the Roaring River Falls (*JG*, 19 October 1915) were unambiguously invested in the erotic connotations of tropical environments and particularly waterfalls. According to Hudson (2001), *A Daughter of the Gods* established the long association" of "eroticism on the silver screen" with Jamaican waterfalls, which since have "often been used as settings for adventurous exploits or romantic, even erotic episodes" (pp. 31–32). As such, the film foreshadowed the equation of the Jamaican landscape with sexuality and hedonism, prefiguring the future "marketing of the Caribbean via imagined geographies of tropical enticement and sexual availability" (Sheller, 2004, p. 31).

A Daughter of the Gods as Early Film Tourism: The Spectacle of Production and Exhibition

During the nine-month filming period, *A Daughter of the Gods* reportedly made a significant impact on Jamaica's economy. The production provided temporary employment for "thousands of Jamaica[n] people" behind and in front of the camera (*JG*, 28 February 1919). According to the *Gleaner*, Brenon had "spent on native labour here over $165.000", including "dressmakers", "an average of 550 people (…) in the Manufacturing and Construction Departments" and "from time to time in the capacity of extra actors, 61.000 local people" (*JG*, 27 April 1916). According to the Jamaican newspaper, the director even established "a special municipality" near Fort Augusta for the thousands of locals he hired during the filming period there (*JG*, 8 October 1915). In addition, the stay of the Fox team at the Myrtle Bank Hotel and their transportation to the various locations across the island created many short-term jobs in the accommodation and transport sectors. While Brenon initially arrived with about 30 staff members, on average his team consisted of over 200 people (*JG*, 27 April 1916).

For the duration of their stay, the Fox team remained in the Myrtle Bank at the Kingston waterfront. Apart from occupying guest rooms, they built workspaces in separate annexes on the property to develop their film recordings (*JG*, 19 October 1915). At the same time, Brenon set up the "headquarters of the Fox Film Company in Jamaica" at Rose Gardens in Kingston (*JG*, 3 February 1916), which reportedly became "the finest outdoor moving picture studio that has ever been built" (*JG*, 19 October 1915). Concerning transport, the company maintained "its own transportation facilities" during their operations at the various locations (*JG*, 27 April 1916). After the filming, the Fox press department even stated that the film

FOX FILM NOTES.

Residents of Port Henderson, Spanish
Town and Districts near
Fort Augusta,

WEATHER PERMITTING,

— ON —

MONDAY, 21st at 6 A.M.

400 Natives!

Men, Women and Children, are offered
employment at Fort Augusta to take
part in the great Fox Film "Daughter
of The Gods."

HERBERT BRENON,
Director General.

Figure 1.4 An advertisement in the *Jamaica Gleaner* of the Fox Film Company offering
temporary employment at Fort Augusta to "400 Natives!"

Source: *JG*, 18 February 1916.

company had used "an entire Caribbean island and all of its population (…) in
the making of the picture" (*Manitoba Free Press*, 10 November 1917). Although
grossly exaggerated, it points to the idea that the location production had a major
employment impact on Jamaica.

In addition, the filming of *A Daughter of the Gods* became a tourist event in
itself, what Ward and O'Regan (2009) call "the spectacle of film production within

the locale" (p. 216). The *Gleaner* reported widely on the location shooting, with that attracting a great deal of interest. According to the newspaper,

> the making of moving pictures is something new in Jamaica, and anything that gives an idea (...) [of] the mode of work will have an attraction for readers anxious to obtain a peep into the arcana of what has always appeared to them to be allied to the marvellous.
>
> (*JG*, 3 February 1916)

Much coverage was devoted to the actors involved in the film production. Their visit created a buzz in Kingston and drew both journalists and ordinary citizens to the hotel where they were staying and rehearsing (*JG*, 1 November 1915). In addition, a New York newspaper described the Rose Gardens studio as the new "show place of the island" and indicated that the property's staff was increased in order to meet the amount of visitors that came by (*New York Morning Telegraph*, 30 January 1916).

Apart from the direct economic benefits associated with the production, the *Gleaner* also devoted much attention to the indirect economic benefits that would result from Jamaica's inclusion in *A Daughter of the Gods*. As mentioned in the introduction, particularly Sullivan repeatedly stressed the publicity value of the film's global exhibition for Jamaica in his *Gleaner* series. He regarded cinema as a new popular medium for tourism promotion:

> We are able nowadays to acquaint the public by projection on the screen with almost every object or scene of interest in the world. (...) We will to all intents and purposes carry back the island of Jamaica (...) by a process of photography.
>
> (*JG*, 28 October 1915)

According to Sullivan, *A Daughter of the Gods* would publicise Jamaica in a way "no amount of advertising could have done" (*JG*, 20 December 1915) due to the great scope of cinema across social classes and national boundaries (*JG*, 28 October 1915). Apart from the tourism promotion through "the presentation of the picture itself", Sullivan argued that the film's publicity campaign would also greatly increase Jamaica's reputation. The American anticipated that in the United States alone, the island's name would appear in over 5,000 newspapers as a result of the film's campaign (*JG*, 20 October 1915).

The expectations about the tourist potential of the film created considerable enthusiasm among Jamaica's colonial officials and business elites. They recognised "the benefit that is sure to accrue to the island from the picture" (*JG*, 4 March 1916) and supported the showcase of "the beauties of our land" to "millions of people all over the world" through "the marvellous Kellerman-Brenon film" (*JG*, 28 October 1915). The Jamaica Tourist Association expressed the hope that *A Daughter of the Gods* would promote Jamaica as a tourist resort for affluent travellers in the United States, the UK and the rest of the world (*JG*, 19 October

THE PARTY OF "MERMAIDS" NOW IN JAMAICA.

Figure 1.5 A photograph of the *Jamaica Gleaner* depicting "the diving girls of the William
Fox Film Company, who will appear as mermaids in the scenes along with Miss
Annette Kellerman," at the Myrtle Bank Hotel in Kingston.

Source: *JG*, 25 September 1915.

1915). Furthermore, they arranged a meeting with Film Fox to see if they could
do anything else "in regards to advertising the island" abroad (*JG*, 26 February
1916). More specifically, they asked if the company could produce "a film to advertise the beauties and attractions of Jamaica in different parts of the United States"
(*JG*, 25 February 1916). Such a travelogue film, they felt, "would be a very good
means of inducing tourists to come here" (*JG*, 25 February 1916). The Fox team
agreed to make "a film typical of Jamaican scenery" for the JTA free of cost (*JG*,
25 February 1916). They also offered, although this time not free of charge, to
organise showings of the travelogue in the United States as "they wanted to do all
they could do to assist in booming the island" (*JG*, 25 February 1916).

Following the world premiere of *A Daughter of the Gods* in New York in October
1916, the film reportedly became an instant "big movie box office success" and
garnered widespread critical acclaim in the United States (*The New York Clipper*,
6 December 1916).[7] The first reviews of *A Daughter of the Gods* in the New York
newspapers were all published in the *Gleaner* and reviews continued to be written

Figure 1.6 An advertisement of *A Daughter of the Gods* in a Canadian newspaper. While the moving picture was clearly promoted as a Kellerman spectacle, no mention is made of Jamaica as the shooting location.

Source: *Manitoba Free Press*, 16 February 1918.

in the following months and even years. The film had a long domestic theatrical run and a gradual international distribution. In addition, Fox rereleased the film in the United States in 1917, 1918 and 1920, each time generating a new round of press coverage. Many of the billings and reviews of the almost three-hour-long film

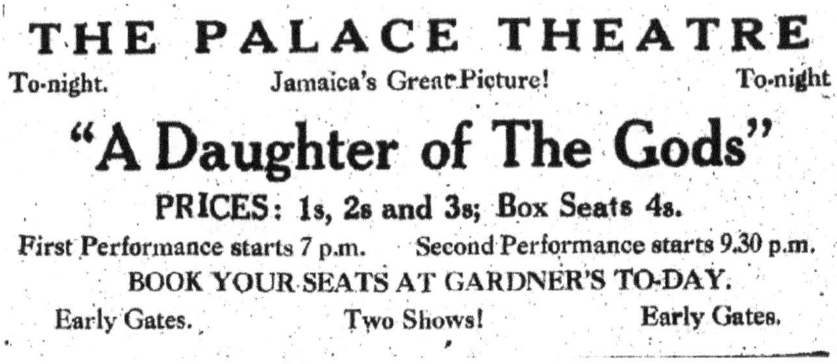

THE PALACE THEATRE
To-night.   Jamaica's Great Picture!   To-night

"A Daughter of The Gods"
PRICES: 1s, 2s and 3s; Box Seats 4s.

First Performance starts 7 p.m.   Second Performance starts 9.30 p.m.

BOOK YOUR SEATS AT GARDNER'S TO-DAY.

Early Gates.   Two Shows!   Early Gates.

Figure 1.7 An advertisement for the first screenings of *A Daughter of the Gods* in Jamaica in March 1919. The film was first shown in The Palace in Kingston, before it went to other venues across the island.

Source: JG, 3 March 1919.

extolled two virtues, first and foremost the virtue of Kellerman's body and secondly the virtue of the exotic setting: "Together with the daring of Miss Kellerman, (…) the natural beauty of the Jamaican seascapes and landscapes makes a picture of great attractiveness" (*New York Herald*, 25 October 1916). Still, most billings and reviews referred to the film's tropical scenery in general terms. They usually did not mention Jamaica as filming location but emphasised the "beautiful pictures to delight the eye of the spectator" (*Variety*, 20 January 1917). However, some critics did mention the location used in *A Daughter of the Gods*, and others not only identified the Jamaican settings, but also described them in highly favourable terms.

In Jamaica, *A Daughter of the Gods* received its premiere in March 1919, well over two years after its original release in the United States. When *A Daughter of the Gods* was finally released in theatres across the island, it reportedly attracted a "mighty crowd" (*JG*, 27 March 1919). While the plot was not always highly appreciated, the film was considered a success from "the spectacular point of view" (*JG*, 28 February 1919). In particular, the *Gleaner* stated that it contained "some very good reproductions of Jamaican scenery" (*JG*, 28 February 1919). In one advertisement published in the newspaper, *A Daughter of the Gods* was even described as "the picture that has won fame through Jamaica's charming scenery" (*JG*, 1 March 1919). It was these kinds of written endorsements that Jamaican tourism stakeholders had hoped to see (more) in American, British and other foreign newspapers when they envisioned the potential indirect benefits through the promotion of Jamaica in the film and the press. At the same time, it reminded them of "the good old days" when Fox Film was "spending money largely on the island" and "some of the local people" experienced "good times", referring to the incidental direct benefits of hosting the major Hollywood runaway production (*JG*, 3 March 1919).

Conclusion

After the initial theatrical run of *A Daughter of the Gods*, it appears that Jamaican tourism stakeholders were disappointed with the outcomes of the location production. The lack of press coverage in the *Gleaner* on the topic after its original release suggests that the film did not have the significant impact in terms of enhancing Jamaica's reputation and increasing the number of tourists as was anticipated. It was only when subsequent foreign film companies travelled to the island that some critical reflections started to appear. The *Gleaner* argued that Jamaica had "lost a fine advertisement" since Brenon had not kept his promise that he would "set forth on the moving picture screen that the scenes (…) had been taken here" (*JG*, 4 February 1920). Indeed, the name of the island was eventually excluded from the film's credits as well as most of the advertisements of the Fox Film Company. On top of that, the travelogue film that the Fox team agreed to produce and exhibit for tourist promotion purposes seemingly never materialised. When the English Film Company visited Jamaica in 1920, to make an actuality film on "all phases of the colony's life" for "educational purposes", the *Gleaner* argued that it would be "the first time" that Jamaica would be filmed in this way, and that it would constitute "a far better advertisement" than *A Daughter of the Gods* (*JG*, 4 February 1920). Furthermore, a year later, when the next "American moving picture concern" came to the island to make another drama, *Love's Redemption* (1921), it was praised that, this time, the Jamaican places where the scenes were shot were "named on the screen" (*JG*, 30 November 1923).

Still, the production of *A Daughter of the Gods* took the awareness of the tourism potential of location filming as well as the collaboration between the film and tourism industries in Jamaica, to a new level. In one of the articles that Sullivan wrote for the *Gleaner*, the moving picture was mistakenly but aptly referred to as an "ad venture", with a space between "ad" and "venture" (*JG*, 20 December 1915). In "early, racist silent cinema" (Denzin, 2002, p. 21), *A Daughter of the Gods* did not only set the tone for "colonialist adventure films" (Dyer, 1997, p. 155) taking place in the British Empire, but also for the future of advertising ventures between the film and tourism industries in Jamaica and other European and American colonies. As early as the mid-1910s, the island's interwoven history of film, tourism and empire took off with Brenon's "million dollar miracle" (*JG*, 28 February 1919). From that moment onwards, the call to establish "a British Hollywood within the Empire" (*JG*, 7 August 1925) or at least "some sort of tropical Hollywood" (*JG*, 16 November 1925) that would encourage American and British filmmakers as well as tourists to visit Jamaica became part of the island's imperial tourist consciousness – and with that the envisioned road of its "chances of development" (*JG*, 4 February 1920) in the century to come. By connecting and integrating different histories and countries, the comparative history of early (empire) cinema both widens and deepens the understanding of the global workings of leisure imperialism. The film tourist practices and discourses of Hollywood's *A Daughter of the Gods* in British colonial Jamaica demonstrate the interconnections between film, tourism and empire in the early twentieth century – and with that, the potential of a transnational comparative perspective within film (tourism) historiography.

Notes

1 An earlier version of this chapter was published as E. Martens (2020). *A Daughter of the Gods* (1916): Film, tourism and empire on location in Jamaica. Special issue on 'Comparative histories of moviegoing'. *TMG Journal for Media History,* 23(1/2), p. 1–40.

2 As newspaper reports of the *Jamaica Gleaner* are the primary source materials for this study, the abbreviation *JG* is used when quoting or paraphrasing from these reports. When referring to other newspapers and magazines, their full title is provided.

3 In general, postcolonial cinema historiography consists of recognising and revealing Euro-American cinemas of the late nineteenth century and the first half of the twentieth century as "mass media tool[s] of European [and American] imperial projects" that "helped shape, enforce, and naturalize the relationships between hegemonizing groups and their 'dominated' others" (Ponzanesi & Waller, 2012, p. 1). Rethinking many of the popular films of early cinema, late silent cinema and classical sound cinema from a postcolonial perspective, these new film histories not only revised old film histories but also old colonial histories. The body of films that has been – and still has to be – rethought became coined as *empire cinema*. Already in the early 1970s, Richards (1973) defined empire cinema as "films which detail the attitudes, ideals and myths of British Imperialism" (p. 2). In the following years, the definition of empire cinema got extended to include the national cinemas of other empires, notably the United States, which emerged as an imperialist power by the early twentieth century. At present, empire cinema is most closely associated with the 1930s and 1940s British and American films "promoting ideologies of popular imperialism" (Chapman, 2006, p. 814). As Chowdhry (2000) notes, empire cinema is "a term now accepted for both the British as well as Hollywood cinema made mainly during the 1930s and 1940s, which projected a certain vision of the empire in relation to its subjects" (p. 1). Indeed, with the proliferation of the imperial adventure genre in the early sound period, "the heyday of the classical cinema of Empire was undoubtedly the 1930s" (Chapman, 2006, p. 814). However, already prior to this period, popular cinema proved itself to be a significant cinema of empire. Over the years, scholars of early cinema have carried out important work to better understand the imperial workings of film production, distribution and exhibition at the end of the nineteenth and the beginning of the twentieth century (see e.g., Burns, 2013; Kirby, 1997; Grieveson & MacCabe, 2011; Grieveson, 2018; Peterson, 2013).

4 Despite the numerous copies that existed during its theatrical run, no surviving copies of *A Daughter of the Gods* are known, making it one of the most important lost films of the silent era.

5 Although *A Daughter of the Gods* is widely recognised as the first film production in the world with a budget over one million US dollars, there is some debate about the size and composition of the film's budget. Several sources state that the "million dollars of expenditure" also included "the cost of advertisement" and the actual production costs were about US$850,000 (*JG*, 27 April 1916). Other sources suggest that the budget was merely used for promotional purposes, that is, to bill *A Daughter of the Gods* as "William Fox's million dollar picture" (*Variety*, 20 January 1917).

6 *Neptune's Daughter* marked the first of several popular feature-length aquatic fantasy pictures starring Australian swimming champion Annette Kellerman, who was often described in the press as "the most beautifully formed woman in the world" (*JG*, 31 August 1915). In these pictures, Kellerman controversially pushed the boundaries of

nudity in silent cinema. With her performance in *A Daughter of the Gods*, Kellerman allegedly became the first female Hollywood star to appear nude on the silver screen.

7 *A Daughter of the Gods* reportedly earned well over US$1 million at the domestic box office, which allegedly made it "the greatest box office movie ever made" (*The New York Clipper*, 20 December 1916). Thus, despite the high production costs, the early blockbuster film already made back its million dollar budget through its domestic run only. The film's worldwide box-office gross is unknown, but newspaper reports from other countries suggest that *A Daughter of the Gods* was a box office hit abroad as well.

References

Ahmed, S. (2000). *Strange encounters: Embodied others in post-coloniality*. London: Routledge Books.

Altman, R. (2004). *Silent film sound*. New York: Columbia University Press.

Biltereyst, D., & Meers, P. (2006). New cinema history and the comparative mode: Reflections on comparing historical cinema cultures. *Alphaville: Journal of film and screen media, 11*, p. 13–32.

Black, C. (1991). *The history of Jamaica*. Second edition. Essex, Kingston: Longman Caribbean.

Bruner, E. (1989). Of cannibals, tourists, and ethnographers. *Cultural Anthropology, 4*(4), p. 438–445.

Bruno, G. (2002). *Atlas of emotion: Journeys in art, architecture and film*. London: Verso.

Burns, J. (2013). *Cinema and society in the British empire, 1895–1940*. New York: Palgrave Macmillan.

Cham, M. (1992). Introduction: Shape and shaping of Caribbean cinema. In *EX-ILES: Essay on Caribbean cinema*. Trenton: Africa World Press, p. 1–43.

Chapman, J. (2006). Cinemas of empire. *History Compass, 4*(5), p. 814–819.

Chowdhry, P. (2000). *Colonial India and the making of empire cinema: Image, ideology and identity*. Manchester: Manchester University Press.

Crick, M. (1989). Representations of international tourism in the social sciences: Sun, sex, sights, savings, and servility. *Annual Review of Anthropology, 18*, p. 307–344.

Davis, P. (1996). *Darkest Hollywood: Exploring the jungles of cinema's South Africa*. Athens: Ohio University Press.

Denzin, N. (2002). *Reading race: Hollywood and the cinema of racial violence*. London: SAGE Publications.

Erdman, A. (2004). *Blue vaudeville: Sex, morals and the mass marketing of amusement, 1895–1915*. Jefferson: McFarland.

Fellow, A. (2013). *American media history*. Boston: Wadsworth.

Gibson, E., & Firth, B. (2005). *The original million dollar mermaid: The Annette Kellerman story*. Crows Nest: Allen & Unwin.

Greene, B. (2018). *Book and camera: A critical history of witches in American film and television*. Jefferson, NC: McFarland.

Grieveson, L. (2018). *Cinema and the wealth of nations: Media, capital, and the liberal world system*. Oakland: University of California Press.

Grieveson, L., & MacCabe, C. (Eds.). (2011). *Empire and film*. London: British Film Institute.

Griffiths, A. (2002). Wondrous difference: Cinema, anthropology, and the turn-of-the-century visual culture. New York and Chichester: Columbia University Press.

Gunning, T. (1996). 'Now you see it, now you don't': The temporality of the cinema of attractions. In Abel, R. (Ed.), *Silent film*. London: Athlone Press, p. 71–84.

Hambuch, D. (2015). Caribbean cinema now: Introduction. *Imaginations: Journal of Cross-Cultural Image Studies, 6*(2), p. 4–9.

Haswell, J., Haswell, R., & Blalock, G. (2009). Hospitality in college composition courses. *College Composition and Communication, 60*(4), p. 707–727

Hudson, B. (2001). *The waterfalls of Jamaica: Sublime and beautiful objects*. Kingston, Jamaica: The University of the West Indies Press.

Jutel, T. (2004). *Lord of the Rings*: Landscape, transformation, and the geography of the virtual. In Bell, C., & Matthewman, S. (Eds.), *Cultural studies in Aotearoa New Zealand: Identity, space and place*. Oxford: Oxford University Press, p. 54–65.

Kirby, L. (1997). *Parallel tracks: The railroad and silent cinema*. Durham, NC: Duke University Press.

Koszarski, R. (1994). *An evening's entertainment: The age of the silent feature picture, 1915–1928*. Berkeley: University of California Press.

Leotta, A. (2011). *Touring the screen: Tourism and New Zealand film geographies*. Bristol: Intellect.

Maltby, R. (2011). New cinema histories. In Maltby, R., Biltereyst, D., & Meers, P. (Eds.), *Explorations in new cinema histories: Approaches and case studies*. Malden: Wiley-Blackwell, p. 3–40.

Martens, E. (2018). The history of film and tourism in Jamaica. In Bandau, A., Brüske, A., & Ueckmann, N. (Eds.), *Reshaping glocal dynamics of the Caribbean: Relations and disconnections*. Heidelberg: Heidelberg University Publishing, p. 193–215.

Musser, C. (2004). Historiographic method and the study of early cinema. *Cinema Journal, 44*(1), pp. 105.

Perez, L. (1974). Aspects of underdevelopment in the West Indies. *Science and Society, 37*(4), p. 473–480.

Peterson, J. L. (2005). Travelogues. In Abel, R. (Ed.), *Encyclopedia of early cinema*. Abingdon and New York: Routledge, p. 639–643.

———. (2013). *Education in the school of dreams: Travelogues and early nonfiction film*. Durham, NC: Duke University Press.

Ponzanesi, S., & Waller, M. (2012). Introduction. In Ponzanesi, S., & Waller, M. (Eds.), *Postcolonial cinema studies*. London: Routledge, p. 1–16.

Popple, S., & Kember, J. (2004). *Early cinema: From factory gate to dream factory*. London: Wallflower Press.

Quilley, G., & Kriz, K. D. (2003). Introduction: Visual culture and the Atlantic World, 1660–1830. In Geoff Quilley and Kay Dian Kriz (Eds.), *An economy of colour: Visual culture and the Atlantic World, 1660–1830*. Manchester: Manchester University Press, p. 1–14.

Richards, J. (1973). *Visions of yesterday*. London: Routledge & Kegan Paul.

Ruoff, J. (2006). *Virtual voyages: Cinema and travel*. Durham, NC: Duke University Press.

Rush, A. S. (2011). *Bonds of empire: West Indians and Britishness from Victoria to decolonization*. New York: Oxford University Press.

Sheller, M. (2003). *Consuming the Caribbean: From Arawaks to zombies*. London: Routledge.

———. (2004). Natural hedonism: The invention of Caribbean islands as tropical playgrounds. In D. T. Duval, (Ed.), *Tourism in the Caribbean: Trends, development, prospects*. London: Routledge, p. 23–38.

Shohat, E., & Stam, R. (1994). *Unthinking Eurocentrism: Multiculturalism and the media*. Reprinted edition (2002). London: Routledge.

Strachan, I. (2002). *Paradise and plantation: Tourism and culture in the Anglophone Caribbean*. Charlottesville: University of Virginia Press.

Thompson, K. (2006). *An eye for the tropics: Tourism, photography, and the framing of the Caribbean picturesque*. Durham, NC: Duke University Press.

Tucker, H., & Akama, J. (2020). Tourism as postcolonialism. In T. Jamal and M. Robinson (Eds.), *The SAGE handbook of tourism studies*. London: SAGE Publications, p. 504–520.

Ward, S., & O'Regan, T. (2009). The film producer as the long-stay business tourist: Rethinking film and tourism from a Gold Coast perspective. *Tourism Geographies: An International Journal of Tourism Space, Place and Environment, 11*(2), p. 214–232.

2 More than just Home of Middle Earth

The History of Film Tourism in Aotearoa New Zealand

Alfio Leotta

Introduction

This chapter examines the evolution of film tourism in Aotearoa New Zealand (NZ) and demonstrates that its prominence is the result of the country's remote geographic location and colonial history. Following the release of Peter Jackson's *The Lord of the Rings* (*LOTR*) trilogy in the early 2000s, New Zealand became one of the most popular film tourism destinations in the world. A few years later, *The Hobbit* trilogy (2012–2014) reignited interest in the 'New Zealand, Home of Middle Earth' tourism campaign. Hobbiton, a major set in both trilogies, has since become one of the most visited tourist attractions in the country. While many studies have explored the characteristics of so-called Middle Earth tourism and its impact on the country's economy, society and identity (e.g. Tzanelli 2004; Roesch 2009; Leotta 2011), this chapter considers the broader relationship between New Zealand, film and tourism, arguing that the versatility of the New Zealand landscape, as well as the settler construction of this landscape as "transposable otherness" (Jutel 2004: 55), paved the way to its branding as Home of Middle Earth. While certain stakeholders and sectors of New Zealand society have embraced film tourism and the economic benefits associated with it, various scholars and commentators, particularly from an Indigenous Māori viewpoint (Te Punga Somerville 2009), have criticised the cultural consequences of the country's conflation with fantasy lands such as Middle Earth.

This chapter will deploy a historiographical approach to map the relationship between film and tourism in New Zealand. The first part of the chapter will examine the way in which European artistic conventions of the nineteenth century, particularly the sublime and the picturesque, shaped certain representations of the New Zealand landscape, which in turn informed the country's tourism imagery. The second part will explore how the government's involvement in the development of the film industry from the 1970s and particularly the 1990s, with the international success of Jane Campion's *The Piano* (1993), was informed by an attempt to leverage feature films to put the country on the global map. Finally, the chapter will discuss the enduring legacy of film tourism in New Zealand in the twenty-first century. Since the early 2000s, discourses about the economic potential of such

DOI: 10.4324/9781003320586-4

tourism had a significant influence on local policymaking. From this point of view, the recent efforts of the New Zealand government to attract and retain major international productions such as Amazon's *Rings of Power* (2022) or the *Avatar* sequel (2022), should be understood, at least in part, as an attempt to replicate the success of *LOTR* tourism. The cultural consequences of film tourism, however, have been contested, as the superimposition of new layers of meaning onto Aotearoa New Zealand threatens Indigenous ways of understanding and interacting with the land.

The Sublime and the Picturesque: The Early History of Film and Tourism in New Zealand

The histories of film and tourism are deeply interconnected, since both cultural activities provide different, but overlapping answers to the modern desire for temporal and spatial mobility (Leotta 2011). One of the most evident manifestations of the relationship between film and tourism is represented by the phenomenon of film-induced tourism. Based on Beeton's definition of film-induced tourism as "visitation to sites where movies and TV programmes have been filmed as well as to tours to production studios, including film-related theme parks" (2005: 11), film tourism could be more loosely defined as the tourist activity associated with the screen industries. Recently, an increasing number of studies have examined the actual benefits and effects of film tourism. Most of the literature has focused on the increase in visitor numbers to destinations depicted in films and TV series (e.g. Riley and Van Doren 1992; Tooke and Baker 1996; Hudson and Ritchie 2006).

The close interconnection between film and tourism can be traced back to early cinema's obsession with exotic views and travel images (see Chapter 1 by Martens). More specifically, such relationship is better understood in the context of the Western industrial and colonial expansion in the nineteenth and twentieth centuries (Gunning 2006). The travel images that were so popular at the turn of the twentieth century brought distant places closer and made them ready for consumption. Western photographers and spectators took possession of the world through pictures of it, while colonialism played an essential role in the commodification of 'foreign views'. Locating the social origin of the photographic medium within the reproduction of certain forms of power, Batchen argues that "power inhabits the very grain of photography's existence as a modern Western event" (1997: 202). The complex relation between visual media, travel and colonialism has been particularly important to the history of New Zealand cinema as it played a crucial role in shaping local film aesthetics. Shortly after the emergence of cinema, many early European film pioneers visited New Zealand and the wider South Pacific to capture images of remote and exotic locations for Western audiences. European production companies such as Pathé Freres sent camera operators to Aotearoa to produce films that capitalised on the spectacle offered by the country's natural landscape (Babington 2007). Similarly, in the 1920s and 1930s, a number of international directors chose New Zealand to make films based on Māori legends (Blythe 1994; Leotta 2011). Such films, usually conceived for international audiences, often featured shots of scenic attractions and scenes of Māori life.

Due to the limited market and expensive cost of film technology, in the first half of the twentieth century, most New Zealand films were produced or funded by state agencies. The New Zealand governments of this period regarded film mainly as a tool of national publicity. From this point of view, New Zealand was a pioneer in the use of film as a means for tourism promotion. Since 1917, moving pictures were produced by the Department of Tourism and Health Resources and mainly targeted tourists from the United States, UK and other British colonies (NZOYB 1976). By 1930, the Government Publicity Office, responsible for government filmmaking activities, was transferred from the Internal Affairs Department to the Tourism Department (Dennis 1981: 8). More importantly, the use of film for tourism promotion by early New Zealand governments was underpinned by a deeply rooted settler culture, which framed the landscape within established European conventions (Pound 1983). In his work on early New Zealand landscape painters, Francis Pound demystifies the myth of an unmediated representation of New Zealand nature, claiming that access to the pure and original land is impossible (Pound 1983). The very act of seeing is an act of possession and displays an unequal relationship of power between the seer and the seen. Early European settlers were confronted with an alien environment which they represented and tamed using familiar artistic conventions of the time.

More specifically, the New Zealand landscape was framed by the canons of the sublime and the picturesque that in turn fed the local tourism industry. As defined by Burke, the sublime is that which excites the emotion of self-preservation (1998: 79). The favourite object of the sublime is nature, and particularly mountains, oceans, sky and storms, while its major qualities are terror, darkness, the superior power of nature and infinity. The aim of the sublime is to produce a sort of pleasurable vertigo and delightful horror reflecting man's sensation of smallness in the face of nature's majesty. Andrews claims that since the end of the eighteenth-century British travellers stormed the countryside, and indeed the whole of Europe, seeking picturesque and sublime sites. He states that the peculiar nature of 'picturesque tourism' was "to find scenery which resembled familiar paintings or poetic descriptions" (1989: 76). Picturesque tourism implies a tension between the search for new and thrilling experiences and the familiarity of traditional artistic practices. The aesthetic experience of the picturesque and the sublime resonates with the dialectic between risk and safety that characterises tourism. Early New Zealand travel films, produced to promote the country to both tourists and settlers, inherited the traditional canons of visual art and established a blueprint that informed subsequent cinema produced in Aotearoa.

The establishment of the Publicity Office in 1921 corresponded with the commencement of regular government film production, mainly scenic short documentary pieces and travelogues. During the mid-1920s, the government approach to cinema began to be marked by the creation of national propaganda. The transfer in June 1930 of the Publicity Office from the Department of Internal Affairs to the Department of Tourism and Health Resorts clearly indicated the establishment of a stricter relationship between national propaganda and the country's tourist promotion. As Dennis points out, during the 1920s, the directive to the

Publicity Office was to empty the films of people in order to prevent "them from being dated by changes in fashion" (1993: 9). The depopulated narratives of the travelogues produced by the Publicity Office allowed only the representation of empty landscapes, occasional tourists or timeless Māoriland. In this sense, scenic views of the fjords or the Southern Alps were timeless and literally interchangeable. For example, the very same footage of alpine landscape was used in several tourism films of this period including *Happy Altitudes in New Zealand Southern Alps* (1933), *Romantic New Zealand* (1934) and *New Zealand Charm: A Romantic Outpost of the Empire* (1935).

The focus on empty landscapes, however, served another important purpose, namely the idea that the natural and social environment was prone to the colonial enterprise. New Zealand landscape, in particular, was represented as raw material that could be materially and imaginatively processed and consumed by the settler, tourist or spectator. New Zealand tourist imagery of the early twentieth century was underpinned by the government's attempt to link tourism and immigration. In particular, the mythology of a scenic wonderland and the depiction of an untouched 'Māoriland' were deployed to attract potential settlement by what a 1930s New Zealand Trade Commissioner defined as "goodtype Europeans" (Taylor in Hillyer 1997: 15). The projection of a favourable image of New Zealand constructed for both tourism promotion but also to attract immigration and further trade was continued by the National Film Unit (NFU), a production company established by the government in 1941.

The representation of untouched rural landscapes implicitly suggested that the country was a good terrain for the colonial enterprise. In turn, the national reception of travel films made in New Zealand contributed to forging the national myth of the country as a scenic wonderland. Referring to the role played by influential travel films such as *Glorious New Zealand* (1925) and *Romantic New Zealand* (1934) in shaping national identity, Hillyer points out that "our institutional memory is created through a series of dehistoricised present moments – a view, a product, a holiday 'snapshot'" (1997: 20). The few white (Pākehā) New Zealanders depicted in *Romantic New Zealand*, for example, are presented as either responsible for the civilising improvements made to the land, or at play on ski slopes and rivers. They first tamed the land, making it a safe and pleasurable tourist playground that they can now finally enjoy. Again, in the words of Hillyer: "the Pākehā of Romantic New Zealand are both eternal tourists and themselves proud pioneers" (1997: 18).

Landscape for Export: *The Piano* as 100% Pure New Zealand

In the 1970s, a number of New Zealand filmmakers and intellectuals began lobbying the government for a state-funded institution responsible for supporting the production of local feature films (Shelton 2005). Convinced by the local and international success of films such as *Sleeping Dogs* and *Off the Edge*, the national government established in 1978 the New Zealand Film Commission (NZFC), which had the task of funding and supporting local feature film productions having "significant New Zealand content" (Shelton 2005: 24). In their influential documentary about New Zealand cinema, *Cinema of Unease* (1995), Sam Neill and Judy Rymer argued that the feature films produced in Aotearoa between the 1970s and 1990s were characterised by the centrality of the rural landscape. Similarly, several critics have stressed the structural importance of landscape in New Zealand feature films. As Bob Harvey, the Mayor of Waitakere City, puts it, "for many years New Zealand film production was without major facilities and studios were unknown. Sets were difficult, so location was everything, both an asset and a challenge" (Harvey and Bridge 2005: 17). Horrocks even argues that "in almost all New Zealand films the physical landscape makes its presence strongly felt not only as scenic background, but as an influence shaping the lives of the characters" (1989: 102). Lindsay Shelton, the first film commissioner responsible for the marketing of films funded by the NZFC, claimed that in the early years of the commission, promoting New Zealand films often meant promoting the country as a whole (Shelton 2005).

At a time in which many international viewers only had a vague knowledge about Aotearoa, its people or even its geographical location, New Zealand films of the 1980s, 1990s and 2000s contributed to put the country on the global map. *The Piano* (1993) was the first New Zealand film to achieve worldwide popularity, gaining a vast success both among the public and critics, winning a Palme d'Or, three Oscars and several other prestigious international awards. The panoramic views of Karekare beach, the most prominent location of *The Piano*, became the distinctive mark of New Zealand for film viewers all around the world. According to Wevers, the landscape in the film functioned as "an authenticating context for a narrative which encoded 'New Zealand' to foreign audiences" (1994: 1). Viewers were struck by the uniqueness of the dark West Coast beaches and the dense, sub-tropical bush. The distinctiveness that characterised the film locations did not alienate Western viewers as it was framed within the familiar Western canon of the sublime. According to Shelton, even before the Oscars, the film was creating interest in New Zealand with particular benefits for tourism and export (Shelton in NZPA 1994: 24). After the Oscar success, the Trade Development Board and the NZFC used stills from *The Piano* in a brochure about international invest-ment in New Zealand. Similarly, in 2001, increasingly aware of the global profile films achieve and create, Tourism New Zealand used an image inspired by *The Piano* in its international advertising campaign '100% Pure New Zealand'. The representation of New Zealand that emerges from *The Piano* conforms, in fact, to the marketing strategy adopted by the national tourism board and in particular its focus on landscape. Indeed, Tourism New Zealand has targeted the "interactive traveler" looking for "new experiences that involve engagement and interaction" and demonstrating "respect for the natural, social and cultural environment" (TNZ 2004: 1).

However, *The Piano* did not merely impact upon tourism and tourist images of the film locations. The representation of a landscape 'for export' has deeply influenced New Zealand's cultural self-image, particularly in subsequent local features. Films such as Lee Tamahori's *Once Were Warriors* (1994), Harry Sinclair's *Topless Women Talk About Their Lives* (1997) and Niki Caro's *Memory and Desire* (1999) implicitly or explicitly commented on the role of film and tourism in the process of landscape commodification. These films acknowledged the global resonance of the image of New Zealand landscape constructed by *The Piano* and other national films. The opening sequence of *Once Were Warriors*, for example, shows an idyllic New Zealand country landscape that fills the frame. However, as the camera gradually zooms back, it reveals a gritty urban setting, within which the rural landscape appears only on an advertising billboard. The opening of the film does not simply oppose the tourist imagery of the country with the harsh city life of South Auckland; it is also an ironic reference to the plethora of New Zealand films that begin with establishing shots of the country's natural beauty (Martens 2006: 75; 2012: 11; Turner 1999: 131; McDonald 1998: 81).

The Tourist Spin-Off of *Lord of the Rings*: New Zealand's Conflation with Middle Earth

Since the launch of the '100% Pure New Zealand' campaign in 1999, Tourism New Zealand has tried to capitalise on the possibilities of non-conventional publicity tools, particularly filmed tourism. As Croy points out, "this image building and promotion process effectively utilises limited financial resources by using other groups' resources to provide the images and then creating association to New Zealand" (2004: 7). Research commissioned by Tourism New Zealand at the beginning of the 2000s identified the country as rich in four assets: landscape, people, adventure and culture (Morgan et al. 2003: 292). The tourist authorities consequently designed a new promotional strategy that positioned New Zealand as "an adventurous new land and an adventurous new culture on the edge of the Pacific Ocean" (Piggott in Morgan et al. 2003: 292). The essence of the New Zealand brand, as conceived by TNZ, is landscape, and in particular a landscape imbued with sophisticated, innovative and spirited values, which allows tourists to express themselves through activities and experiences. Landscape plays a crucial role in tourism as a function of commodification, which orientates space towards the selling of tourist destinations and experiences.

Landscape was one of the major selling points of *LOTR*. The films used some of the country's most famous tourist sites, such as the Southern Alps, Queenstown and Mount Ruapehu, and producer Barry Osborne admitted that "throughout, we picked the most spectacular appropriate locations we could find" (cited in Duncan 2002: 101).[1] According to Butler, Jackson's virtuoso camera movements, which represent the key aesthetic component of the *LOTR* trilogy, reinforced the idea of an omnipotent observer while highlighting the scenic beauty of the country (2007: 162). The trilogy provided New Zealand with an unprecedented level of international exposure and led to serendipitous opportunities for the local economy in the form of significant flows of film tourism. Since the release of the *LOTR* trilogy, New Zealand has become one of the most popular film tourism destinations in the world. A 2002 study commissioned by Tourism New Zealand revealed that 95% of the international visitors to the country were aware that *LOTR* had been filmed there, and that for 9% of the respondents *LOTR* was one the reasons that prompted their visit (NFO NZ cited in Croy 2004; see also Statistics New Zealand in Leotta 2011; Bowler 2012). The *LOTR* tourism spin-off led to dozens of local tour operators around the country offering Middle Earth related products, some of which specifically targeting hard-core fans and foreign tourists. Tourism New Zealand and Air New Zealand played a crucial role in the launch of the 'New Zealand, Home of Middle Earth' campaign, creating a wealth of promotional material based on the two 'Ring franchises'. New Zealand incorporated *LOTR* into its national identity to the extent that the films have become an acknowledged component of the country's heritage.

The association between New Zealand and Middle Earth seemingly benefited from the alignment between the representation of landscape in the *LOTR* movies and the pre-existing tourism marketing strategies deployed by the local authorities. Tourism New Zealand's promotional strategy seamlessly converged with the representation of the natural environment in the *LOTR* films. One of the main challenges faced by Frodo and Bilbo in their journeys is the interaction with a landscape that plays an active role in narrative terms. The conflation of the fantasy land with the real one was also predicated on the pre-existing cinematic and touristic imagery of the country. As Smith-Rowsey points out, "in Jackson's films the presumption of familiarity with medieval Europe is counter-balanced, to a significant extent, by a presumption of unfamiliarity with New Zealand itself" (2007: 142). The alignment between New Zealand's tourism imagery and *LOTR* was further strengthened by the development of co-marketing strategies. Fans' interest in the films' locations led to New Line's awareness of the great potential of developing a marketing strategy that would involve local tourist stakeholders. The government of the time was responsive to this idea and it allocated millions of dollars to leverage the association between the country and *LOTR*. Pete Hodgson was appointed as associate minister to manage the involvement of the New Zealand government in *LOTR*-related activities. Hodgson was quickly nicknamed 'The Minister of the *Lord of the Rings*', a designation that demonstrates the importance of the project in the governmental agenda (Thompson 2007: 311).

Invoking Baudrillardian notions of hyper-reality and simulation, the conflation between New Zealand and Middle Earth has attracted the interest of many cultural theorists who explored the power dynamics at play in this convergence. Jutel, for example, points out how in *LOTR* New Zealand becomes a "space of radical dépaysement" (2004: 60). Jutel translates dépaysement as 'out-of-nationess' claiming that this notion "represents the simultaneous attraction of geographic defamiliarisation, and the separation of the lost homeland" (2004: 60). Building upon Jutel, Kavka and Turner (2009) argue that the use of New Zealand landscape as a transposable otherness for export should be explained by the logic of the settler franchise, which in turn is a legacy of the colonial history of the country. More specifically, the *LOTR* production might be taken as a cinematic synonym for the political logic of settler societies: "For the enterprise of settler New Zealand to succeed it has had to remove the place from its peoples and history" (Kavka and Turner 2009: 231) – not unlike the production of the film itself. Similarly, referring to international fantasy productions made in Aotearoa, Perry inverts the commonplace of a 'unique New Zealand landscape', arguing that New Zealand is the perfect location for such productions as "a low wage country that otherwise exists only as just such a scenically diverse, but temporally and spatially indeterminate, fiction" (1998: 106). By removing the place from its history and people, *LOTR* enhanced New Zealand international fame as exporter of transposable exotic settings. From this point of view, the versatility of the New Zealand landscape could be read as a legacy of the country's colonial past. While in the nineteenth century, New Zealand's main economic and cultural function was the production of raw material to be produced and consumed elsewhere, in the twentieth century and into the twenty-first century, New Zealand is still producing raw material, exotic views and transposable landscapes for the centre of the empire, now represented by Global Hollywood.

In the wake of the impact of *LOTR* on New Zealand tourism, a number of local stakeholders tried to capitalise on the potential tourist spin-off associated with films shot in New Zealand. *Whale Rider* (2002), directed by Niki Caro and based on the book of the same name by Witi Ihimaera, generated international awareness of, and interest in, New Zealand's scenic beauty. Witi Ihimaera himself, who was also an executive producer of the film, claimed that he hoped that the film would make people want to go to the places where the story was filmed (Matthews 2003: 20). Tourism New Zealand was responsible for funding some of the costs of the American premieres and Tourism Minister Mark Burton himself was present at the pre-release screening for the US travel media. Shortly after the release of the film, Whale Rider Tours was established by some locals in the Gisborne region, where the film was shot. Similarly, *The Last Samurai* (2003) a Hollywood production shot in New Zealand yet set in Japan, generated hopes of a '*LOTR*-effect' in the Taranaki region, where the movie was filmed. Once again Tourism New Zealand attempted to capitalise on the international success of the film by promoting the film locations, while locals offered to the film's main location, the Samurai Village. Shortly after the release of *Whale Rider* and *The Last Samurai*, the locations featured in both movies reportedly benefited from a brief surge in tourism interest. However, the long-term effect of these films on New Zealand tourism seems to have been negligible (TNZ 2006).

The release of *The Hobbit* trilogy between 2012 and 2014, however, renewed international interest in Middle Earth tourism. New Zealand's Ministry of Business, Innovation and Employment estimated that international spending in the Matamata district, the home of Hobbiton, more than tripled (Kloeten 2014). Unlike the tourism associated with *Whale Rider* and *The Last Samurai*, Middle Earth tourism was able to stand the test of time. Hobbiton, which was rebuilt using materials for *The Hobbit* films, has become one of the most popular tourist destinations in the country, attracting a gradually increasing number of tourists and reaching a record of nearly 650,000 tourists in 2019, the year before the emergence of COVID-19. It remains difficult to pinpoint a single factor to explain the enormous popularity of Middle Earth related film tourism. It has been argued that the style and narrative of Jackson's films, characterised by an aesthetic of spectacle, a photorealistic representation of Middle Earth and an emphasis on maps and mobility might have been responsible for activating the viewers' tourism imagination. At the same time,

The *LOTR* and *The Hobbit* films undoubtedly benefited from their worldwide distribution and marketing: the two trilogies were watched by hundreds of millions of viewers, grossing more than US$3 billion each. Apart from the global popularity of Tolkien's novels, which had already led to the development of vast fan communities, the success of both franchises could be explained by the significant investment by the producers. Finally, the partnership between the film producers and the New Zealand government was renewed for the release of *The Hobbit* films and led to even more extensive co-marketing opportunities, including the premiere of *An Unexpected Journey*, the redevelopment of Hobbiton and the launch of various promotional videos and fan competitions.

However, tourism associated with *LOTR* and *The Hobbit* was arguably also the result of New Zealand's geographical, social and cultural predisposition. For many viewers before these films, New Zealand was a place as imaginary as Middle Earth. Epithets like 'down under' or the 'edge of the world' conveyed the idea of a land kept pristine and unspoiled due to its distance from the First World. Wright argues that New Zealand, a country that can be seen as remote and at the very edge of the earth, has an affinity with the realms of imagination, an affinity that aligns it particularly with the fantasy vision of J.R.R. Tolkien (Wright 2000: 52). Barker and Mathijs noticed that, for most overseas *LOTR* audiences, the New Zealand depicted in films is an abstract, distanced and ideal location, which naturally embodies goodness and purity (2007: 108). For these viewers, New Zealand represented a perfect destination for a pilgrimage of self-rediscovery. As such, New Zealand as Middle Earth, inheriting the colonial fantasy of the settler culture, also provides metropolitan audiences with an exotic 'there'. There is an eerie resonance between New Zealand as Middle Earth and the colonial rhetoric of New Zealand as 'Britain of the South Seas'. New Zealand has been culturally constructed as a rural idyll and as an extension of the British countryside. Both the 'New Zealand Home of Middle Earth' campaign and the colonial construction of the country as a Britain of the South Seas conceived and defined New Zealand as what Foucault (1986) would call a 'heterotopia' – a place which is simultaneously present and absent. The title of Bilbo's book, *There and Back Again*, that features at the beginning of *The Hobbit* trilogy, is in this respect particularly significant. The films were constructed as a travel narrative whose destination is a utopian elsewhere at the border between fantasy and reality.

Concluding Remarks

The myth of Aotearoa New Zealand as a scenic wonderland continues to shape New Zealand's national identity and to influence the creative vision of home-grown filmmakers. The visual treatment of New Zealand's landscape in the films directed by contemporary filmmakers such as Jane Campion, Peter Jackson and Taika Waititi is imbued with the aesthetic of the picturesque and the sublime that characterises New Zealand's cinematic tradition. The post-2000 conflation between Middle Earth and New Zealand has seemingly been beneficial to various stakeholders both within the tourism and film industries. First, it contributed to redefining the international reputation of New Zealand as a film-friendly destination. By removing the place from its history and people, *LOTR* enhanced New Zealand's international fame as exporter of transposable exotic settings, such as standing in for Japan in *The Last Samurai*. Since *LOTR*, international filmmakers have made use of the companies and facilities owned by Peter Jackson, contributing to their financial profitability and technological development.[2] However, the Middle Earth films could also be seen as accounts undergirded by colonial histories and contemporary fantasies and anxieties about race relations. A number of scholars have pointed out how the conflation between New Zealand and Middle Earth constructs an imaginative geography in which the importance of Māori culture is diminished and partly silenced. More specifically, the toponymical identification of certain New Zealand locations with fictional Middle Earth places such as Hobbiton or Rivendell risks erasing Indigenous history and understanding of the land (Te Punga Somerville 2009).[3]

Since the production of *The Hobbit* trilogy, hopes for a boost of film tourism in New Zealand have mainly been focused on two large international productions, James Cameron's *Avatar* film franchise and Amazon's *LOTR* television spin-off, *Rings of Power*. In 2013, Cameron and 20th Century Fox, the studio that owns the *Avatar* franchise, signed a Memorandum of Understanding with the New Zealand government, which entitled the filmmakers to a 25% rebate on qualifying expenditure – 5% more than the standard amount offered to large film productions. The government offered this extra incentive at the condition that the producers would develop a marketing plan to promote "New Zealand as 'the place where Pandora was brought to life' or similar, with a view to maximising New Zealand's international brand" (Beehive 2013). Since the signing of the MOU, Cameron featured in Tourism New Zealand's promotional material; however, the release of *Avatar*'s first sequel *The Way of the Water* (2022) failed to bring significant returns to tourism. Similarly, the New Zealand government offered Amazon an estimated US$100 million incentive to film *Rings of Power* in New Zealand (RNZ 2022). Tourism New Zealand played a crucial role in the negotiations because of the perceived importance of maintaining and reinforcing New Zealand's international brand as 'Home of the Lord of the Rings'. Despite the strong investment of the government, however, the series failed to provide New Zealand the same international visibility it gained by Peter Jackson's films, possibly in part due to the criticism surrounding its alleged lack of fidelity to Tolkien's original texts. Furthermore, in 2022 Amazon announced that the production of the TV series would be moved to the UK. Amazon's decision was met by negative reactions in New Zealand, where many commentators expressed concern about the impact of the move on Middle Earth Tourism.

In the early 2020s, the future of film tourism in New Zealand seems uncertain. More broadly, the cultural identity of the country remains fluid and contested. Since the late 1990s, international productions such as *LOTR*, but also *Xena: Warrior Princess*, *The Last Samurai* and *Avatar*, transformed New Zealand into what Jutel calls a 'transposable otherness', a blank canvas fit to suggest a variety of locations and historical time periods. Future developments in film tourism in New Zealand might provide an indication of the way in which the country's cultural identity will develop in the twenty-first century. It is to be seen how the growing prominence of Indigenous culture as well as an increasing awareness of the uniqueness of the country's history might impact New Zealand tendency to rely on texts and narratives conceived elsewhere to define its own (film) heritage and identity.

Notes

1 In some cases, the reference to government-produced travel films is explicit. The opening sequence of *LOTR: The Two Towers* (2002), for example, presents a spectacular aerial view of the Southern Alps, also known as the Misty Mountains. Set against the epic soundtrack of the film, the sequence bears a striking resemblance to the final scene of *This Is New Zealand*, a three-screen film produced by the NFU, which Jackson watched in 1971 as a ten-year-old boy (MacDonald 2014). The last scene of the film, which features a bird's-eye view of the Southern Alps accompanied by the climatic build of Sibelius' *Karelia Suite*, became the most famous part of this travelogue. Like *This Is New Zealand*, the opening sequence of the *Two Towers* emphasised the sublime 'naturalness' of the local landscape contributing to increase the international media exposure of the country.
2 The significant economic impact of *LOTR* tourism has arguably also enhanced Jackson's political leverage in New Zealand. During the 2010 Hobbit labour dispute that opposed the Film Producers and the Actors Equity Union, the government's decision to side with Jackson was influenced, among other things, by the potential economic return of *The Hobbit*'s tourist spin-offs. In order to retain Jackson's production in New Zealand, the government offered an increased US$ 20 million tax concession and US$ 13 million to offset marketing costs conditional on marketing New Zealand by developing various promotional material.
3 In addition, it has been argued that the orcs and the Uruk hai, most of whom were played by Polynesian actors, articulated the settler dread of the savage indigenes (Kavka and Turner, 2009: 232). By contrast, the Hobbits display clothes, accents and a tendency to walk barefoot that equate them with displaced Britons or, in other words, Pākehā.

References

Babington, B. 2007. *A History of the New Zealand Fiction Feature Film*. Manchester: Manchester University Press.
Barker, M. and Mathijs, E. 2007. "Seeing the Promised Land from Afar." *How We Became Middle Earth,* edited by A. Lam and N. Oryshchuk. Zurich: Walking Tree Publisher: 107–28.
Batchen, G. 1997. *Burning with Desire: The Conception of Photography*. Cambridge, MA: MIT Press.
Beehive. 2013. "Memorandum of Understanding Relating to the Making of the New *Avatar* Films." 13 December. www.beehive.govt.nz/sites/default/files/Memorandum_of_Unders tanding_New%20Avatar_Films.pdf
Beeton, S. 2005. *Film-Induced Tourism*. Buffalo, NY: Channel View Publications.
Blythe, M. 1994. *Naming the Other: Images of the Maori in New Zealand Film and Television*. Metuchen, NJ: The Scarecrow Press.

Bowler, Kevin. 2012. "Leveraging the Benefits of Film Tourism." *Tourism New Zealand*. 12 June. www.tourismnewzealand.com/tourism-news-and-insights/tourism-insights/leverag ing-the-benefits-of-film-tourism/

Butler, D. 2007. "One Wall and No Roof Make a House: The Illusion of Space and Place in Peter Jackson's *The Lord of the Rings*." In *How We Became Middle Earth*, edited by A. Lam and N. Oryshchuk. Zurich: Walking Tree Publisher: 149–68.

Croy, W. G. 2004. "The Lord of the Rings, New Zealand, and Tourism: Image Building with Film." Working Paper Series of the Department of Management. Melbourne: Monash University.

Dennis, J. [ed.]. 1981. *The Tin Shed*. Wellington: Clive Sowry Editions.

Duncan, J. 2002. "Ringmasters." *Cinefex*, 89: 64–131.

Foucault, M. 1986. "Of Other Spaces." *Diacritics*, 16: 22–7.

Gunning, T. 2006. "The Whole World within Reach: Travel Images without Borders." In *Virtual Voyages*, edited by J. Ruoff. London: Duke University Press: 25–41.

Harvey, B. and Bridge, T. 2005. *White Cloud, Silver Screen*. Auckland: Exisle Publishing.

Hillyer, M. 1997. *We Calmly and Adventurously Go Travelling: New Zealand Film, 1925– 35*, Master's Thesis. Auckland: University of Auckland.

Horrocks, R. 1989. "The Creation of a Film Feature Industry." In *Te Ao Marama*, edited by S. Toffetti and J. Dennis. Torino: Le Nuove Muse: 100–3.

Hudson, S. and Ritchie, J. R. B. 2006. "Film Tourism and Destination Marketing: The Case of Captain Corellis Mandolin." *Journal of Vacation Marketing* 12(3): 256–68.

Jutel, T. 2004. "*Lord of the Rings*: Landscape, Transformation, and the Geography of the Virtual." In *Cultural Studies in Aotearoa*, edited by C. Bell and S. Matthewman. Auckland: Oxford University Press: 54–65.

Kavka, M. and Turner, S. 2009. "This Is Not New Zealand: An Exercise in the Political Economy of Identity." In *Studying the Event Film*, edited by H. Margolis, S. Cubitt, B. King and T. Jutel. Manchester: Manchester University Press: 230–38.

Kloeten, N. 2014. "Hobbit Triples Tourist Spend." *Stuff*. 28 November. www.stuff.co.nz/ business/industries/63638178/hobbit-movies-triple-tourist-spend

Leotta, A. 2011. *Touring the Screen: Tourism and New Zealand Film Geographies*. Bristol: Intellect.

———. 2016. *Peter Jackson*. London: Bloomsbury.

Martens, E.S. 2006. Once Were Warriors*: The Aftermath–The Controversy of* OWW *in Aotearoa New Zealand*. Amsterdam: Aksant Academic Publishers.

———. 2012. "Maori on the Silver Screen: The Evolution of Indigenous Feature Filmmaking in Aotearoa/New Zealand." *International Journal of Critical Indigenous Studies* 5(1): 2–30.

Matthews, P. 2003. "The Making of *The Whale Rider*: Turning a New Zealand Classic into an International Hit Movie." *New Zealand Listener*, 1 February, 18–25.

McDonald, L. 1998. "Film as a Battleground: Social Space, Gender Conflict and Other Issues in 'Once Were Warriors'." In *The Changing Field*, edited by P. Cleave. Palmerston North: Campus Press: 73–89.

Morgan, N. J., Pritchard, A. and Piggott, R. 2003. "Destination Branding and the Role of the Stakeholders: The Case of New Zealand." *Journal of Vacation Marketing* 9(3): 285–99.

Neill, S. and Rymer, J. 1995. *New Zealand Cinema: Cinema of Unease*. BFI Century of Cinema Series [Videorecording].

NZOYB. 1976. *Tourism: The Invisible Export. New Zealand Official Yearbook*. Wellington: Department of Statistics.

NZPA. 1994. "Piano Gives NZ World Springboard." *New Zealand Herald*. 24 March: 24.

Perry, N. 1998. *Hyperreality and Global Culture*. London: Routledge.

Pound, F. 1983. *Frames on the Land: Early Landscape Painting in New Zealand*. Auckland: Collins.

Riley, R. and Van Doren, C. S. 1992. "Movies as Tourism Promotion: A Pull Factor in a Push Location." *Tourism Management* 13: 267–74.

RNZ. 2022. "*Lord of the Rings*: Unveiling a Different Middle Earth." 23 July. www.rnz.co.nz/news/national/471482/lord-of-the-rings-unveiling-a-different-middle-earth

Roesch, S. 2009. *The Experiences of Film Location Tourists*. Bristol: Channel View Publication.

Shelton, L. 2005. *The Selling of New Zealand Movies*. Wellington: Awa Press.

Smith-Rowsey, D. 2007. "Whose Middle-Earth Is It?: Reading *The Lord of the Rings* and New Zealand's identity from a Globalized, Post-Colonial Perspective." In *How We Became Middle Earth*, edited by A. Lam and N. Oryshchuk. Zurich: Walking Tree Publisher: 129–45.

Te Punga Somerville, A. 2009. "Asking that Mountain: An Indigenous Reading of *The Lord of the Rings*?" In *Studying the Event Film*, edited by H. Margolis, S. Cubitt, B. King and T. Jutel. Manchester: Manchester University Press: 249–58.

Thompson, K. 2007. *The Frodo Franchise: The Lord of the Rings and Modern Hollywood*. Berkeley, CA: University of California Press.

TNZ. 2004. "Interactive Travellers: Who are They?" [Online]. Available: www.newzealand.com/travel/library/y42301_23.pdf [Accessed 7 July 2006].

TNZ. 2006. "*The Last Samurai* New Zealand Film Location." [Online]. Available: www.newzealand.com/travel/destinations/regions/taranaki/last-samuraifeature/last-samurai-feature-home.cfm [Accessed 12 July 2006].

Tooke, N. and Baker, M. 1996. "Seeing Is Believing: The Effect of Film on Visitor Numbers to Screened Locations." *Tourism Management* 17: 87–94.

Turner, S. 1999. "Once Were English." *Meanjin* 58(2): 122–40.

Tzanelli, R. 2004. "Constructing the 'Cinematic Tourist': The 'Sign Industry' of The Lord of the Rings." *Tourist Studies*, 4 (1): 21–42. doi:10.1177/1468797604053077

Wevers, L. 1994. "A Story of Land: Narrating Landscape in some Early New Zealand Writers or: Not the Story of a New Zealand River." *Australian and New Zealand Studies in Canada*, 11 June: 1–11.

Wright, A. 2000. "Realms of Enchantment: New Zealand Landscape as Tolkienesque." In *New Zealand: A Pastoral Paradise?*, edited by I. Conrich and D. Woods. Notthingam: Kakapo Books: 52–59.

3 Where East Becomes West

Imaginative Geographies of Eastern Europe in Popular Indian Cinema

Krzysztof Stachowiak

Introduction

Local, regional, and national film commissions are constantly trying to attract film and media investment to their locations, often intending to boost tourism as a key aspect of the new international division of cultural labour (cf. Miller et al., 2005). Places become part of a region's creative capital, attracting investors by offering organisational and financial facilitation. Meanwhile, film producers are also, or perhaps especially, interested in the on-location shooting because the unique local landscapes provide an authenticity that enhances the film's storyline. Increasingly, the importance of a location is being integrated into a broader political and economic context (interest in the location must be sustained and managed), and its attractiveness is becoming dependent on its adaptability for different creative purposes (Goldsmith & O'Regan, 2005, p. 8).

The use of foreign locations in film production has been the subject of much academic research and discussion, especially concerning Hollywood (Coe, 2000; Elmer & Gasher, 2005; Foster et al., 2015; Gasher, 2002; Scott, 2002, 2005). International locations outside the United States have been positioned in these debates as sites of economic, cultural, and technological transactions that are closely linked to Hollywood's status as a powerful global media industry. Studies of other film industries have been sporadic or piecemeal in this regard and most often refer to an analysis of how specific places and spaces are transformed through film production taking place in these locations (see, for example, Carl et al., 2007; Hubbard, 2002; Kim, 2010; Stachowiak & Stryjakiewicz, 2018; Young & Young, 2008) or through film tourism (Beeton, 2005, 2010; Riley et al., 1998). There is a lack of analysis in the literature on other major film industries in relation to the use of foreign locations, and India, being the largest in terms of the number of films produced, provides a good starting point to fill this gap.

The choice of locations for popular Indian film productions tends to be eclectic, given their narrative and storytelling styles (Hogan, 2008). These locations often include iconic monuments or landscapes that provide a picturesque backdrop to song sequences but also those regions or parts of the world that provide a sense of novelty and exoticism. These qualities made specific locations more popular with film crews than others, as they offered almost everything a blockbuster film

DOI: 10.4324/9781003320586-5

should (Willis, 2003). Many studies exist evaluating the scope and role of Indian filmmaking in Western Europe, especially in the UK (Krämer, 2016), Switzerland (Gyimóthy, 2015, 2018; Janta & Hercog, 2023), Italy (Cucco & Scaglioni, 2014), Portugal (Lourenço, 2017), Germany (Krauss, 2012), Netherlands (Nanjangud & Reijnders, 2020), Iceland (Nanjangud & Reijnders, 2022) or Finland and Sweden (Sunngren-Granlund, 2023).

However, little is known about the situation in Eastern Europe. In the second decade of the twenty-first century, Indian filmmakers have become increasingly interested in Eastern Europe. The region gained in popularity mainly at the expense of Western European countries. This was due, among other things, to lower production costs, the high level of development of Eastern European countries (no longer so different from Western Europe) or financial incentives for international film productions (Mitric, 2021; Parvulescu & Hanzlík, 2022). Kozina and Jelnikar (2023), using the case of the Central European country of Slovenia, note that the comparative advantages of the country as a destination for Indian filmmakers are affordable filming locations, a great variety of landscapes displaying beautiful natural scenery as well as cultural heritage, skilled technicians, and talented performers. However, Slovenia has only acted as a backdrop, mainly for 'song and dance' scenes. Therefore, this chapter aims to take a more detailed look at how Eastern European landscapes appear in Indian films made there.

Despite the cultural and landscape diversity of Europe, for Indian filmmakers, Eastern Europe can appear as a kind of monolith, not much different from Western Europe (Urbanc et al., 2023). In this sense, Eastern European regions represent a general notion of 'the West'. Using the concept of imaginative geographies, this chapter will take a closer look at the visuality and figurative value of Eastern European landscapes in Indian cinema. In doing so, it will examine to what extent these images are stereotypical, which could suggest reverse orientalism. This reflection will be supported by an analysis of Indian films shot in Poland, which provides empirical evidence of how Eastern European space and landscapes are used in Indian film and to what extent this use is similar to or different from previous Indian productions in Western Europe.

Mapping cinematic imaginaries: Indian film's representation of Europe and the construction of imaginative geographies

Depictions of foreign locations, encompassing both people and environments, cultures, and interpretations of the natural world, serve as expressions of the wishes, dreams, and apprehensions of those who create them. These representations also reflect the power dynamics between the creators and the subjects they depict as 'different'. Imaginative geographies, a term used to describe such depictions, refer to the mental pictures of the world and its diverse inhabitants that aid a group in establishing its sense of identity. These representations are cultural and carry significant emotional and ideological significance. In human geography, examining imaginative geographies treats these depictions with importance (Driver, 2014; Gregory, 1995). Images, influential forces in shaping individuals and perceptions of the world, also influence the world. As a result, imaginative geographies blur the boundaries between what is commonly regarded as the 'objective' reality and the domain of the 'fictional'. In essence, they are considered real not because they precisely mirror the world but because they have mirrored and strengthened people's imaginative constructs of the world in tangible and practical ways.

The Palestinian American literary critic Edward Said originally proposed the concept of 'imaginative geographies"' in his influential critique of Orientalism. Said (1978) elucidates how Western colonial powers perceived non-Western cultures, particularly the 'Orient', as unchanging and primitive. This perception served a crucial purpose, as it allowed colonial powers to establish their own identities in contradistinction to these 'others' by positioning themselves as advanced, dynamic, and sophisticated. This binary opposition between the civilised American or European and the primitive native became pivotal in colonial dominance struggles, whose echoes persist in contemporary times. Said argued that imaginative geographies are intricately linked to political-economic power and the unequal social relations underpinning racism.

Said's (1978) focus on observing, looking, and visuality brought attention to the cultural construction of perception. While Said's critique of Orientalism heavily employed visual metaphors, the imaginative geographies at the core of his analysis were primarily expressed through text. Nonetheless, scholars have been attracted to both textual and visual representations within human geography, encompassing art forms like film and photography, shedding light on viewing as an embodied and profoundly sensory practice (Schwartz & Ryan, 2020). These imaginative geographies are not regarded as solely the outcome of cognitive processes. As cultural constructs, their images are infused with elements of fantasy, desire, and the unconscious. Within these images lie evaluative comparisons, akin to what Said (1978) referred to as a "poetics of space", through which locations acquire 'figurative value'. Said argued that these figurative values not only contribute to the creation of "otherness" but also play a complex role in shaping the identity of the observing subject in an intricate dialectical relationship. Imaginative geographies thus sustain images of 'home' as well as 'abroad', 'our space' as well as 'their space'.

Film has the remarkable capacity to generate intricate and immersive 'worlds' within its cinematic narrative. As Yacavone (2015) pointed out, films are not just passive representations of reality but are active creators of distinct realms. These 'film worlds' encompass many elements, including visual aesthetics, narrative structures, and thematic underpinnings, all working in harmony to craft a coherent and immersive universe within the confines of the screen. This underscores that cinema is not merely a medium for storytelling but an art form that crafts parallel realities for viewers to engage with through its various cinematic techniques and artistic choices. Film can be, therefore, regarded as a medium for 'world-building' or 'worldmaking', offering audiences the chance to step into and experience these created realities. Thus, it becomes part of a process that more or less consciously imposes representations of places, people, and cultures over other existing or potential representations of these subjects (Hollinshead et al., 2009; Stachowiak, 2023).

In this context, the imaginative geographies of Indian film productions in Europe discussed here may be the result of specific perceptions of Europe by Indians. Research on Indians' perceptions of Europe indicates that, for most of them, the continent is essentially a foreign, exotic place with tourist attractions to which only a privileged stratum of society has access (Jain, 2019, p. 95). In India, there is a considerable deficit of information about Europe due mainly to mutual indifference and neglect of contacts, a permanent lack of mutual knowledge, and the fact that "Europe is still marginal in the Indian collective memory" (Goddeeris, 2011, p. 7). From the point of view of film audiences in India, Europe (or the West more broadly) is as much unknown as it is different, not only in the sense of tourism but also culturally. According to Krämer (2016, p. 53), the representation strategy of portraying European spaces in Indian cinema can be interpreted as "postcolonial revenge". This is particularly evident in Britain mainly due to colonial ties. This allows filmmakers to allude to issues such as patriotism or national pride, especially when they transform Britain into an Indian space on the screen and thus perform a "counter-hegemonic discursive appropriation or reverse colonialism" (Krämer, 2016, p. 55). This simplification can be seen as a form of "reverse orientalism" (Gabriel & Wilson, 2021). In this context, Europe is portrayed exotically and uniformly to facilitate storytelling and conform to the broader theme of presenting the West as something distinctly different.

Research method

Grounded in the concept of imaginative geographies, which explores the intricate interplay between cultural constructs, geographical imaginaries, and cinematic representations of locations, this chapter will further explore the way in which Eastern European locations are used in Indian film. A content analysis of a sample of Indian films shot in Eastern Europe was carried out to deconstruct the representation of these landscapes and to identify any prevalent stereotypes or tropes. This allowed for a nuanced understanding of how Indian filmmakers perceive and depict Eastern European locations, shedding light on the intricate interplay between film production, cultural identity, and geographic imaginaries within the context of the Indian film industry's expansion into this region. In order to do so, the methodological approach was based on a previous photo analysis to identify the geographical imagination of landscapes (Urbanc et al., 2021), which was further developed for film analysis by Kozina et al. (2021).

The study followed systematic steps, presented in Figure 3.1. It was part of the international project 'FilmInd: The Indian film industry as a driver of new socioeconomic connections between India and Europe' (https://filmind.philhist.unibas.ch/en/). The data collection process, that is, the selection of scenes and the gathering of information about them, was carried out by a team of researchers. The analysed films were shot in Poland, which is the largest Eastern European country. It should be noted that Poland should not be considered as a representative example of the entire Eastern European region here. Nevertheless, as the largest and most diverse country in terms of landscape and an area of great interest to Indian film crews,[1]

(1) Film selection procedure

Films identified in IMDb database
Criteria:
Form: film
(Co)production country: India
Filming location: Poland
Year of production: 2006-2018

↓

Relevent films: 24

↓

Filters:
1. Public availability in streaming media (such as YouTube, Facebook or Vimeo)
2. Popularity (films rated by at least 1,000 viewers in IMDb database)

↓

Final pool of 10 films

↓

(2) Scenes selection

Of all the scenes in the 10 films examined, a total of 138 scenes filmed in Poland were identified and included in the further study

(3) Variables development

Variables describing the characteristics of scenes relating to geographical characteristics such as visible landscape elements and socio-cultural characteristics such as the characters' behaviour, local cultural symbols, and the scene's relevance to the film's plot.

↓

(4) Data extraction

Creation of an audiovisual database containing full films, clips of analysed scenes, and image stills.

↓

(5) Scenes analysis

The researchers analysed each scene regarding the variables highlighted in step 3, assigning predefined values and categories to each scene. The characteristics of the scenes were then described using descriptive statistics (frequencies, percentages, mean) and crosstabulations.

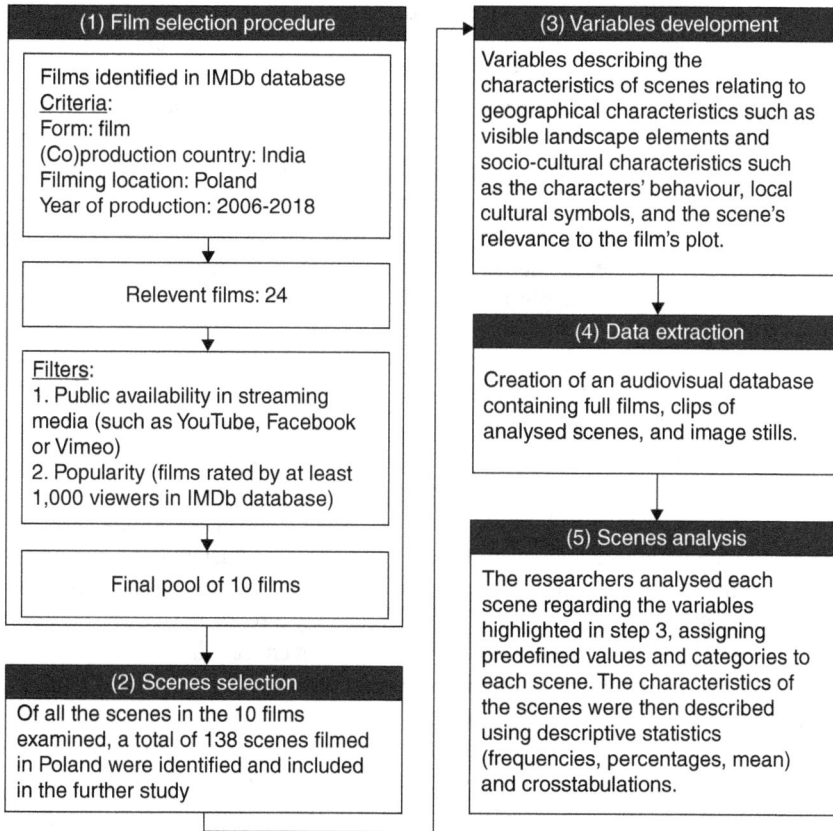

Figure 3.1 Research procedure.

Source: Own elaboration.

with 24 Indian films shot in Poland up to 2018, it can serve as a proxy for Eastern European landscape. The first Indian film shot in Poland was *Fanaa* in 2006, so the investigation included films shot since 2006. The study was conducted in 2019–2020, therefore, all films released in the period 2006–2018 were considered. From all 24 films shot in Poland during this period, the ones that had the greatest impact, measured by popularity, were selected (number of ratings on IMDb). Therefore, the empirical part of the study is based on a sample of ten films released between 2006 and 2018.

Figurative value of Polish landscapes in Indian films

This and the next section of this chapter delve into the realm of imaginative geographies as portrayed in cinematic narratives. These two sections present the results of an empirical analysis of scenes of Indian films made at least in some parts of Poland, thus showing the local landscape, objects, or places on the screen. This analysis has been annotated with commentary and reference to the literature to immediately capture similarities or differences to representations of other places (mainly Western European countries). The films analysed contained a total of 138 scenes. Their length varied depending on the scene's nature – the more dynamic ones, such as musical or action scenes, changed location more often, while static scenes, such as conversation scenes, were more extended. In the course of the analysis, 93% of the locations were identified by the researchers, that is, the exact name or location of the place where the scene was shot, which allowed them to pinpoint the precise role that the depicted locations played in the film.

To illustrate the role of Polish landscapes, an 'exposure ratio' was proposed, which is the ratio of the length of scenes shot in Poland to the entire length of the film (Table 3.1). This ratio shows what part of the film takes place in Polish settings. For example, in *Bangistan* (2015), over 26% of the film time was shot in Poland. This results in a quarter of the film exposing the Polish landscape, hence its exposure to the viewer is the highest. However, *Bangistan* was a film that was not very well received by the audience, which may not have translated into the memorability of the location. A high exposure rate (around 15%) came with commercial blockbusters like *Faana*, *Kick*, and *Mersal*, which may already have a promotional effect. In the last mentioned film, one of the songs shot entirely in Gdansk, a city in the north of Poland, received 94 million views on YouTube (as of 5 January 2023). The dominant landscape type appearing in the videos was the urban landscape, which dominated 81% of the scenes (Figure 3.2). It should be noted that non-urban landscapes were associated not only with rural areas and agriculture but also with mountainous areas, which appeared visually appealing, particularly in the 'song and dance' scenes. The most prominent elements on screen were historic parts of cities and historic buildings, which appeared in almost 70% of the scenes shot in the urban landscape (Figure 3.3).

It is worth noting that the popularity of Indian cinema has moved beyond the diaspora and its presence is no longer essential. Indian productions in Europe are made in places where there is no Indian community. Even in countries with a large Indian diaspora, such as the UK, many Indian films have used local settings exclusively or almost exclusively for their visual qualities. Krämer (2016, p. 50) points to the example of the film *Mohabbatein* (2000), in which university buildings in Oxford replace a fictional Indian public school. The architectural solemnity of the Gothic buildings is not only visually impressive but also contributes to the portrayal of the fictional setting as grand, austere, and emotionally hostile. The buildings, although non-Indian, have been used to represent an Indian boarding school. Similar situations can be found in films shot in Poland. In the opening scene of *Kick* (2014), the modern space of the centre of Warsaw, depicted in distinctly cool colours, is a visual representation of the main character's sadness and

Table 3.1 Main features of Indian films and scenes shot in Poland

Film data					Scenes shot in Poland		
Title[a]	Year	Director	Rating[b] (No. of votes)	Runtime (minutes)	Number of scenes/shots	Total length of scenes/shots (minutes)	Exposure ratio[c] (%)
24	2016	Vikram K. Kumar	7.8 (22 810)	164	16	15	9.4
Andhadhun	2018	Sriram Raghavan	8.2 (95 833)	139	2	5	3.8
Bangistan	2015	Karan Anshuman	4.5 (1 349)	135	26	36	26.6
Fanaa	2006	Kunal Kohli	7.1 (33 663)	168	9	24	14.4
Fitoor	2016	Abhishek Kapoor	5.4 (4 979)	131	3	5	4.1
Kick	2014	Sajid Nadiadwala	6.0 (70 934)	146	7	22	15.4
Mersal	2017	Atlee	7.5 (35 833)	172	36	25	14.3
Saguni	2012	N. Shankar Dayal	4.8 (1 185)	148	35	5	3.6
Shaandaar	2015	Vikas Bahl	3.6 (5 985)	144	3	8	5.8
Shivaay	2016	Ajay Devgan	6.1 (10 875)	153	1	2	1.2

Source: Own elaboration.

[a] Films in which the 'song and dance' scene was filmed in Poland are shown in bold.
[b] Weighted average vote at IMDb database (www.imdb.com/).
[c] Total length of scenes shot in Poland/film runtime.

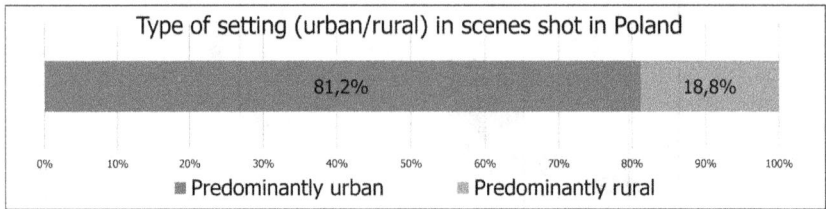

Type of setting (urban/rural) in scenes shot in Poland

81,2%   18,8%

0%   10%   20%   30%   40%   50%   60%   70%   80%   90%   100%

■ Predominantly urban   ■ Predominantly rural

Saguni (2012)

Bangistan (2015)

Figure 3.2 Type of setting in scenes shot in Poland.

Source: Own elaboration.

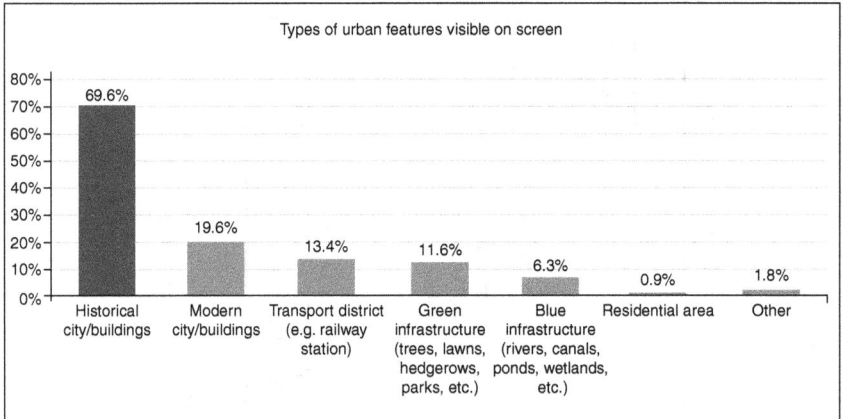

Types of urban features visible on screen

69.6%

19.6%

13.4%

11.6%

6.3%

0.9%

1.8%

Historical city/buildings   Modern city/buildings   Transport district (e.g. railway station)   Green infrastructure (trees, lawns, hedgerows, parks, etc.)   Blue infrastructure (rivers, canals, ponds, wetlands, etc.)   Residential area   Other

Figure 3.3 Types of urban features visible on screen.

Source: Own elaboration.

despondency. This exemplifies that in many Indian films, the realism of the setting is secondary to its picturesqueness, atmosphere, or symbolism. Often, the scenery does not play a significant role in the film's diegesis, making it irrelevant whether the filming location is in the UK, Poland, or another country.

Location unawareness and exoticism

Indian filmmakers often perceive Eastern Europe as a monolithic entity akin to Western Europe, despite the vast cultural and landscape diversity that Europe encompasses (cf. Kozina & Jelnikar, 2023; Stachowiak & Balcerak, 2022; Sunngren-Granlund, 2023; Urbanc et al., 2023). This simplification can be attributed to cinematic convenience and a broader trend of representing 'the West' as a symbol of exoticism and modernity in Indian cinema (Krämer, 2016). Filmmakers may view both Eastern and Western Europe as part of this broader Western construct, simplifying the complex mosaic of European cultures and regions for storytelling purposes. This perception not only serves narrative expediency but also reflects audience expectations, as Indian viewers may have limited exposure to the nuances of European regions (see case studies in Stachowiak et al., 2023). Therefore, imaginative geographies play here a prominent role. Consequently, Eastern European settings in Indian films may be employed to evoke a sense of the exotic and cosmopolitan, aligning with the overarching theme of the West as an emblem of otherness. While this simplification has its cinematic merits, it also underscores the need for more nuanced portrayals and cultural exchange between India and Europe to better represent the diversity of European regions.

Unawareness of the location in which the film takes place was quite evident in the examined scenes shot in Poland. Only in 4% of scenes are there clear indications of what kind of place it is, such as a visible name of the place or a distinctive city skyline or landmark. This leaves the viewer uninformed of the location in which the action takes place. What was more meaningful was that in some scenes, the locations were digitally transformed or pretended to be other locations. From a quantitative point of view, however, such scenes were few (only 5%). Yet, their significance for the film's plot was considerable, as they created the impression that the action was taking place in Western European cities. In *Fitoor* (2016), the city of Cracow pretended to be London, while in *Mersal* (2017), although the film was shot in Poland, the entire action of the film took place in France. Hence, Polish cities (Gdansk, Poznan, Rzeszow) pretended to be French, mainly Paris. In such cases, the impression of Frenchness was created by the set design: flags, inscriptions in French, or brands. Digital modification of the depicted landscape was also used. The most blatant example of such manipulation is the landscape collage, which consists of pasting not only a characteristic object to indicate the place (in this case it was the Eiffel Tower) but also elements of other cityscapes – in this case, it was the Chicago skyline (Figure 3.4).

Another kind of location unawareness was the use of spaces and landscapes as a picturesque backdrop for song sequences that constitute a kind of love fantasy. As Gopal and Moorti (2008) note:

> The song and dance sequence, then, is the single most enduring feature of popular Hindi cinema, although song-dance is hardly unique to Hindi film. From the earliest days of sound, song-dance has characterised the commercial product in other Indian languages, including Tamil, Telugu, Marathi, and Bengali.
>
> (p. 1)

Rzeszów (Poland) as Paris (France)   *Mersal* (2017)

Figure 3.4 An example of a 'landscape collage' in the film *Mersal* (2017).

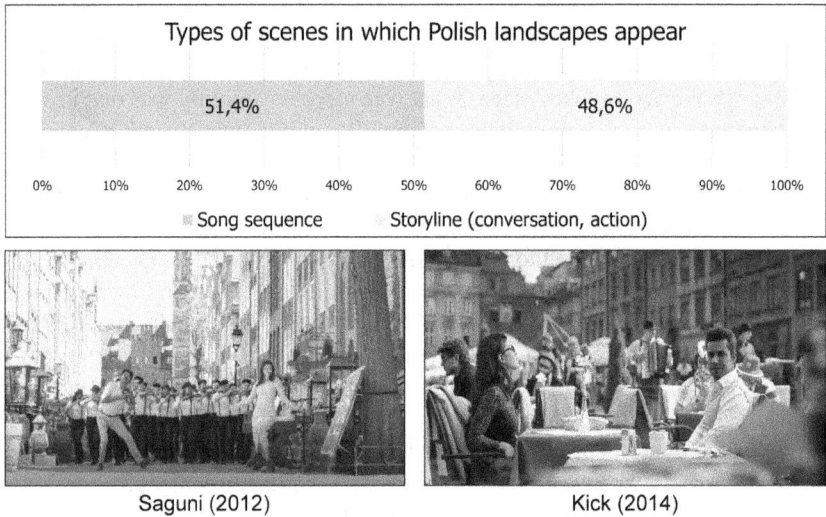

Saguni (2012)   Kick (2014)

Figure 3.5 Types of scenes in which Polish landscapes appear.

Source: Own elaboration.

More than half of the scenes shot in Poland were musical scenes (Figure 3.5), meaning that Polish landscapes were not only an attractive backdrop for song and dance scenes, but also became part of the storyline.

In the musical sequences, the film jumps from one setting to another (often foreign) location, which is meant to provide a scene that is a 'mindscreen' of the

The Tatra Mountains (southern Poland)   City of Cracow (southern Poland)

The Table Mountains (south-western Poland)   Ogrodzieniec Castle (south-central Poland)

Figure 3.6 Location hopping in the 'song and dance' scene of the film *Saguni* (2012).

Source: Own elaboration.

protagonist and to indicate that the audience is witnessing the thoughts of one or more characters. An example of this is the song in the film *Saguni*, whose imagery includes several Polish locations (Figure 3.6). Hogan (2008, pp. 165–166) points out that this technique is used particularly when depicting love fantasies in Indian films. Usually, the exact setting of the fantasy sequence, and therefore the question of whether the shooting location was in Switzerland, the UK, or Poland, is irrelevant to the film's meaning. What matters instead is the picturesqueness and impressive scenery of the location and its potential exotic appeal to Indian audiences. However, the landscapes and especially the buildings depicted in such fantasy sequences usually allow for at least a rough identification of the geographical and cultural sphere of the filming location. Even unidentified European objects and places thus trigger the notion of the 'West' and (from an Indian perspective) foreign traditions. This notion becomes more concrete when Europe serves as the setting for entire plots or even films.

Fostering imaginative geographies: The role of film commissions and destination marketing

The role of creating imaginative geographies of Polish and European landscapes is played not only by Indian filmmakers but also by local film commissions and destination marketing organisations. Along with Indian cinema's interest in Europe, there was also a growing interest in Indian cinema in Europe, and it was not limited to the diaspora (Krämer, 2016). Productions made in the second decade of the twenty-first century increasingly turned to settings exotic to Indians, which was one ingredient in the recipe for a commercially successful project. The various system incentives used by Western and Eastern European countries were introduced to attract filmmakers to new locations (Orankiewicz, 2022).

Apart from the United Kingdom, one Western European country that was very popular with Indian producers and location scouts was and still is Switzerland. The focus on Switzerland as a filming location for Indian film producers is justified for several reasons. Cooperation in the film industry between India and Switzerland was established decades ago, with the first film being shot in 1964. The conflict in Kashmir sparked the first adventures of Bollywood directors who sought mountainous landscapes outside India, and Hindi directors based in Mumbai became interested in filming in alpine settings (Gyimóthy, 2015). Indian producers came to Poland in the early twenty-first century for a similar reason. While it may be unlikely that Poland can replicate the success of Switzerland – famously known as a more film-friendly and less bureaucratic location (Janta & Hercog, 2023) – more recent cases show a link between Indian film settings and the overseas travel choices of Indian tourists (Josiam et al., 2020), with the media comparing Poland and Switzerland in this regard.

As previously mentioned, the first Indian film to be partly shot in Poland was *Fanaa* (2006). It starred the great Bollywood stars Aamir Khan and Kajol in the lead roles. The snow-capped Tatras and traditional highland homes replaced the Kashmiri landscapes in which part of the film is set (Figure 3.7). Although *Fanaa* was a commercial success, it did not increase Poland's popularity among Indian film producers. Only a few years later, after a complete hiatus, the Polish promotional campaign began to bear fruit (Balcerak et al., 2023). Three institutions were particularly helpful in bringing Indian filmmakers to Poland: The Polish Tourist Organisation, Film Polska Productions, a private film proxy company geared towards cooperation with India, and the Cracow Festival Office. The last mentioned is responsible for promoting the city of Cracow. The Polish Tourist Organisation has been targeting Indian film producers since 2009, and in 2012,

Figure 3.7 Snow-capped peaks of the Polish Tatra mountains pretending to be Kashmir in the film *Fanaa* (2006).

launched the 'I like Poland' campaign focused on the Japanese, Chinese, and Indian tourism markets and again included film producers. Attracting the producers of the film *Kick* (2014) was part of this effort, and to make it work, the Polish Tourism Organisation offered the filmmakers a financial incentive.

In addition to incentives, film commissions offer assistance in choosing a location and suggest characteristic elements of native culture, which then have a chance to appear on the screen. This was the case with the film *Bangistan* (2015), which was shot mainly in Cracow, an important cultural centre of Poland. In the musical scenes of the film, Polish dancers appear in traditional Cracow costumes, dancing and singing Polish folk dance, characteristic of the city and region of Cracow.[2] The presence of such elements was possible thanks to the promotional activity of the local film commission.[3] However, the image created as a result of it is typical of the tourist imaginary. In this way, a tourist and marketing narrative is woven into the feature film, and the film itself begins to serve as a specific platform for tourism promotion. This is the goal of local film commissions, but in this way, local destination marketing organisations participate in replicating a simplified and stereotyped image of a given place and creating imaginative geographies (cf. Pan & Hsu, 2014).

Conclusions

The exploration of imaginative geographies in cinematic narratives, particularly within the context of Indian films shot in Poland and Europe, exposes an interplay of landscape, symbolism, and cultural identity. These films utilise landscapes figuratively, akin to Said's (1978) "poetics of space", imbuing locations with symbolic and figurative values that contribute to the construction of 'otherness' and shape the observer's identity.

The portrayal of Eastern Europe in Indian cinema, examined in this chapter, reveals a complex interplay between cinematic convenience, audience expectations, and the broader trend of representing 'the West' as an exotic and modern entity. Indian filmmakers often depict Eastern Europe as a monolithic entity, simplifying the region's rich cultural and geographical diversity. This simplification can be seen as a form of 'reverse orientalism', where Eastern Europe is exoticised and homogenised for the convenience of storytelling and to align with the overarching theme of the West as an emblem of otherness.

The chapter highlighted imaginative geographies, where Polish landscapes and locations are often transformed or misrepresented for cinematic purposes. Whether Polish cities pretending to be French or British, or historical landmarks used as backdrops for song sequences, the focus remains on creating picturesque and exotic settings rather than accurately representing Europe's geographical and cultural nuances. Furthermore, the chapter highlights the role of Indian filmmakers, local film commissions, and destination marketing organisations in shaping imaginative geographies. While these collaborations offer incentives and assistance in choosing locations, they also tend to reinforce simplified and stereotyped images of the places involved. In essence, the cinematic representation of Eastern Europe in Indian films reflects a complex interplay of cultural perceptions, economic interests, and storytelling convenience, underscoring the multifaceted nature of reverse orientalism in contemporary cinema. This might have consequences for industries relying on such perceptions such as tourism. Indian filmmakers' simplification of Eastern Europe, aligning it with Western Europe and using it as an exotic backdrop, contributes to a skewed perception of the region for Indian audiences. This perception of Europe as exotic and other reinforces stereotypes and limited understanding of European diversity.

At the same time, the use of Eastern European landscapes in Indian films has become increasingly prevalent in recent years. While the use of European landscapes in Indian films has been a recurring theme for decades, the choice of Eastern European landscapes over Western European landscapes has raised questions regarding the reasons behind this trend. The most likely reason for the preference for Eastern European landscapes in Indian films is the relative affordability of filming in these locations, which Balcerak et al. (2023) confirmed for Poland and Kozina and Jelnikar (2023) for Slovenia. Filmmakers often choose cost-effective locations and offer production incentives; Eastern European countries have been known to offer such incentives. The lower cost of filming in these locations allows filmmakers to create grandeur and a sense of escapism for their Indian audience without incurring exorbitant production costs. Examples of Polish cities pretending to be Western European cities confirm the greater ease of filming in Eastern Europe. At the same time, Eastern European landscapes in Indian films are unique in terms of locations. Unlike Western European landscapes, which have been extensively filmed and are well-known to the Indian audience, Eastern European landscapes are relatively unexplored and offer a sense of novelty. The unfamiliarity of these landscapes allows filmmakers to create a sense of mystery and exoticism, which adds to the film's appeal (see more in Kozina & Jelnikar, 2023; Stachowiak & Balcerak, 2022; Urbanc et al., 2023).

Additionally, the use of Eastern European landscapes in Indian films can also be attributed to the changing cultural dynamics in India. India's growing interest in Eastern European culture and its people has led to a greater appreciation of the landscapes and architecture of these countries (cf. Kozina & Jelnikar, 2023). Poland could have taken the lead in enticing Bollywood to shoot films, soap operas, commercials, and periodic melodrama in Polish locations – palaces, parks, castles, monuments, gardens, and old cities and towns (Mishra, 2018). However, Poland sporadically appears in Indian films as Poland. Most often, it either imitates Western landscapes or forms a generic but exotic backdrop to the events depicted in the film.

Acknowledgements

The research presented in this chapter was carried out under the project 'FilmInd – The Indian film industry as a driver of new socio-economic connections between India and Europe', financed by the Polish National Science Centre (project no. 2017/27/Z/HS4/00039). The author would like to thank Malwina Balcerak and Marcin Adamczak for their contribution to the analysis of the data.

Notes

1 Media reports that Poland is becoming hugely popular as a shooting location for Bollywood movies and is set to replace Switzerland, which was the all-time favourite, see more at https://themigrationbureau.com/2017/04/16/poland-bollywoods-new-swit zerland/ (accessed 16 January 2023).
2 www.rmf24.pl/kultura/news-gwiazda-bollywood-dla-rmf-fm-powinniscie-zobaczyc-bangistan,nId,1483817 (accessed 20 October 2023).
3 https://wiadomosci.wp.pl/najazd-indyjskiego-kina-na-polske-efekt-bollywood-podbije-krakow-6027728060850817a (accessed 20 October 2023).

References

Balcerak, M., Stachowiak, K., & Adamczak, M. (2023). Production culture, interpersonal relations and the internationalisation of filmmaking industry: The case of Indian film productions in Poland. In K. Stachowiak, H. Janta, J. Kozina, & T. Sunngren-Granlund (Eds.), *Film and place in an intercultural perspective: India-Europe film connections* (pp. 159–178). Routledge. https://doi.org/10.4324/9781003293347-12

Beeton, S. (2005). *Film-induced tourism*. Channel View Publications. http://site.ebrary.com/id/10110145

Beeton, S. (2010). The advance of film tourism. *Tourism and Hospitality Planning & Development, 7*(1), 1–6. https://doi.org/10.1080/14790530903522572

Carl, D., Kindon, S., & Smith, K. (2007). Tourists' experiences of film locations: New Zealand as 'Middle-Earth'. *Tourism Geographies, 9*(1), 49–63. https://doi.org/10.1080/14616680601092881

Coe, N. M. (2000). On location: American capital and the local labour market in the Vancouver film industry. *International Journal of Urban and Regional Research, 24*(1), 79–94. https://doi.org/10.1111/1468-2427.00236

Cucco, M., & Scaglioni, M. (2014). Shooting Bollywood abroad: The outsourcing of Indian films in Italy. *Journal of Italian Cinema and Media Studies, 2*(3), 417–432. https://doi.org/10.1386/jicms.2.3.417_1

Driver, F. (2014). Imaginative geographies. In P. Cloke, P. Crang, & M. Goodwin (Eds.), *Introducing human geographies* (3rd ed., pp. 234–248). Routledge.

Elmer, G., & Gasher, M. (2005). *Contracting out Hollywood: Runaway productions and foreign location shooting*. Rowman & Littlefield.

Foster, P., Manning, S., & Terkla, D. (2015). The rise of Hollywood East: Regional film offices as intermediaries in film and television production clusters. *Regional Studies, 49*(3), 433–450. https://doi.org/10.1080/00343404.2013.799765

Gabriel, S. P., & Wilson, B. (Eds.). (2021). *Orientalism and reverse Orientalism in literature and film: Beyond East and West*. Routledge. https://doi.org/10.4324/9781003105367.

Gasher, M. (2002). *Hollywood North: The feature film industry in British Columbia*. UBC Press.

Goddeeris, I. (2011). *EU–India relations*. Leuven Centre for Global Governance Studies.

Goldsmith, B., & O'Regan, T. (2005). *The film studio: Film production in the global economy*. Rowman & Littlefield.

Gopal, S., & Moorti, S. (2008). *Global Bollywood: Travels of Hindi song and dance.* University of Minnesota Press.

Gregory, D. (1995). Imaginative geographies. *Progress in Human Geography, 19*(4), 447–485. https://doi.org/10.1177/030913259501900402

Gyimóthy, S. (2015). Bollywood-in-the-Alps: Popular culture place-making in tourism. In *Spatial dynamics in the experience economy* (pp. 170–185). Routledge.

Gyimóthy, S. (2018). Transformations in destination texture: Curry and Bollywood romance in the Swiss Alps. *Tourist Studies, 18*(3), 292–314. https://doi.org/10.1177/1468797618771692

Hogan, P. C. (2008). *Understanding Indian movies: Culture, cognition, and cinematic imagination.* University of Texas Press.

Hollinshead, K., Ateljevic, I., & Ali, N. (2009). Worldmaking agency–worldmaking authority: The sovereign constitutive role of tourism. *Tourism Geographies, 11*(4), 427–443. https://doi.org/10.1080/14616680903262562

Hubbard, P. (2002). Screen-shifting: Consumption, 'riskless risks' and the changing geographies of cinema. *Environment and Planning A, 34*(7), 1239–1258. www.envplan.com/abstract.cgi?id=a3522

Jain, R. K. (Ed.). (2019). *Changing Indian images of the European Union: Perception and misperception.* Springer. https://doi.org/10.1007/978-981-13-8791-3_7.

Janta, H., & Hercog, M. (2023). Contemporary Indian film productions in Switzerland. In K. Stachowiak, H. Janta, J. Kozina, & T. Sunngren-Granlund (Eds.), *Film and place in an intercultural perspective: India-Europe film connections* (pp. 118–135). Routledge. https://doi.org/10.4324/9781003293347-10

Josiam, B. M., Spears, D. L., Dutta, K., Pookulangara, S., & Kinley, T. (2020). Bollywood induced international travel through the lens of the involvement construct. *Anatolia, 31*(2), 181–196. https://doi.org/10.1080/13032917.2020.1749349

Kim, S. (2010). Extraordinary experience: Re-enacting and photographing at screen tourism locations. *Tourism and Hospitality Planning & Development, 7*(1), 59–75. https://doi.org/10.1080/14790530903522630

Kozina, J. (Ed.). (2021). *Comparative report on the use of European locations by the Indian film industries.* ZRC SAZU.

Kozina, J., & Jelnikar, A. (2023). Slovenia as a new contender in attracting Indian filmmakers within the context of Central and Eastern Europe. In K. Stachowiak, H. Janta, J. Kozina, & T. Sunngren-Granlund (Eds.), *Film and place in an intercultural perspective: India–Europe film connections* (pp. 136–158). Routledge. https://doi.org/https://doi.org/10.4324/9781003293347-11

Krämer, L. (2016). *Bollywood in Britain: Cinema, brand, discursive complex.* Bloomsbury Publishing.

Krauss, F. (2012). Bollywood's circuits in Germany. In A. G. Roy (Ed.), *The magic of Bollywood: At home and abroad* (pp. 295–317). SAGE Publications India.

Lourenço, I. (2017). Bollywood in Portugal: Watching and dancing practices in the construction of alternative cultural identities. *Etnográfica, 21*(1), 175–202.

Miller, T., Govil, N., McMurria, J., Maxwell, R., & Wang, T. (2005). *Global Hollywood 2.* British Film Institute.

Mishra, K. D. (2018). Poland's enchanted Bollywood. *Journal of European Popular Culture, 9*(2), 119–129.

Mitric, P. (2021). The policy of internationalisation of East European film industries: East–West co-productions 2009–2019. *Studies in Eastern European Cinema, 12*(1), 64–82. https://doi.org/10.1080/2040350X.2020.1800184

Nanjangud, A., & Reijnders, S. (2020). Cinematic itineraries and identities: Studying Bollywood tourism among the Hindustanis in the Netherlands. *European Journal of Cultural Studies*, *25*(2), 659–678. https://doi.org/10.1177/1367549420951577

Nanjangud, A., & Reijnders, S. (2022). On the tracks of musical screenscapes: Analysing the emerging phenomenon of Bollywood Filmi-song tourism in Iceland. *Tourist Studies*, *22*(2), 175–199. https://doi.org/10.1177/14687976221090728

Orankiewicz, A. (2022). The role of public support for the film industry – An analysis of movie production incentives in Europe. *Research Papers of Wroclaw University of Economics and Business*, *66*(2), 90–104.

Pan, S., & Hsu, C. H. C. (2014). Framing tourism destination image: Extension of stereotypes in and by travel media. In F. Hanusch & E. Fürsich (Eds.), *Travel journalism: Exploring production, impact and culture* (pp. 60–80). Palgrave Macmillan. https://doi.org/10.1057/9781137325983_4

Parvulescu, C., & Hanzlík, J. (2022). Beyond postsocialist and small: Recent film production practices and state support for cinema in Czechia and Romania. *Studies in European Cinema*, *19*(2), 129–146. https://doi.org/10.1080/17411548.2020.1736794

Riley, R., Baker, D., & Doren, C. S. V. (1998). Movie induced tourism. *Annals of Tourism Research*, *25*(4), 919–935. https://doi.org/http://dx.doi.org/10.1016/S0160-7383(98)00045-0

Said, E. W. (1978). *Orientalism*. Vintage Books.

Schwartz, J. M., & Ryan, J. R. (Eds.). (2020). *Picturing place: Photography and the geographical imagination*. Routledge.

Scott, A. J. (2002). A new map of Hollywood: The production and distribution of American motion pictures. *Regional Studies*, *36*(9), 957–975. https://doi.org/10.1080/0034340022000022215

Scott, A. J. (2005). *On Hollywood: The place, the industry*. Princeton University Press.

Stachowiak, K. (2023). From real to reel and back again: Multifaceted relations between film and place. In K. Stachowiak, H. Janta, J. Kozina, & T. Sunngren-Granlund (Eds.), *Film and place in an intercultural perspective: India-Europe film connections* (pp. 16–38). Routledge. https://doi.org/10.4324/9781003293347-3

Stachowiak, K., & Balcerak, M. (2022). Egzotyczna Europa. Eskapady popularnego kina indyjskiego na Stary Kontynent w XXI w. (Exotic Europe. The escapades of popular Indian cinema to the Old Continent in the 21st century). *Kwartalnik Filmowy (Film Quarterly)*, *120*, 46–66.

Stachowiak, K., Janta, H., Kozina, J., & Sunngren-Granlund, T. (Eds.). (2023). *Film and place in an intercultural perspective: India-Europe film connections*. Routledge.

Stachowiak, K., & Stryjakiewicz, T. (2018). The rise of film production locations and specialised film services in European semi-peripheries. *Hungarian Geographical Bulletin*, *67*(3), 223–237. http://ojs3.mtak.hu/index.php/hungeobull/issue/view/111

Sunngren-Granlund, T. (2023). Film productions as a part of the regional identity in rural areas–the case of the Nordic countries. In K. Stachowiak, H. Janta, J. Kozina, & T. Sunngren-Granlund (Eds.), *Film and place in an intercultural perspective: India–Europe film connections* (pp. 179–192). Routledge. https://doi.org/10.4324/9781003293347-13

Urbanc, M., Fridl, J., & Resnik Planinc, T. (2021). Landscapes as represented in textbooks and in students' imagination: Stability, generational gap, image retention and recognisability. *Children's Geographies*, *19*(4), 446–461. https://doi.org/10.1080/14733285.2020.1817333

Urbanc, M., Gašperič, P., & Kozina, J. (2023). Portraying European landscapes in Indian films. In K. Stachowiak, H. Janta, J. Kozina, & T. Sunngren-Granlund (Eds.), *Film and place in an intercultural perspective: India–Europe film connections* (pp. 63–82). Routledge. https://doi.org/https://doi.org/10.4324/9781003293347-6

Willis, A. (2003). Locating Bollywood: Notes on the Hindi blockbuster, 1975 to the present. In J. Stringer (Ed.), *Movie blockbusters* (pp. 255–268). Routledge.

Yacavone, D. (2015). *Film worlds: A philosophical aesthetics of cinema.* Columbia University Press.

Young, A. F., & Young, R. (2008). Measuring the effects of film and television on tourism to screen locations: A theoretical and empirical perspective. *Journal of Travel & Tourism Marketing, 24*(2–3), 195–212. https://doi.org/10.1080/10548400802092742

4 The Unfulfilled Potential of the Radio Archives

Designing a Prototype to Uncover Hidden Stories in the Landscape

Rebecca Nedregotten Strand

Introduction

An elderly woman relives her worst trauma. With a trembling voice, she describes a rockfall causing a giant wave that killed 41 persons on 7 April 1934. Accompanying her are the living images of the Norwegian village where it happened, nearly 90 years later. A quiet place with steep mountainsides rising from a narrow fjord, the sun mirroring itself in the water. This scene is part of the documentary series *Landet frå lufta (Norway from a Bird's Eye View)*, which I made while working in the Norwegian Broadcasting Corporation (NRK) (NRK, 2016). All the footage is aerial, and the soundscape was constructed in post-production. Using the NRK's radio archive for this purpose, I noticed that something interesting happened when the sound of the past met the present-day landscape. The contrasting effect of the woman's dramatic story and the tranquillity of the place decades later made her words more impactful – awakening the imagination and igniting a new interpretation of the landscape in the gap between then and now. Later, I left the NRK to become an academic researcher and discovered something else. The archives were no longer easily available to me. People outside of the corporation only get limited access.

The digitisation of cultural heritage has led to a paradigm shift that affects how and to whom it can be disseminated. According to Jin and Liu (2022: 7):

> The use of technology, such as computers, the Internet and new media in the digitization of cultural heritage has changed the initially simple object relations of cultural heritage and composed a complex object relation involving heritage owners–heritage managers–governments–experts–audiences and other stakeholders.

One of these stakeholders are the national public broadcasters. Many, like the NRK, have been the only broadcasters in their countries with the right to broadcast radio nationally for decades. They have published millions of clips, documenting people's stories, small and large events in places all over the world. These are now made available for private listening in public archives, but downloading and re-use for non-commercial purposes is restricted. Broadcasting archives represent a

DOI: 10.4324/9781003320586-6

global resource that can be utilised to reveal stories that are hidden from view and provide a deeper understanding of our surroundings. The recordings are part of our mutual, cultural history and can be reconnected with the locations they are linked to if implemented in low-threshold digital location-based storytelling that is now available for anyone via smartphones (Revill et al., 2020). Could the radio archives be utilised to benefit society more if access is given for use in location-based dissemination of cultural heritage?

This chapter explores the unfulfilled potential of the public service broadcasters' radio archives as a tool for disseminating cultural heritage on location. It describes how the method of media design is deployed to develop the prototype *Pastfinder*, a web application that uses archive sound from the NRK to augment a protected landscape with a rich history. The concept is transferable to archives of other broadcasters with a similar legacy and aims to exemplify what can be gained by transferring its content out of the past and into the present. Nyre (2009: 14) argues the need for researchers to take on the role as developers of media that challenges the role of the main industry actors and potentially have positive influence on society: "If media technology is not considered a central arena of the struggle for power and influence, then the potential for change in the mass media will be thrown out with the bathwater." The main research questions for this study are: How can radio archives be utilised and made more relevant by implementing them in location-based technology to disseminate cultural heritage? Why is this specific type of sound suitable for this purpose? And could it potentially represent a new form of media tourism?

Much research has been done on how media tourism impacts local communities, and how tourists experience it (van Es et al., 2021). This project aims to explore the concept further by using the method of media design to develop an actual prototype that might represent a new form of media tourism. Earlier research projects such as Bederson (1995), Hight (2005), Bolter et al. (2006) and, more recently, Sikora et al. (2018) and Tsepapadakis and Gavalas (2023) have designed prototypes for location-based audio-augmented reality dissemination. *Pastfinder* differs from these by making use of an already existing resource, the radio archive.

The background section gives a brief introduction to public service broadcasting, the digitisation and accessibility of their archives, and the specific location chosen for this study. In the theoretical section, connections between humans and place, location-based cultural heritage, and the concepts of aura (Benjamin, 2008), and 'imaginative heritage' (Reijnders, 2021) are explored. Further, the text describes the method of media design and the development of the prototype. In view of the theoretical framework, the analysis discusses how using historic sound clips in location-based dissemination of cultural heritage can contribute to realising the unfulfilled potential of the archives and potentially become a new form of media tourism.

Background

The term 'public service broadcasting' in this text broadly refers to corporations that are non-commercial and publicly funded, with a democratic duty to transmit to everyone and the intention to ensure that what is produced has some sort of societal value (Gramstad, 1989; Moe & Syvertsen, 2009). This model was first introduced in 1926 with the establishment of the British Broadcasting Corporation (Gramstad, 1989) and later adapted in different forms by many European countries, Japan (Itoì et al., 1978), Australia, Canada, and South Africa (Moe & Syvertsen, 2009). The degree of intervention and level of public funding varies and thereby also the broadcaster's position in society (Moe & Syvertsen, 2009). The NRK, one of the world's oldest active public service broadcasters, is modelled after the BBC and has produced radio since 1933. In 2020, its funding became tax-based; prior to this, every household paid annual licence fees. As a national public service broadcaster, the NRK were obligated to make content from all over Norway. They had, and still have, reporters working across the nation, and several regional broadcasting stations airing locally while also delivering content for the national broadcasts.

Digitisation has made it possible to store these broadcasts in public archives on a large scale (Revill et al., 2020). More than 1.9 million audio clips are available for private listening through the Norwegian National Library's website (Nasjonalbiblioteket, n.d.). A vast majority of these are recordings from the NRK, as they had a monopoly until the 1980s. Metadata with keywords related to locations, events, dates etc. is used to navigate the archive and find the relevant material. Even if copyrights are cleared for non-commercial purposes, getting access to download and permission to re-use the sound is rather complex. One must contact the NRK, pay them for the time they use to download and send the file and, if the sound is to be made publicly available, there is an annual licence fee to be paid per minute published. Accessibility is limited and relatively costly if making use of other public service broadcasters' archives, such as the Swedish, Sveriges Radio (Kungliga Biblioteket Publik Verksamhet, 2023), and the BBC (BBC, 2023a and 2023b; Revill et al., 2020: 292). Similar issues of costs and accessibility are problematised by Misek (2022). His film *A History of the World According to Getty Images* critically addresses the power of international image banks, such as Getty Images.

One of the many places featured in the NRK's radio archive is Dovrefjell, the mountain plateau chosen as the location for this study. It plays a prominent role in the birth of Norway as a nation and its struggle for independence (Steinsland, 2014) and is deeply affected by human activities. It's an iconic place, referred to upon the signing of the first Norwegian constitution in 1814: The delegates held each other's hands and declared to be "United and faithful until Dovre falls" (Steinsland, 2014). Thousands of tourists visit the place each year (Dovrefjell nasjonalparkstyre, 2021). The attraction lies not only in Dovrefjell's outstanding nature but also in experiencing what Reijnders (2016) refers to as 'places of the imagination'. Dovrefjell features in Viking sagas, myths, literature, theatre, and films. The first humans arrived after the last Ice Age, following the wild reindeer. Prior to the reformation, pilgrims crossed the mountain plateau on their way to St. Olav's shrine in Nidaros. At the end of the sixteenth-century, the Kingsroad was built, making it possible to travel with a four-wheel horse-drawn carriage. In more recent history, human encroachments like a railway, Norway's main road, mining, and a military shooting field have dominated the landscape. In the 2000s, industrial remnants of both the mine and shooting field were removed as a part of the largest nature restoration project in Norwegian history (NINA, 2021). This area is now part of the Dovrefjell-Sunndalsfjella National Park.

These types of complex stories related to place can be found all over the world. They are hard to grasp when hidden from sight, but if revealed, could contribute to a more profound understanding of our surroundings. In the contemporary land-scape, they are not always easily spotted, but in the radio archives, many of them still exist. Reconnecting the audio to the places they once were recorded in, and/or whose stories they told, could help trigger our imaginations and visualise their past.

Places, meaning, and memory

Human site-specific stories can be traced back more than 45,000 years. In 2017, scientists discovered what was considered the world's earliest known representa-tional work of art; a cave painting in Lang Tedongnge in Indonesia (Brumm et al., 2021). However, most human traces in the landscape are fading, erased, or replaced. We need our imagination to reimagine them. According to van Es et al. (2021: 11), humans have the capacity to mentally construct places without being there, but also to envision how places may have looked in the past. This point is also argued in Jin and Liu (2022), who state that "the simulation and simulacrum of cultural heritage in virtual worlds open up a wider space for the visual imagination".

Thirty years ago, Ryden (1993) envisioned three-dimensional mapping of landscapes. Although geographical maps show topography and physical elements, his idea was the design of maps that could convey how the landscape has been influenced by the people living in it. Ryden suggested the term 'chorographer' for a person mapping and interpreting the invisible landscape, infusing it with meaning. The means to realise this concept came with the rapid development of digital media. In this vein, Sample (2014: 73) writes: "Location-aware apps could prove to be the means to create such maps of the invisible landscape. I have a vision of infusing geosocial applications with a sense of depth, with meaning that goes beyond the check-in." Through what Farman (2014) refers to as "creative misuse", artists, authors, and scholars have contributed to develop content for mobile phones way beyond their intended use: "The result is often a deeper sense of place and a stronger understanding of our position within that place" (Farman, 2014: 5).

Location-based dissemination of cultural heritage is one example of these types of "misuse", with several research projects utilising emerging technologies to inter-pret history. Liestøl (2009) has developed the mobile augmented reality platform *SitSim*, which reveals historic sites and events in urban, rural, and even submerged landscapes (Liestøl et al., 2021). It provides visitors with technical aids for experi-encing past events and present-day landscape simultaneously. This combination is something users have found rewarding, and user-tests have shown the concept to be suitable for both formal and informal learning situations (Liestøl & Morrison, 2014: 210). Oppegaard and Adesope (2013) designed an interactive location-based educational application for Fort Vancouver National Park. External evaluation found it to be an effective resource for learning but with a need for increased level of interactivity (Oppegaard & Adesope's, 2013: pp. 101–103). McGookin et al.'s (2019) cultural heritage application *Explore* was designed to disseminate a site with complex layers of heritage to users whose main purpose for visiting wasn't necessarily to explore cultural heritage. Their research revealed a need for more studies on how cultural heritage can be communicated "as a secondary tangential activity" (McGookin et al., 2019: 208).

In a study of narrative practices on cultural heritage, López Salas (2021: 9) analyses eight heritage sound walks and observes that they are "heavily aligned with the visual as the user is both a listener and an observer, but with an emphasis on sound". Bederson (1995) introduced the term audio-augmented reality (AAR), and Knowlton et al., (2002) developed an AAR-based dissemination concept for a warehouse-district in Los Angeles with the project, *34 North 118 West*. It brought a railroad community back to life with the use of soundscapes and narrations by actors playing the roles of former residents. The space was "inhabited with the sonic ghosts of another era" (Knowlton et al., 2002). Ghosts are also central to "the Oakland project", where users visiting a cemetery listen to the stories of those buried there (Bolter et al., 2006). *Pastfinder* is also intended to evoke the past with the use of sound. It's closely linked with Hight's (2005) concept of 'narrative archaeology', where instead of digging up the past physically, one uses locative audio to reveal what is hidden beneath the surface. Hight saw location-based media as a form of art that would develop together with the emerging locative technologies opening the opportunities for 'locative narratives': "Narrative, history, and scientific data are fused in landscape, not a digital augmentation, but a multi-layered, deep and malleable resonance of place" (Hight, 2006: 2).

In the field of media tourism, Reijnders (2016: 112–113) emphasises that memory's role is significant and crucial for how we experience "places of the imagination." While he refers to media tourism triggered by popular culture such as films and books, I argue that using the radio archive to disseminate cultural heritage is a way of evoking memory and the imagining of places through a different type of media. Sound helps conjure these imaginations, and the authenticity of an archive clip produced for radio serves a different purpose in doing so than new recordings describing previous events. In his renowned essay "The Work of Art in the Age of Mechanical Reproduction'', first published in 1936, Benjamin (2008) describes the aura of an original artwork and the loss of this when technology made it possible to reproduce them. Bolter et al., (2006) provide an interesting analysis of Benjamin's arguments in view of the later development of digital media. They agree that aura, the feeling of authenticity and awe we experience when for instance seeing an original piece of art as opposed to reproductions of the same artwork, is in a permanent state of crisis. However, they argue that mixed reality "should be more effective at conveying aura than other media forms" (Bolter et al., 2006: 29). Its hybrid form allows for experiencing the virtual and the physical simultaneously. Benjamin (2008) described aura as the sense of something that is unattainable and distant even if you get very close to it. Bolter et al. (2006) state that the aura of the actual location, in their case the Oakland Cemetery, is sensed and possibly enhanced "by building a sense of distance-through-proximity" because one is exposed to content that augments the place. A similar argument is found in Jin and Liu (2022), who reflect on how virtual reality (VR) allows for more fluidity when it comes to disseminating cultural heritage. When sites are freed from the physical reality of their current form, users can move through them in different eras without the limitations of time and space. They conclude that: "In the age of virtual reality, the aura, which Benjamin believed had dissipated in the age of mechanical production, is revived" (Jin & Liu, 2022: 9).

Radio archives contain large amounts of original recordings of people living in, describing, and experiencing a place. They have a different and possibly stronger aura than new recordings telling the same stories. The content is diverse and multi-faceted, containing a range of different representations of place, from the factual to the fictional. Introducing the term 'imaginative heritage', Reijnders (2021) argues for a more holistic approach to heritage studies and a broader involvement by the stakeholders that contribute to media tourism. He writes that

> the concept of imaginative heritage can be used as an important instrument by local governments. On the one hand, it allows them to stimulate new media tourism initiatives that align with other types of local heritage. On the other hand, it will help them unlock "forgotten" forms of heritage from the past.
>
> (Reijnders, 2021: 21)

Media producers such as the public service broadcasters are one potential contributor. If allowing for more extended use of their archives, in, for instance, location-based cultural heritage dissemination, they could contribute to what Reijnders suggests: "stimulate a new form of media tourism initiatives that align with other types of local heritage" (Reijnders, 2021: 21). Giving visitors more knowledge of their surroundings and the history that shaped them.

Designing the prototype

The prototype *Pastfinder* is a low-threshold technical solution where curated clips from the NRK's radio archive are used to augment a vulnerable landscape. It has been developed using the method of media design (Fagerjord, 2012) to research how radio archives can be utilised and made more relevant by implementing them in location-based technology to disseminate cultural heritage. The aim is also to demonstrate why this specific type of sound is well suited for this purpose and explore if this type of use might represent a new form of media tourism.

Nyre, (2009:11) argues that media researchers should use their knowledge to contribute to better public communication, calling it a "moral duty". Also, that what is designed serves a "public purpose" and that researchers should have "a max-imally conscious approach to the ethical and cultural implications of the solutions they are making" (Nyre, 2014: 90). Methodically, media design is influenced by the concept 'research by design' with the researcher taking on a "dynamic research perspective" (Fagerjord, 2012; Sevaldson, 2010). Instead of asking what is there, one explores the potential of what might be and is faced with "wicked problems" that can't be answered until one has "found the solution" (Sevaldson, 2010: 17). The researcher takes on the role as designer to produce knowledge while constantly reflecting on the process, thereby contributing to developing the practice further (Sevaldson, 2010). *Pastfinder* is a password-protected web application made with the software WordPress and the open-source map framework Leaflet (Sevincer, 2023). Key elements like curation of sound clips, location for each specific clip, visual design, and user interface evolves through a series of iterations and regular evaluation. This includes desktop development, a user test of an initial desktop version (Osen, 2022), and three periods of field work.

Firstly, the location was chosen. Dovrefjell is a popular tourist destination due to its spectacular nature and role as a place of 'imaginative heritage' (Reijnders, 2021). It's also the realm of the wild reindeer (Flemsæter et al., 2013). They are considered an endangered species and Norway has an international responsibility to protect them. Efforts are being made to canalise people to areas where there is little risk of disturbing the herd. One of these areas is the Kingsroad and Pilgrim's Route across the Hjerkinnhø hill, mostly located within a protected landscape (Lovdata, 1974). Humans have been present here since the end of the Ice Ages, resulting in a complex system of traces, symbols, signs, and monuments. This seven-kilometre trail is a popular walk among both tourists and locals. For these reasons, it was decided that the route is well suited to explore how radio archive sound can be used for augmentation. Much of the cultural heritage that is referred to in the clips is physically linked to this area.

The clips and specific placing of each of them (POI) were chosen through a curation process. A broad search using words related to Dovrefjell was conducted in the radio archive of the Norwegian National Library. From a total of 49 clips, ten were selected and purchased using the following criteria: information value, copyrights cleared, entertainment value, variation in themes, quality of sound, and not containing misinformation or breaking the press' code of ethics. The ten POI along the route are meant to be experienced when walking from north to south. They are marked with a flower icon on the interactive map. Clips have been shortened in length to a maximum of 00:03:30, but not edited in any other way. In the following section three clips are presented and analysed in view of the study's purpose.

Revealing what is hidden

Mobile technology enables us to add elements to the real world, with the goal of enhancing the richness of it. It allows for new ways in which media tourists can memorise and explore "places of the imagination" on location (Reijnders, 2016) With the designer applying Ryden's (1993) role of *chorographer,* the prototype aims to infuse the landscape with meaning (Farman, 2014). Sound clips are chosen to work as AAR with the intention of contributing to the experience of walking the already established route. In Figure 4.1, the route is shown in blue with POI marked with title.

1   *The Vinje Memorial* (see Figure 4.2) is located next to an engraved slate depicting two persons, a royal carriage, a gate, and a Norwegian text that reads: "Once again I see the same mountains and valleys. Here, Aasmund Olavson Vinje and Malene from Foldale took a rest in July 1860." Without context, the place gives little meaning; the monument is located off the main trail, and as it's hidden from view by vegetation, you could easily walk past without noticing it. The narrative asks the listener to picture the odd couple depicted on the slate, author Vinje, and the local elderly woman, Malene, following the King's ceremonial crossing of Dovrefjell in 1860, a journey described in Vinje's book *Ferdaminne fraa Sumaren 1860* (Vinje, 1996). His writings are considered as one of the highlights in Norwegian literature (Sandberg, 2023). He is one of many contributors to Dovrefjell's 'imaginative heritage' (Reijnders, 2021). The archive sound was recorded on location for the anniversary in 1960 when the monument was revealed and celebrated, while today it appears to be forgotten. There is not even a sign indicating where to find it. The AAR attempts to reawaken its fading past, contributing to the users' knowledge and understanding of the landscape.

2   *Shooting field and mine* describes the area at a time when both the military shooting field and industrial mine were active. From the POI, there is a broad view of the landscape, the narrative indicates which way to look and describes how today's view contrasts with what the two persons in the clip saw in 1976. In it, a reporter and the owner of a local hotel discuss the development in the area since the 1920s. The interviewee describes the mine, military installations, and a radio link station, while the reporter complains that the mountain plateau is losing its uniqueness. They predict that the industrial 'monster' one day will be demolished. Those visiting today see a landscape where almost all traces of industry are erased due to the nature restoration project (NINA, 2021), while this clip explicitly reveals what is hidden. There is an illusion of something pristine and untouched, when the truth is that humans have controlled and altered this place for centuries. If some of these epochs are removed from memory, we might not learn from previous mistakes. This area is now part of the Dovrefjell-Sunndalsfjella National Park, while a new shooting field has been established somewhere else (Hammer, 2023).

3   In the final POI, *Until Dovre falls*, the narration tells of Dovrefjell's symbolic role through history. In the archive clip, one hears part of a military officer's speech to Norwegian soldiers in the air force training base, Little-Norway in Canada, on the Norwegian National Day, 17 May 1943. He refers to the afore-mentioned signing of the first Norwegian constitution. The NRK broadcasted from both the UK and United States during the war, because German forces had taken control over the NRK in Norway (Dahl & Dybing, 1978). Hence, the officer's words had a motivational purpose for those 500,000 persons risking punishment by listening to illegal broadcasts. This POI has a more abstract approach to the landscape than the others. It's an example of how media can be

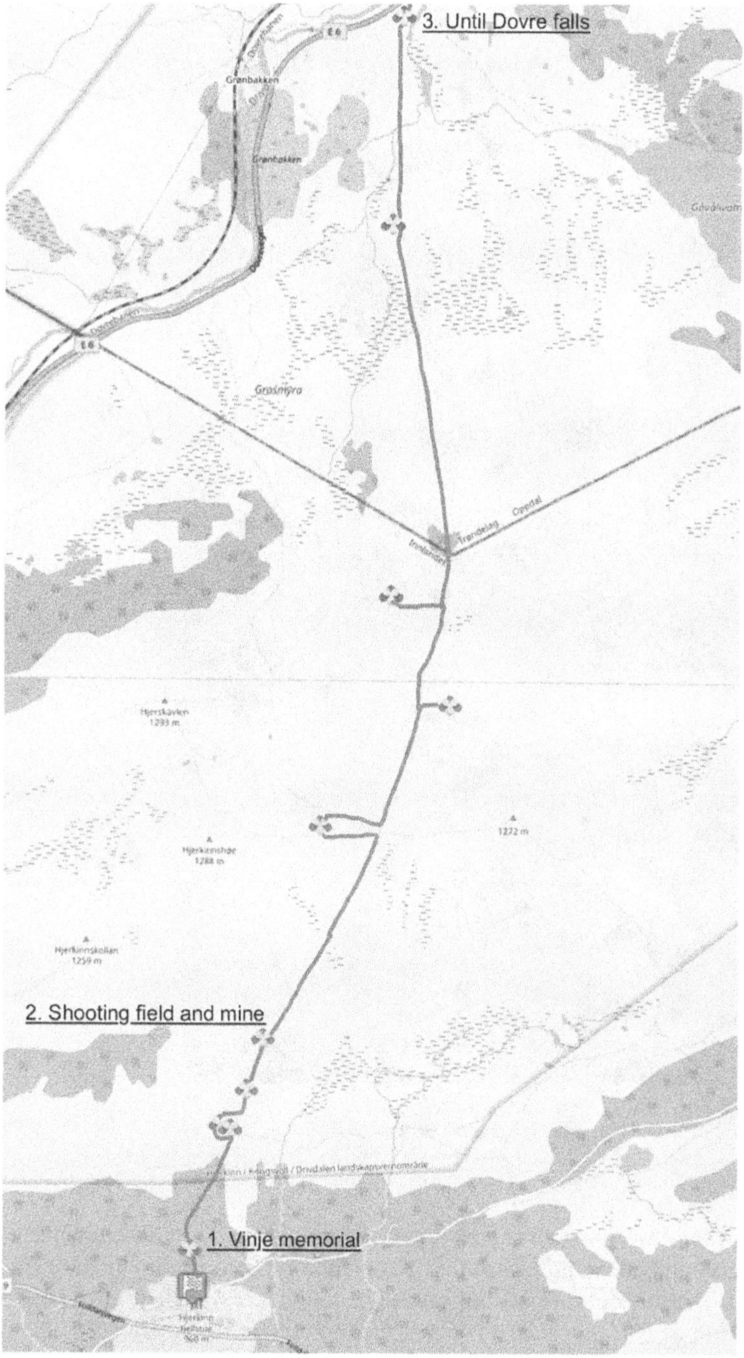

Figure 4.1 Screenshot of *Pastfinder* with route marked and three POI highlighted (slightly distorted map as the figure is a merge of two screenshots). Photo: Strand, R. N.

Figure 4.2 The Vinje memorial. Photo: Strand, R. N.

a tool for propaganda and how a place is used rhetorically to build a sense of national unity. Dovrefjell is not just a mountain plateau, it's a symbol of Norway itself. The roots of this symbolism can be found in the ancient myths depicting the birth of Norway, and in sagas of the Norwegian kings written by Icelandic historian and poet, Snorri Sturluson (1178–1241) (Steinsland, 2014). One of them tells of King Harald Fairhair, said to have been raised by a Jotun called

Dovre. His story is told at one of the other POI. These types of mythifications explain some of Dovrefjell's seemingly never-ending part in the shaping of the nation. It's also used with sarcasm and irony in Henrik Ibsen's *Peer Gynt* from 1867 and still plays a symbolic role as the home of mythic creatures in modern productions like the comical fantasy movie *Trolljegeren* (2010), and the Netflix film *Troll* (2021).

According to Reijnders (2016), memory plays a key role in how we experience 'places of the imagination'. As these examples are meant to highlight, the content of each POI is chosen to augment the existing landscape and the radio archive clips are key to achieving this. Their function is to reveal hidden stories that are linked to both physical cultural heritage and the 'imaginative heritage' of the area. It represents a hybrid form of dissemination, evoking memories, triggering the imagination, and linking us more closely to the past and its place.

Out of the vaults and into society

This chapter has explored how allowing for use of archive sound in location-based dissemination could contribute to better understanding of our surroundings and deeper knowledge of place, thereby playing a societal role beyond mere entertainment (Nyre, 2014).

A radio archive contains the voices of people who have lived in, experienced, and described places. Modern technologies give us the tools to bring them back there and implementing them in cultural heritage dissemination could be a purposeful way of doing so. It's, of course, possible to write and record new narratives to disseminate the same content, but as the text aims to argue, something would be lost on the way – the aura of the original and how it affects us. This need for authenticity also means that the concept of using the radio archive to disseminate cultural heritage on location has limitations that can be criticised. Clips cannot be translated from their original language and thereby made available to all. It's possible to add a voice-over translating the archive sound, but if the users don't understand the original clip, they are likely to miss out on some of its authenticity. One is also at the mercy of what has actually been recorded and preserved and therefore cannot tell just any story. In addition, one must be aware that these archives are not objective or completely reliable sources. While researching the archive, I have found information to be both incorrect and missing from the metadata, such as names or places being misspelled or left out. One must tread carefully and be aware of how the changing contexts and time might influence how recordings are perceived today. They should be interpreted critically and in view of what time they were recorded, who was interviewed, and why. Including a narrative is one way of informing users of such factors, allowing for the archive to be used despite its faults.

The study has attempted to demonstrate the potential for this type of dissemination to work as a different form of media tourism, creating more locally mediated experiences. Recent events such as the global COVID-19 pandemic, and the effects of climate change, show how travel patterns can suddenly change, and force or motivate us to find destinations closer to home. van Es et al. (2021: 7) predict that: "In this increasingly mobile world, people will search for new and inventive ways of creating a sense of home, a sense of belonging." Making use of the radio archives represents one such potential invention. As is the case with Dovrefjell, there are many destinations where tourism is a potential threat to vulnerable nature – and cultural heritage. Efforts are made to direct travellers to specific areas and motivate them to avoid others. Jin and Liu (2022) argue that a more creative approach to how cultural heritage is disseminated, and combining it with VR technology represents possible solutions to these challenges. A version of *Pastfinder* that makes it possible for users to visit Dovrefjell via VR headsets is developed to explore how the radio archive will function if implemented in remote experiences. There are multiple ways in which the broadcasters' archive material could be reused. They are both cultural heritage in their own right and a tool for disseminating cultural heritage in our surroundings. But for their unfulfilled potential to be realised, those who hold the key to these memory-vaults need to be less restrictive in terms of who can access them and how. Researchers might contribute to this by making use of emerging technologies to do further research on how historic broadcasting material can be used globally in media tourism to benefit society more.

References

BBC (2023a). Researching the BBC Archives. BBC. Retrieved 14 February 2023 from www.bbc.co.uk/archive/researching-bbc-archives/zrqpwty

BBC (2023b). What's in the BBC Archives? BBC. Retrieved 14 February 2023 from www.bbc.co.uk/archive/whats-in-the-bbc-archives/zmvvxyc

Bederson, B. (1995). Audio augmented reality: A prototype automated tour guide. In *Conference on Human Factors in Computing Systems* (pp. 210–211). ACM.

Benjamin, W. (2008). *The Work of Art in the Age of Mechanical Reproduction* (J. A. Underwood, Trans.) [Essay]. Penguin (1936).

Bolter, J. D., MacIntyre, B., Gandy, M., & Schweitzer, P. (2006). New media and the permanent crisis of aura. *Convergence (London, England)*, *12*(1), 21–39. https://doi.org/10.1177/1354856506061550

Brumm, A., Oktaviana, A. A., Burhan, B., Hakim, B., Lebe, R., Zhao, J.-X., Sulistyarto, P. H., Ririmasse, M., Adhityatama, S., Sumantri, I., & Aubert, M. (2021). Oldest cave art found in Sulawesi. *Scientific Advances*, *7*(3). https://doi.org/10.1126/sciadv.abd4648

Dahl, H. F., & Dybing, M. (1978). *"Dette er London": NRK i krig 1940–1945*. Cappelen.

Dovrefjell nasjonalparkstyre (2021). Besøksstrategi for verneområdene på Dovrefjell. www.nasjonalparkstyre.no/uploads/files_dovrefjell/Forvaltningsplan/2021-08-Besoksstrategi-Dovrefjell-ENDELIG-m-Informasjonsplan.pdf

Fagerjord, A. (2012). Design som medievitenskapelig metode. *Norsk medietidsskrift*, *19*(3), 198–215. www.idunn.no/nmt/2012/03/design_som_medievitenskapelig_metode

Farman, J. (2014). *The Mobile Story: Narrative Practices with Locative Technologies*. Routledge.

Flemsæter F., Rønningen K. & Holm F. E. (2013). Dovrefjells moralske landskap. Villrein, ferdsel og inngrep i Dovre-Rondane-regionen, Norsk senter for bygdeforskning.

Gramstad, S. (1989). Kringkasting i folkets teneste: ei innføring i allmennkringkasting (Vol. 17). Møre og Romsdal distriktshøgskule.

Hammer, E. (2023). Regionfelt Østlandet in *Store norske leksikon* at snl.no. Retrieved 28 August 2023 from https://snl.no/Regionfelt_%C3%98stlandet

Hight, J. (2005). Narrative archaeolgy. *Media in Transition 4: The Work of Stories, 4* (Cambridge, MA: May 6–8, 2005). https://web.mit.edu/comm-forum/legacy/mit4/papers/hight.pdf

Hight, J. (2006). Views from above: Locative narrative and the landscape. *Leonardo Electronic Almanac 14*(7–8), 1–9.

Itoì, M., Shiono, H., Kataoka, T., & Nagatake, S. (1978). *Broadcasting in Japan* (Vol. 8). Routledge.

Jin, P., & Liu, Y. (2022). Fluid space: Digitisation of cultural heritage and its media dissemination. *Telematics and Informatics Reports*, *8*, 100022. https://doi.org/10.1016/j.teler.2022.100022

Knowlton, J. S., N.; Hight, J. (2002). *38 North 118 West*. Retrieved 28 August 2023 from http://34n118w.net/

Kungliga Biblioteket Publik Verksamhet. (2023). Avgift bruk av filer fra Sveriges Radioarkiv. Personal email to R. N. Strand, 6 February 2023.

Liestøl, G. (2009). Augmented reality and digital genre design – Situated simulations on the iPhone. *IEEE International Symposium on Mixed and Augmented Reality – Arts, Media and Humanities*, 29–34. https://doi.org/doi:10.1109/ISMAR-AMH.2009.5336730

Liestøl, G., Bendon, M., & Hadjidaki-Marder, E. (2021). Augmented reality storytelling submerged. Dry diving to a World War II Wreck at ancient Phalasarna, Crete. *Heritage*, *4*(4), 4647–4664. https://doi.org/10.3390/heritage4040256

Liestøl, G., & Morrison, A. (2014). The power of place and perspective. In A. De Souza e Silva (Ed.), *Mobility and Locative Media: Mobile Communication in Hybrid Spaces* (1st ed.). Routledge. https://library.oapen.org/handle/20.500.12657/46453

López Salas, E. (2021). A collection of narrative practices on cultural heritage with innovative technologies and creative strategies. *Open Research Europe*, *1*, 130. https://doi.org/10.12688/openreseurope.14178.1

Lovdata (1974). Forskrift om vernebestemmelser for Drivdalen landskapsvernområde, Sør-Trøndelag, Kongsvoll landskapsvernområde, Hjerkinn landskapsvernområde, Oppland. https://lovdata.no/forskrift/1974-06-21-6

McGookin, D., Tahiroğlu, K., Vaittinen, T., Kytö, M., Monastero, B., & Carlos Vasquez, J. (2019). Investigating tangential access for location-based digital cultural heritage applications. *International Journal of Human–Computer Studies*, *122*, 196–210. https://doi.org/10.1016/j.ijhcs.2018.09.009

Misek, R. (2022). *A History of the World According to Getty Images* [Film]. T. Nilsen, www.ahistoryoftheworldaccordingtogettyimages.com

Moe, H., & Syvertsen, T. (2009). Researching public service broadcasting. In T. Hanitzsch, K. Wahl-Jorgensen, & International Communication Association (Eds.), *The Handbook of Journalism Studies* (pp. 399–413). Routledge.

Nasjonalbiblioteket. (n.d.). Nettbiblioteket NRK Digitalt radioarkiv. Nasjonalbiblioteket. Retrieved 23 August 2022 from www.nb.no/search?mediatype=radio&recordSource=dra.nb.no

NINA. (2021). *Ecosystem restoration*. Norsk institutt for naturforskning. Retrieved 21 April 2021 from www.nina.no/english/Research/RestorationEcology

NRK. (2016). Risiko (Season 1, Episode 4) [TV]. R. N. Strand, Ed. & Trans. *Landet frå lufta*. NRK. https://tv.nrk.no/se?v=DVFJ61004313

Nyre, L. (2009). Normative media research: Moving from the ivory tower to the control tower. *Nordicom Review*, *30*(2), 3–17. https://doi.org/10.1515/nor-2017-0148

Nyre, L. (2014). Media design method. Combining media studies with design science to make new media. *The Journal of Media Innovations, 1*(1). https://doi.org/https://doi.org/10.5617/jmi.v1i1.702

Oppegaard, B., & Adesope, O. (2013). Mobilizing the past for the present and the future: Design-based research of a model for interactive, informal history lessons. *Journal of Teaching and Learning with Technology, 2*(2), 90.

Osen, I., A. (2022). *Facilitating a Digital Route: A Mobile Experience Designed for the King's Road across Dovrefjell.* Volda University College. Unpublished.

Reijnders, S. (2016). *Places of the Imagination: Media, Tourism, Culture.* Routledge.

Reijnders, S. (2021). Imaginative heritage: towards a holistic perspective on media, tourism, and governance. Locating Imagination in Popular Culture: Place, Tourism and Belonging (pp. 19–34). Routledge. https://doi.org/10.4324/9781003045359

Revill, G., Hammond, K., & Smith, J. (2020). Digital archives, e-books and narrative space. *Area (London 1969), 52*(2), 291–297. https://doi.org/10.1111/area.12413

Ryden, K. C. (1993). *Mapping the Invisible Landscape: Folklore, Writing and the Sense of Place.* University of Iowa Press. www.jstor.org/stable/j.ctt20h6sc9

Sample, M. (2014). Location is not compelling (until it is haunted). In J. Farman (Ed.), *The Mobile Story* (pp. 80–90). Routledge. https://doi.org/10.4324/9780203080788-12

Sandberg, K., L. (2023). *Aasmund Olavsson Vinje.* Store norske leksikon. Retrieved 28 August 2023 from https://snl.no/Aasmund_Olavsson_Vinje

Sevaldson, B. (2010). Discussions and movements in design research. *Formakademisk, 3*(1). https://doi.org/10.7577/formakademisk.137

Sevincer, V. (2023). *Teknisk info webapp.* Report. Unpublished.

Sikora, M., Russo, M.,Derek, J., & Jurčević, A. (2018). Soundscape of an archaeological site recreated with audio augmented reality. *ACM Transactions on Multimedia Computing Communications and Applications, 14*(3), 1–22. https://doi.org/10.1145/3230652

Steinsland, G. (2014). *Dovrefjell i tusen år: mytene, historien og diktningen.* Vigmostad Bjørke. Bergen.

Tsepapadakis, M. & Gavalas, D. (2023). Are you talking to me? An audio augmented reality conversational guide for cultural heritage [101797]. [Amsterdam]: https://doi.org/10.1016/j.pmcj.2023.101797

van Es, N., Reijnders, S., Bolderman, L., & Waysdorf, A. (2021). Locating Imagination in Popular Culture: Place, Tourism and Belonging. Routledge. https://doi.org/10.4324/9781003045359

Vinje, A. O. (1996). *Ferdaminne fraa Sumaren 1860* (D. Gundersen, Ed.). Gyldendal. https://doi.org/oai:nb.bibsys.no:999616277194702202URN:NBN:no-nb_digibok_2008061900069

5 Environmental *Imaginaria* in the Age of Extinction

Perspectives from the "Critical Zone"

Rodanthi Tzanelli

Introduction

This chapter explores a trend in the cultural registers of travel, one which relies on the refashioning of local, national and/or regional environments as imaginaries of identity via popular media, such as film, sound/music installations and their extensions in the digisphere. To be more precise, the chapter is not concerned with media forms as such, but the ways mediations of non-human lifeforms are managed in new audio-visual environments via new technologies. Hence, the chapter is a social theorist's perspective on the ways the said environments enter the realm of the artifice to be taxonomised as living compositions of nature (what I later term 'biomediations'). Such refashioning of lifeforms takes the shape of *imaginaria*, themed natural environments that are pleasing to the prospective visitor's gaze. The theme is based on the semiotic properties of a terrestrial location, that is, a/ the place that the exhibited landscape and/or group of species *allegedly* resides. Popular media are important in this refashioning only insofar as they allow representational centres to semiotically organise the themed environments, suggesting to audiences that we do indeed deal with a self-contained place and culture belonging to a dominion.

This process takes place in global contexts of environmental and social crises that we associate with climate change, climate-induced migrations of humans and other species, but also multispecies extinction. However, my approach to biomediation, which debates the colonisation of imagination through the commercialisation of environments belongs to a group of competing debates regarding the relationship between artifice, imagination and nature in, what Latour (2018, pp. 79–80) dubbed the "critical zone". Designating "climate change/catastrophe" as a transdisciplinary field spanning the humanities and the social sciences, the critical zone is not a pedagogical exercise, a "boutique" methodology on travel aesthetics, or yet another anthropocentric version of critical cosmopolitanism, but a posthuman approach to generative action. It is a way to scope the competing interpretations and experiences that define the relationship between research and public life.

Another perspective would direct our attention away from the conflicts produced in biomediations of nature: it would point out that artistic narratives centring on the

DOI: 10.4324/9781003320586-7

identities of environments tell us stories about planetary futures in which the said environments may not survive, whereas scientific interventions on climate change tend to act to prevent or slow down such processes. It would proceed to stress that these two agential forces, which move in different realms of action (the former in that of creativity and the latter in that of technical progress, utilitarian change and survival), are turned into partners in the production of imaginative installations of environmental enclosures under the threat of extinction. It takes a great deal of work to organise such narratives (i.e., to materialise them in audio-visual forms and sculptural objects to display) and human resources (i.e., the technical skills of directors, sound and image engineers, etc.). However, without the help of scientific findings on climate, and general scientific expertise, which often stretch to the ways these *imaginaria* are constructed (i.e., under particular ecological specifications), the said *imaginaria* would not exist. Otherwise put, then, this is a chapter about the way different worlds of science, imagination and culture come together in a troubled dialogue centring on the making and remaking of the environment as an object of tourist fascination. Although it covers a variety of semantic fields before modernity, contemporary uses of 'fascination' have evolved to a 'projection screen for the imagination of the consumer': an arrangement of aesthetic experiences connected by projective technologies that we find in music-making and filmmaking (Schmid et al., 2011, pp. 4–5).

A literal reference to a copy (Latin root *ima*) of forms of life (Latin suffix *ago*) that have morphed into objects of speculation or amazement as allegedly finished images (i.e., images that have acquired a dominant interpretation; Tzanelli, 2021a, p. 89), environmental *imaginaria* are neither museums nor scientised versions of nature; and yet, they are forced into a symbiosis with both. They can be designated (underwater or inland) territories of immense ecological significance and beauty, or enclosed spaces/buildings in which rare or at-risk (of extinction) species are preserved in an ordered/classificatory fashion; but they can also be mediated versions of the above in documentaries and movies, particularly when(ever) they remain inaccessible to most or all humans or cease to actually exist as territories. Mediated products of them can also be mobilised as exhibition styles in such themed establishments (i.e., museums featuring extinct species, natural parks or botanical gardens), so we may say that audio-visual renditions and organisations of nature persist across different versions of them. The latter observation suggests connections with cinematic tourist playscapes. However, the rootedness of audio-visual representations in mechanical reproductions of life can be deemed troublesome: turning lifeforms into dead object/matter to be admired "from afar", such mediated environmental *imaginaria* are more than leisurely or pedagogical understandings of planet earth's dying ecosystems, or "dark tourism", that is, visitations to sites of disaster and collective memory. They are generators of atmospheres, multisensory feelings concerning the end of the world and our way of coping with it (Haraway, 2016). These feelings produce the very field of multiple interpretations about the causes and consequences of human action on nature.

A critical approach to the development of environmental *imaginaria* in the tourism industry and experience invites a careful analysis of the unpredictable ways in which the Anthropocene affects planetary futures. However, the chapter does not subscribe to the futuristic scenarios regularly performed by scientists (i.e., the de-growing suggestion that dictates lower carbon consumption or the ecological modernisation promised by the implementation of eco-technologies; Urry, 2016, pp. 183–185), who try to slow down the Great Acceleration of extinction; instead, it reviews reactive/responsive *aggregates* of lay/leisurely and expert/scholarly perspectives on this unpredictability. All the perspectives in the critical zone struggle for autonomy and legitimacy because they represent different *imaginaries* of how our planet and its populaces, including human and non-human species and inanimate technologies, can live together. From all these competing interpretations, I debate three dominant ones across academic disciplines and public fora. The first concentrates on how the "deep histories" of human mastery of nature communicate with "flat accounts" concerning the insertion of themed *imaginaria* into the business of tourism and leisure. The second interpretation highlights the role of such *imaginaria* as lay forms of "worldschooling", performative education facilitating terrestrial/virtual encounters with planetary problems and registers of hope. The third considers the activist design of the critical zone: who speaks on behalf of whom in the global controversy. I conclude on the significance of mediated *imaginaria* as forms of *planetary legacy* in their own right in a world of postnational mobilities.

The Thesis on Exchange (En)closure: Questions of Power and Commodity Aesthetics

This thesis concentrates on the implication of environmental enclosures in the formation of natural parks, natural museums and species displays in national and regional economies. This thesis cuts across popular cultural channels populated by digital press agencies, television programmes on ecological themes, but also specific academic debates concerning futurist designs of sustainable travel and consumption (see Tzanelli, 2020; 2021b). The advent of audio-visual technologies in the twenty-first century reintroduced the taxonomical logic of the colonial enterprise of knowledge in academic debates on these enclosures. Now these enclosures are discussed as aspects of a blended biopolitical (the management of all forms of life by institutions) and ecological agenda (the institutional use of forms of life in different domains of human activity – here, tourism consumption – Hollinshead et al., 2009). Because of the continuous coerced non-human species transportations to former colonial centres, where such environmental *imaginaria* are currently infrastructurally most embedded, critical scholarship speaks of a double plundering of territory (a political rendition of belonging) and land (a biological and physical aspect of it) (see also Lapointe, 2021). Representations of forced species migrations that advertise the content of environmental *imaginaria* assists in producing a novel perspective on the ways the planet is damaged: if indeed humans are part of the land/nature which they reside, then mining, crushing and extracting from it damages them as bodies, minds and hearts, ultimately turning them into 'dead' icons (Clark & Szerszynski, 2021, pp. 146–149). In legal studies, such historical approaches posit questions of biopiracy (Brockway, 2002), whereas in tourism marketing they facilitate critical debates on the transformation of branding into the only valid/reasoning mode for global economic growth (Lury, 2004). Between these two poles, we may place the rise of commodity aesthetics in the place of filial networks of heritage and inheritance (including inheriting pristine lands and natural habitats to future generations).

Etymologies alone will not suffice but are indicative of the origins of this movement to blend artifice with nature: "environment" is a Scottish translation of Goethe's *Umbegung* or surrounding, better explicated through Carlyle's, Thomas Reid's and Sir William Hamilton's philosophical elaborations on the increased authority of technology in the (re)formation of natural spaces. Associated with the modernising project of "Lowland Clearances" (the expulsion of Lowlanders from their home territories by estate owners wishing to populate these areas with livestock or crops – see Jessop, 2012), the "environment" is a mark left on histories of the Anthropocene by intersecting socio-political and economic changes in the age of the Enlightenment. A geological age or period commencing with either automobile invention or the Industrial Revolution, during which human activity has been the dominant influence on climate and the environment, the Anthropocene is commonly seen as the era of human domination over all forms of planetary life. However, even "human life" is subjected to hierarchical criteria, with some social and cultural categories deemed to be of lower worth (Indigenous populations or lower classes). From the outset, *en-vironer* (old French) was understood as a *circuit* or *exchange enclosure* (both translations of the term), which could only be assessed under specific circumstances: the nature of (agrarian) exchange of said mastered land/nature for money. Today, the oxymoron's rationale resides in the ways institutions may simplify socio-cultural conceptions of the environmental *imaginarium* via new media representations so as to reach diverse prospective audiences.

Decades ago, Hollinshead (1998) brought these controversies to the house of tourism studies through a discussion of Stephen M. Fjellman's critique of the "legerdemain" (the digital re-enchantment of locations and objects of tourist fascination) and the "entrepreneurial violence" ("the deliberate social-engineering" of histories through pastiche, hybridisation or concealment and omission of is inconvenient to expose to global publics) "by which the Disney companies construct the supreme commodity fairyland at Walt Disney World" (Hollinshead, 1998, pp. 58–59). Whereas Hollinshead's approach to the "distory" or distorted "pastiche" history produced in these postmodern fairylands (1998, p. 81) does not quite explain the violence inflicted upon the environmental inheritance of the transported species, his approach to entrepreneurial violence does, up to a point. This is associated with the removal of the species' "journey" from the themed location's biographical record, which is displayed to visitors. The omission is often reproduced on the themed site's digital advertising, a recurring decision that generates phantoms of hospitality "in situ". As Germann Molz (2018) has argued, contemporary forms of hospitality progressively rely more on both new media and the actual removal of hosting bodies from tourist destinations (i.e., guest-tourists book via Airbnb accommodation and arrive at it to meet hardly ever the host-owners of the property). The multispecies groups residing in environmental *imaginaria* are beginning to resemble these absent human hosts of the global Airbnb lands, completely muted in the context of late capitalism's *exchange enclosures*.

But there are also other guests in such circuits of tourism, who are more disempowered than we may assume: the local managers of these *imaginaria*. Network capitalism has ensured that even the alleged managing institutions of environmental *imaginaria* do not possess entire control over how they can present such sites. Take for example Tropical World in Leeds, UK: a glasshouse display of flora, fauna and their accompanying physical atmospheres from different parts of the "Tropics". Its mission statement is split between the education of future generations (school visits are frequently accommodated to its visitor scheduling) and entertainment for adult visitors. The latter audience pool includes as diverse groups as whole families, visitors to the adjacent Roundhay Park for picnicking, cycling and walking, and festival or concert attendees, who frequently litter the park (Tzanelli, 2021a). There is no aggregate "public attitude" towards Tropical World's mission statement, which is further obscured by the missing narratives of its species' journeys from the "Tropics" to Leeds. On the contrary, the Tropical World's website prioritises attempts to brand the establishment's enterprise via the activities of young *British* generations. Managed by the Leeds Research Council, which has suffered economic losses over Brexit and other socio-political crises (e.g., COVID-19), the Tropical World is the victim of a complex delay in de-colonising: a multicultural Leeds, which includes migrant communities from the obscured "tropical zones"; a country increasingly hit by climate change, which made promises to stand at the forefront of scientific research on such matters; and a blended artistic and scientific imagination, whose staging of questions of heritage and legacy for the public gaze is sabotaged by profit-making forces.

Thus, when we discuss how worlds of tourism enter the battlefields of the critical zone, we must be cautious to not forget that during our actual or virtual visitations to environmental *imaginaria*, we may be encountering a fully updated version of what Herzfeld (2002) once called "cryptocolonialism". In the era of colonisation, even economically disempowered countries that were never colonised suffered the intellectual violence of Western progress. By analogy, today even the so-called developed world suffers the twin effects of a parochial attachment to the "white man's burden" to civilise and an economic disempowerment that drives decisions to exoticise forms of colonised Otherness for a tourist gaze.

The question of theming alone is a problem in knowledge enterprises such as those of Tropical World or the Botanical Gardens in Kew: both "labels" facilitate parochial blends of imperialism, colonialism and nationalist exceptionalism with the ghosting landscapes of the Anthropocene. The latter involve the replacement of the original life-enhancing entanglements that sustained particular plundered lands, with the ordered environments/landscapes in which their immutable species are displayed. Recent attempts to recreate entire environments at Tropical World merit public financing and are supported by international experts. Instead, the institution is forced into the market competition that comes with attracting any visitor (Tzanelli, 2021a).

Such legacies of representation have knock-on effects on scholarly critiques which presume that leisure and travel partake in planetary wrongdoing without paying attention to context. Their critiques assume an inherent complicity between institutional approaches to heritage and new/old forms of environmental mediation of the environment. Even the causes of environmental pollution are reduced to a demonisation of audio-visual production of exotic imaginaries of nature. Undoubtedly, there are links between production practices and pollution, or the exploitation of both impoverished and Indigenous labour in themed resorts and the reduction of environmental Indigenous heritage to audio-visual entertainment in movies, but the relationship between causation and correlation merits qualification.

The Thesis on Transcultural Mobilities: From Tourism to Travel/ Pilgrimage in the Anthropocene

Regardless of whether audio-visual technologies' semiotic function is mobilised by (corporate, national and regional) institutional projects of representing the environment (Neumann & Zierold, 2010, pp. 104–105), environmental *imaginaria* make new worlds of travel also independently from them. Out-of-context transitions from political-economic critiques of *imaginaria* display to the cultural poetics of site-seeing and visitation tend to filter human activity via economic imperatives. To consider culture also independently from economics, one must address the diversity of environmental *imaginaria's* visitor profiles as a blended expert/institutional/public project in its own right (Shelton, 2013). For this chapter's purposes, such concerns are summed up under a question: "how can we accommodate an economy of attention to critical climate futures among the objectives and motivations of visitors to such themed sites?" A similar question is asked today by museologists, scholars studying the ways museums stage their narratives for publics (Witcomb, 2015), however, the climate change angle is often not among their priorities. Producing economies of attention that question the conditions of environmental survival in the Anthropocene does not facilitate a traditional approach to tourism mobilities, as it brings to centre stage the calibration of a new multisensory approach to hospitality.

Thus, first, visitors to such *imaginaria* can hardly be classified as traditional "tourists". Capitalist imperatives may frame visits to such enclosures in terms of rationalised exchange, and "distorical" designers may attempt to dazzle visitors all they want, but to assume that those who visit environmental *imaginaria* lack critical skills or independently acquired knowledge is too far-fetched. The assumption is often the product of scholarship, which refuses to disrupt set roles and relationships (i.e., who is studied and how by whom and on what sort of ethical guidelines) in the inherently transdisciplinary fields of tourism and hospitality studies (Hollinshead et al., 2021). There will be space to explore the role of non-human agents in the narratives of these sites in the following section, as they posit unique challenges to the use of new media technologies and communication codes in hospitality at large.

Suffice it to focus here on such *imaginaria's* human visitors, who tour them without *really* being tourists. I assign them with the term or nomenclature "post-tourist", which I replace thereafter with that of the "traveller/pilgrim of the Anthropocene" for reasons that become apparent below.

The atmospheric qualities of environmental *imaginaria* as physical/territorial enclosures invite the visitors' multisensory affective involvement. New landscapes and species incite curiosity to apprehend and touch. They may also invoke feelings of awe that we associate with the environmental sublime: a difficult to pinpoint feeling stemming from an encounter with nature's power to command death and life (Bell & Lyall, 2002, p. 5). In addition, a visit to such physical enclosures is not for the faint-hearted or indifferent to physical or emotional risk visitor but directed towards the post-tourist thirst for pedagogical and physical adventure (Jansson, 2018). Unlike the Disneyland McWorlds of Fjellmann (Hollinshead, 1998), certain types of environmental *imaginaria* "ask" visitors to wave their rights to personal safety, and they do so voluntarily, because they love to experience the form of "edge" in their lives (Lyng, 2005). From the minimal demand to instantly acclimatise their body to humid environments, or climb up steep rocks, to diving into deep waters to visit orchestrated narratives of local marine lifeworlds, visitors are invited to take in a single breath of the environmental sublime, embrace it as an experience, and possibly relay it to friends and the world on visitor logs, blogs and personal videos.

Relevant themed sites facilitating this blend of ecotourism and post-tourism as voluntary risk-taking are six underwater museums in Granada, the Bahamas, Lanzarote and England. These include hundreds of submerged artefacts/sculptures designed by Jason deCaires Taylor, who is also an experienced scuba diver (Córdoba Azcárate, 2018, p. 12). The eerie sculptural forms, which are intended to problematise the loss of marine life due to global warming, can be visited by those with some experience in scuba diving (Tzanelli, 2020, p. 159). Subsequently, the six museums attract blended tourist visitors, some of whom take the opportunity to dive in organised tours in the warmer waters of the countries hosting these collections. The warmer the zone, the more touristified the museum's location. Another such underwater museum is installed in the Great Barrier Marine Park close to Townsville, Northern Queensland. As examples of ecological art, deCaires Taylor's sculptures are supposed to repopulate rapidly dying reef zones with sealife, but as tourist attractions, they partake in their pollution by the tourist hordes. Two conflicting agendas of post-tourist activity emerge in these at-risk *imaginaria*, but both end up enhancing tourist flows.

Because these pedagogical/experiential projects partake in niche tourism design while also encouraging the photographic and digital reproduction of the sites by visitors, they are integral to the new post-tourist imaginaries of travelling the damaged planet. Visiting these sites in person is economically prohibitive for the poorer tourist, so embodied visitations facilitate a new sensory-aesthetic division of leisure, allowing the upper middle classes and the super-rich to become associated with different ethics, aesthetics and ecologies (Cresswell & Merriman, 2011, pp. 6–7). Such ecologies turn the death drive into symbolic capital while removing disinterested play from leisure in favour of moral betterment. Because the very same *imaginaria* enter a second round of mobilities via social media platforms, their travel span and audience profiles increase and diversify. Enter the new "post-tourist" of the Anthropocene, who can travel virtually anywhere from their home PC, to participate in the aesthetic and emotional drama of nature's death and rebirth staged at the said physical sites.

With these observations in mind, some scholars and publics place the digital mediation of environmental *imaginaria* and their onsite post-tourist activities on a pedagogical continuum. There is a link to add to the use of multimedia representations in new museum sites, which I cannot fully address here. The prevalence of a "cinematics" of travel has become ubiquitous in these environmental museums not just as a cynical marketing style, but also as a way to tell stories about our planet. And there are such attempts ready at hand, such as the one by Greek cinematographer wildlife photographer and freediver, Stefanos Kontos, who managed to stage an Underwater Gallery of sealife and shipwreck objects outside Amorgos, to tell a story about Greek seascapes, flora and fauna. Free of charge and open to the public, including those who want to scuba dive to the *imaginarium's* premises, this initiative addressed blended types of post-tourism: onsite adventurers who wished to experience the marine sublime; publics who could not travel to the island but could watch videos of dive-tours by Kontos and his team to the site; and a meticulously designed website of the whole venture in which anyone with internet connection can learn about the initiative's ecological and aesthetic ethos (Tzanelli, 2020).

Ultimately, the terrestrial and virtual visitors of such projects are pilgrims of the Anthropocene. The term addresses new entanglements between human bodies, new technologies and very old sites of environmental birth and death at risk from climate change. The pilgrims are asked to feel, before fully comprehending formations of nature and matter as worlds in process. Thus, the pilgrims can engage with planetary problems at this particular moment in time. For some of these pilgrims, the moments of visitation to these *imaginaria* can be relived as many times as they wish on their computer screen. And it is this repetition that encourages new techno-anthropological interpretations of the world, in which the adult mind and heart are treated as too young in the grand scheme of things to comprehend phenomena such as the Great Acceleration, the Anthropocene and posthuman ethics, but pliable enough to begin to feel their way through the loss of our times. The thesis on Anthropocenic pilgrimage is based on a version of ontological realism (Barad, 2007): learning/comprehending while travelling, with the auxiliary use of new technologies, or "worldschooling", minus the social class limitations that the original definition purports (Germann Molz, 2021). The approach is directed predominantly to adult audiences to equip them not with practical sensibilities to consider planetary crises that are here to stay.

The Thesis on Multitude versus Biomediation: Unmuting and Resurrecting in the New Critical Zone

As soon as the second thesis promises to decolonise our embodied multisensory competencies, the practice of gazing upon multispecies compositions in environmental *imaginaria* reintroduces questions of power: humans watch, inspect and learn from them as objects of fascination but do not really interact with them reciprocally. This style of interaction challenges the visitors' adventurous educational-tourist pursuits, making them look more like "neo-colonisers" of a planet humans learned to mute in violent ways: through resource extraction, extravagant tourist resort and amusement park development and extreme urbanisation (Stinson et al., 2020). The third body of knowledge reorients the ontological realism of the second, with a warning that what is *now* absent or close to extinction will continue to affect with its ghostly radiance the symbiotic assemblages of what still exists (life).

This is an alternative "deep" reading of microscopic ghosts of geological memory: fungi, algae, bacteria and other organisms whose mark was left on layers in stones and soil. Such ghosts atmospherically visualise death and dying, foretelling further species extinction. The technique of atmospheric amplification is mobilised in Rio de Janeiro's Museum of Tomorrow, an *imaginarium* that does not host environments and species but utopian possibilities and dystopian narratives on climate change, including short films about environmental futures, which favour catastrophism (i.e., scenarios suggesting that the end of all life is inevitable – Hulme, 2017; Tzanelli, 2017). Environmental *imaginaria* stand in this body of knowledge as landscaped witnesses of loss, demanding a stratigraphic (layer-uncovering and reading) investigation of extinction in which different times merge (Barad, 2017a, pp. 105–107) to diffract the troubling coexistence of different species (Barad, 2017b). However, to reiterate by using different phrasing, the idea of theming nature preserves the ethos of the romantic tourist (Urry, 1990) who attempts to forage memory through diary-keeping while resiliently touring the new foraging routes of the Anthropocene (Tsing, 2015). To facilitate a true decolonisation of the foraging gaze, the science of staging environmental *imaginaria* must use the travel styles of new media recording only as a gateway to the art of multispecies listening (Tsing, 2015, pp. 17–25).

This seems a far cry from conventional sociological approaches to environmental activism, only if we deny that we experience a "reciprocal capture" in the Anthropocene (Stengers, 2010, p. 35): because we live in an unfolding event in which our mode of existence does not transcend that of other species we must learn to work and care with/for them (Rose, 2017, p. 52). Another mode of "capture" implicated in the production of environmental *imaginaria* tends to regress to old-fashioned romantic attachments to death as an identity-making form of sacrifice to allow each nation or region partaking in the touristification of its landscapes to label its beautiful landscape atmospheres as biological properties. This is commonly known as "biomediation" and happens via techniques of moving image production and digital advertising that dazzle audiences (Sheller, 2009; Thrift, 2010). Biomediation in environmental *imaginaria* is a technique of using the properties of physical exhibits to designate the uniqueness of the region from which they come (Thacker, 2004). However, "uniqueness" stands as a form of capital prestige that facilitates the region's global networking (Larsen & Urry, 2008). This mode of capture bleeds into the politics of the Anthropocene, because it associates the display of muted non-human species to visitors with the coercion of the disenfranchised to contribute to climate catastrophe (Latour, 2018, p. 83): poor workers who are asked to contribute to the very production of environmental *imaginaria* by culling their own land for corporate tourism business, and impoverished regions of the Global South who have to commercialise its unique biological properties to survive in global markets (Córdoba Azcárate, 2020). To this, one must add the exponential global growth of tourism, especially in heritage hotspots, which leads to crowding, further pollution and the displacement of localities in favour of development.

The third thesis invites the production of new multispecies worlds of imagination. Firstly, it calls for the organisation of assemblages of action against climate injustice, in which scholars, local publics and international organisations have equal voice but accountability proportional to the group's implication in climate wrongdoings (Haldrup, 2020; Sheller, 2020). Indigenous groups are seen as advocates not of "land" or "landscape", but the closest to a mediator of the rights of those beings (animals, plants, birds, bugs) that speak "asemic" languages humans cannot understand (Pyyhtinen, 2022). There is a call that in the maintenance of environmental *imaginaria* these new formations should also include Indigenous scientific experts with deep knowledge on the displayed environments. Such experiments have been conducted in Brazil, in areas of poverty and neglect, such as those resided by the *quilombos*, communities composed of peoples of African, Indigenous and European descent. *Quilombos* constructed independent societies outside the plantation system. Because the territories they reside today have developed into tourist destinations akin to those of environmental *imaginaria*, justice movements in them were concentrated on nominating Indigenous experts as their land's spokespersons (Guerrón Montero, 2020). In Sri Lanka, the institution of Ceylon Tea as a heritage attraction points in the same direction, with Tea Museums such as those in Hantane, Kandy, standing as environmental *imaginaria* in the place of the actual tea plantations. Notably, activist discourse on the environment and workers' rights is mostly mediated via the writings of travel bloggers, rather than fully organised activist groups (Jayathilaka, 2020).

Therefore, the new resurgent formations of the critical zone must call for transparency in the ways environments and human labour are *re*-presented, including how they *feature* in media formats and platforms. This transforms audio-visual technologies into worldmaking tools (Jensen, 2022) that facilitate the ontological freedom of what is included in environmental *imaginaria* from what corporate objectives may support, which is mostly profit-orientated (on posthumanism and freedom, see Hollinshead et al., 2021). As an emancipatory project, the third thesis must mobilise imaginaries of freedom and justice. The question is how to calibrate them.

Conclusion: Environmental *Imaginaria* as Ex-Samples of *Planetary* Legacy

It is my contention that the competing *bodies* of knowledge discussed in this chapter do not exist outside entanglements of scientific, scholarly, popular and artistic imagination, as Latour (2018) styles his version of the critical zone. Imagination is only contingently tied to systems of production and consumption, so its power to positively affect futures belongs to what he terms the system of endangering terrestrials, aggregates of human agents, animate beings and actors with some capacity to react to the contexts in which they live (Latour, 2018, pp. 82–83). The real challenge is our current inability to operationalise imagination as a faculty not exclusive to humans – something returning the chapter's inquiry to the original meaning of the word as aggregates of human/non-human bodily movement that generate images of life. This is just the start of an experimental journey – for, if indeed, we should stop talking about systems of production and consumption and draft instead a just system for endangering terrestrials, environmental *imaginaria* must not change their function but their creative directionality. By this I mean that their mission statement must not enclose natural otherness but its victimisation, thus introducing honest reconciliation statements in its premises. Such a process would not turn colonial legacies into the endangerment's focus but use them as a preamble to a "Great Deacceleration" of posthuman memory.

Practically speaking, existing environmental *imaginaria* should start using audio-visual fixtures into confessional tools: mechanisms of slowing down the process of thinking about the role of land and earthly life as common heritage, rather than national or regional property. A just system of engendering terrestrial life should facilitate, above all, awareness of shared risks, environmental interdependencies and a new relational posthuman consciousness (Ateljevic, 2008, pp. 188–289). Notably, the place of new media fixtures is not essential but auxiliary in this project. Our current "lack" is not in technologies but in the arts of multispecies communication.

References

Ateljevic, I. (2008). Transmodernity: Remaking Our (Tourism) World? In *Philosophical Issues in Tourism*, edited by J. Tribe, pp. 278–300. Channel View Publications.

Barad, K. (2007). *Meeting the Universe Halfway. Quantum Physics and the Entanglement of Matter and Meaning*. Duke University Press.

Barad, K. (2017a). No Small Matter: Mushrooms Clouds, Ecologies of Nothingness, and Strange Typologies of Specetimemattering. In *Arts of Living on a Damaged Planet*, edited by A. Tsing et al., pp. 103–120. University of Minnesota.

Barad, K. (2017b). Troubling Time/s, Ecologies of Nothingness: On the Im/possibilities of Living and Dying in the Void. In *Eco-Destruction*, edited by M. Fritsch et al. Fordham University Press.

Bell, C., & Lyall, J. (2002). *The Accelerated Sublime*. Praeger.

Brockway, L.H. (2002). *Science and Colonial Expansion*. Yale University Press.

Clark, N., & Szerszynski, B. (2021). *Planetary Thought*. Polity.

Córdoba Azcárate, M. (2018). Fuelling Ecological Neglect in a Manufactured Tourist City: Planning, Disaster Mapping, and Environmental Art in Cancun, Mexico. *Journal of Sustainable Tourism, 27*(4), pp. 503–521.

Córdoba Azcárate, M. (2020). *Stuck with tourism*. University of California Press.

Cresswell, T., & Merriman, P. (2011). *Geographies of Mobilities: Practices, Spaces, Subjects*. Routledge.

Germann Molz, J. (2018). Discourses of Scale in Network Hospitality: From the Airbnb Home to the Global Imaginary of "Belong Anywhere." *Hospitality & Society, 8*(3), pp. 229–251.

Germann Molz, J. (2021). *The World is Our Classroom*. New York University Press.

Guerrón Montero, M. (2020). Legitimacy, Authenticity, and Authority in Brazilian Quilombo Tourism: Critical Reflexive Practice Among Cultural Experts. *Tourism, Culture & Communication, 20*(2–3), pp. 71–82.

Haldrup, M., Samson, K., & McGowan, M.K. (2020). Toxic Climates: Earth, People, Movement, Media. *Performance Philosophy, 5*(2), pp. 252–59.

Haraway, D.J. (2016). *Staying with the Trouble*. Duke University Press.

Herzfeld, M. (2002). The Absent Presence: Discourses of Cryptocolonialism. *South Atlantic Quarterly, 101*(4), pp. 899–926.

Hollinshead, K. (1998). Disney and Commodity Aesthetics: A Critique of Fjellman's Analysis of "Distory" and the "Historicide" of the Past. *Current Issues in Tourism, 1*(1), 58–119.

Hollinshead, K., Ateljevic, I., & Ali, N. (2009). Worldmaking Agency–Worldmaking Authority: The Sovereign Constitutive Role of Tourism. *Tourism Geographies, 11*(4), pp. 427–443.

Hollinshead, K., Suleman, R. & Nair, B.B. (2021). Trilogy of Strategies of Disruption in Research Methodologies: Article 1 of 3: The Unsettlement of tourism Studies: Positive Decolonization, Deep Listening, and Dethinking Today. *Tourism Culture & Communication, 21*(2), pp. 143–160.

Hulme, M. (2017). *Weathered*. Sage.

Jansson, A. (2018). Rethinking Post-Tourism in the Age of Social Medias. *Annals of Tourism Research, 69*, pp. 101–110.

Jayathilaka, K.G. (2020). The Worldmaking Agency of the Sri Lankan Travel Blogger. *Tourism, Culture & Communication, 20*(2–3), pp. 117–127.

Jensen, O.B. (2022). Re-designing World-Making and Mobilities in the Techno-Anthropocene, Conference Presentation. København, Denmark.

Jessop, R. (2012). Coinage of the Term Environment: A Word without Authority and Carlyle's Displacement of the Mechanical Metaphor. *Literature Compass, 9*(11), pp. 708–720.

Lapointe, D. (2021). Tourism Territory/Territoire(s) Touristique(s): When Mobility Challenges the Concept. In *Progress in French Tourism Geographies*, edited by M. Stock, pp. 105–111. Springer.

Larsen, J., & Urry, J. (2008). Networking in Mobile Societies. In *Mobility and Place* edited by J.O. Baerenholdt, B. Granås and S. Kesserling, pp. 89–101. Ashgate.

Latour, B. (2018). *Down to Earth*. Polity.

Lury, C. (2004). *Brands*. Routledge.

Lyng, S. (2005). *Edgework*. Routledge.

Neumann, B., & Zierold, M. (2010). Media as Ways of Worldmaking: Media-Specific Structures and Intermedial Dynamics. In *Cultural Ways of Worldmaking*, edited by V. Nünning and A. Nünning, pp. 103–118. De Gruyter.

Pyyhtinen, O. (2022). Lines that Do Not Speak: Multispecies Hospitality and Bug-Writing. *Hospitality & Society, 12*(3), pp. 343–359.

Rose, D.B. (2017). Shimmer: When All You Love Is Being Trashed. In *Arts of Living on a Damaged Planet: Ghosts and Monsters of the Anthropocene*, edited by A Lowenhaupt Tsing, H.A. Swanson, E. Gan and N. Bubandt, pp. G51–G63. Minnesota Press.

Schmid, H., Sahr, W.-D., & Urry, J. (2011). Cities and Fascination. In *Cities and Fascination*, edited by H. Schnid, W.-D. Sahr and J. Urry, pp. 1–15. Routledge.

Sheller, M. (2009). The New Caribbean Complexity: Mobility Systems, Tourism and Spatial Rescaling. *Singapore Journal of Tropical Geography, 30,* 189–203.

Sheller, M. (2020). Mobility Justice. In *Handbook of Research Methods and Applications for Mobilities*, edited by M. Büscher, M. Freudendal-Pedersen, S. Kesselring, and N. Grauslund Kristensen, pp. 11–20. Edward Elgar Publishing.

Shelton, A. (2013). Critical Museology: A Manifesto. *Museum Worlds: Advances in Research, 1*(1), pp. 7–23.

Stengers, I. (2010). *Cosmopolitics I.* University of Minnesota.

Stinson, M., Grimwood, B., & Caton, K. (2020). Becoming Common Plantain: Metaphor, Settler Responsibility, and Decolonizing Tourism. *Journal of Sustainable Tourism, 29,* pp. 234–252.

Thacker, E. (2004). *Biomedia.* University of Minnesota Press.

Thrift, N. (2010). Understanding the Material Practices of Glamour. In *The Affect Theory Reader*, edited by M. Gregg and G.J. Seigworth, pp. 289–308. Duke University Press.

Tsing, A. (2015). *The Mushroom at the End of the World.* Princeton University Press.

Tzanelli, R. (2017). *Mega-events as Economies of the Imagination.* Routledge.

Tzanelli, R. (2020). *Magical Realist Sociologies of Belonging and Becoming.* Routledge.

Tzanelli, R. (2021a). Colonial Heterotropics and Global Heritage Aesthetics in Roundhay's Tropical World, Leeds. In *Experimental Museology*, edited by M. Achiam, et al., pp. 83–89. Routledge.

Tzanelli, R. (2021b). *Cultural (Im)mobilities and the Virocene.* Edward Elgar.

Urry, J. (2016). *What Is the Future?* Polity Press.

Witcomb, A. (2015). Toward a Pedagogy of Feeling: Understanding How Museums Create a SPACE for Cross-Cultural Encounters. In *The International Handbooks of Museum Studies*, edited by A. Witcomb and K. Message, pp. 321–344. Wiley.

Part 2

Visiting Places of the Imagination

Fandom, Experience and Affects

6 The Everyday Tourist

Traveling the Theater of the Mind in the Wake of Permacrisis

Christine Lundberg, Vassilios Ziakas, and Kristina N. Lindström

Introduction

The present age is characterized by high instability and uncertainty, identified by the recently coined term 'permacrisis' (*Forbes*, 2023). It is described as a continuous period of turmoil and multiple crises, which call for a rethinking of conventional business models and modalities. In a 2023 *Forbes* article, digital transformation was proposed as key to counter the effects of permacrisis, which entails a shift towards creating resilient business that is pressure tested for disruptions. This means developing new business models and modus operandi that allow for agility and adaptability through the use of technology. Tourism is without doubt one of the sectors affected most by the permacrisis. According to McKinsey and Company (2023), metaverse could be the next evolution for the tourism industry. The metaverse allows for collective experiences, where digital and physical worlds converge, creating completely new, hybrid, immersive experiences, also referred to as 'extended realities' (augmented and virtual reality). McKinsey and Company's (2023) estimated value for the travel sector to be over $20 billion by 2030 and as such, they paint a financially strong picture for a touristic metaverse. However, they also emphasize that today, its primary relevance, is for the events and the meetings, incentives, conferences, and exhibitions (MICE) sector and they raise concerns about users' data protection and virtual rights to locations. While the move towards digital forms of tourism would logically take place independently from permacrisis, it appears that its turbulent conditions have significantly accelerated digitalization along with the restructuring of broader socio-economic systems and sectors (Tzanelli, 2021). As a result, it is critical to ask what the constitution of tourism is in today's society, what tourism means (or what tourism is like), if it is defining quality up to now, the physical travel, is de-emphasized or put aside.

This chapter looks at the case of popular culture as a special entanglement of affective connections, building digital, online communities that elevate virtual travel to a real-like embodiment of the world and open new possibilities for its reconstruction. The idea of the 'Theater of the Mind' is here employed, a jargon first used in radio drama (Verma, 2013) and later in the context of role-playing games, recognizing the link between the mind and media through audience imagination, facilitating reconstructions of travel to places experienced through popular

DOI: 10.4324/9781003320586-9

culture expressions. Given the definition of popular culture as what people do in their everyday life (Fedorak, 2018), it is here put forward that virtual travel in itself is a new and hybrid expression of popular culture. On this basis, the emergence of 'everyday tourist' as a new type of traveler is identified that transcends traditional definitions of tourism. The purpose of this chapter is to better understand the evolving digital-driven transformation of tourism and the pivotal role of popular culture as sublimation of everyday life, which challenges not only the conventional perceptions of destinations and tourist-systems as fixed constructs but also the traditional dichotomies between fantasy–reality, everyday–extraordinary, and mobile–sedentary. The analysis operationalizes a mobilities perspective to explore the nature of structural change and its implications for the redistribution of power in the tourism (production) system. It is concluded that 'everyday tourism' constitutes a form of 'popular culture world-making' that may act as an antidote to the deleterious effects of permacrisis, but attention is also needed on preventing and/or mitigating its potential dark side.

Tourism Mobilities and Popular Culture

The notion of mobilities is fundamental in tourism through its focus on the connection of tourists and destinations, primarily through corporeal mobilities, but also through the mobilities of organic and induced images of the destination (i.e., virtual tourism). Media mobilities and virtual travel have always been a strategically important part of destination marketing, in the same way as news media and popular culture consumption have evoked or created reluctance to visit places portrayed, sometimes out of control of destination management (Lundberg & Lindström, 2020). Miami in the 1980s is one 'pre-digitalization' example of a destination capitalizing on the iconic TV series *Miami Vice*, portraying Miami together with a novel integration of contemporary pop and rock music and stylish or stylized visuals. Nevertheless, parallel to the success of the story of the Miami-based undercover detectives Crocket and Tubbs and their fighting against drug cartels in the paradise-like environment of Miami, the darker image of a society suffering from drugs and drug-related crimes reported about in the news media, blurred with the fictional images (Morgan & Pritchard, 1998). Consequently, the crucial intertwined relationships between different categories of mobilities existed long before digitalization became a key driving force of society.

However, over the years, the development of advanced information and communication technologies have created complex mobilities production and consumption patterns, further challenging the notion of how people and places are connected in the tourism system (e.g., Jansson, 2020; Reijnders et al., 2022) and the tourism system per se (Lindström, 2019; Lundberg & Lindström, 2020). In the wake of the pandemic, the traction of virtual tourism through popular culture consumption accelerated and replaced the ability to go and visit places 'in real life'. Hence, the crucial role of technology in reshaping the structure of tourism and driving tourism mobilities is of paramount role for the understanding of relationships between tourism and popular culture placemaking (Verma et al., 2022). In addition to being a cornerstone of tourism, applying the theoretical notion of mobilities prevents a limited understanding of tourism as a sedentary and non-ordinary activity, but as an everyday activity blurring and intersecting with other categories of mobilities, for example virtual travel to places through popular cultural consumption (Cohen et al., 2015; Edensor, 2007; Haldrup & Larsen, 2010; Hall, 2005).

Popular culture plays a crucial role as a place image-maker and there is a myriad of examples where destination marketers use popular culture to strategically promote awareness and create anticipations among audiences to visit a place during and after, for example, watching screen productions, reading books, or listening to music. Consequently, one can argue that the tourism system/supply chain has evolved/converged as a consequence of the development of digitalization (e.g., streaming services, social media, virtual reality, augmented reality, artificial intelligence) and the complex web of conflation of popular culture/tourism mobilities in the wake of this development. Norway is one of many examples of a country where the business of physical tourism mobilities increasingly intertwine with the business of on-screen production and thus the business of virtual tourism. The blockbuster productions of *Mission: Impossible* (*Fallout* and its sequel *Dead Reckoning*) with Tom Cruise as the leading actor made the decision to film advanced action scenes in Norway. A combination of the beautiful Norwegian mountain and fjord landscape, easily accessible through advanced transportation infrastructure and the support of the Norwegian state, reinforced this decision (Innovasjon Norge, 2022). The business of consuming places, physically and virtually, in the wake of on-screen productions, has evolved as an important part of tourism and destination development, restructuring the tourism production system and the power relations within it (Lundberg & Lindström, 2020). Complex intersectoral collaborations between the tourism and the film industries as well as blurred policy and governance landscapes at various levels challenge the traditional understanding of tourism mobilities, its impacts and the governance strategies that need to be put in place to balance business growth and protection of natural landscapes and local communities.

This development can be understood through the lens of the intellectual roots of mobilities in sociology and the so-called new mobilities paradigm. It springs out of a reaction in social science towards the lack of engagement and simplified understanding of the systematic movement of people, for example, for leisure and pleasure (Sheller & Urry, 2006; Urry, 2000, 2007). The paradigmatic shift is both a backlash to sedentarist theories and an extension of the 'spatial turn' in the social sciences, stressing the interpretation of space and spatiality of human life. As pointed out by Sheller (2004), mobilities take place through the mobilization of locality and the rearrangements of the materiality of place. The notion of mobilities is here understood as a phenomenon encompassing the movement of people, objects, and ideas, across different spectrums of movement. Building on the seminal work by Sheller and Urry (2006) and their theoretical pillars underpinning the so-called new mobilities paradigm in the social sciences, there are a number of crucial factors to take into account to understand the underlying premises of tourism mobilities in the wake of digitalization and permacrisis. Places that are connected to each other will have consequences for how people perceive them.

Mobile sociotechnical networks shrink distances, making things and people closer to each other in advanced and crucial ways. Consequently, popular culture consumption/production is not an objective transmission of place images but converging the meaning of place as it connects people to places and shapes how they relate to them through fictional stories. Places are physically and subjectively shaped by people, capital, material, and immaterial things. It is therefore important to pay attention to the distribution of power as well as the winners and losers in the placemaking machinery of popular culture tourism. Mobilities are dynamic, complex adaptive systems, on the one hand, striving for stability (path-dependency), and on the other hand willing to change when stability is disrupted. Hence, it is of importance to both pay attention to popular culture tourism production/consumption path-dependency and the disruptions to patterns to better understand this phenomenon (Sheller & Urry, 2006, 2016).

Adding to the complexity, advanced and rapid digitalization mean new, sometimes blurred, relations between physical travel and different forms of mediated tourism (e.g., virtual tourism). Media and the general communication landscape have gone through similar development in the wake of the rapid digitalization. This results in the tourist having an opportunity to engage in multiple (digital) formats and platforms. Single media consumption (and production) is, thus, replaced with complex webs of transmedia cultural circulation (Freeman & Gambarato, 2018; Jansson, 2020). At first sight, it is tempting to define this as an increasingly blurred media landscape, obstructing the understanding of virtual tourism mobilities. However, as pointed out by Jansson (2020), one must be sensitive to new distinctions evolving through transmedia cultural circulation. Jansson (ibid.) proposes a typology aiming at explicating "how the normalization of transmedia in society contributes to further de-differentiation along the lines of the post-tourism thesis" (p. 392). The typology contains a triple articulation of de-differentiation entailing the transmedia tourist. First, transmedia further extend tourist practices and attitudes in everyday life (ubiquitous transmedia tourist); second, transmedia accentuate the de-differentiation between 'home' and 'away' (decapsulated transmedia tourist); and third, transmedia generate and extract digital audience data (streamable transmedia tourist). This epitomizes the emergent configuration of virtual travel as an ontological mode of action challenging conventional perceptions of tourism, reality, and the world.

Irrealism and Virtual Travel

Drawing upon mobilities theory, Tzanelli (2020) examined websurfing of cultures and landscapes to suggest that such virtual travel should be considered as a novel secular form of metamovement. This type of touristic activity produces multiple versions of reality (world versions), both in conjunction with corporate internet design and independently from technocratic control. It also produces online "travel" communities reorganizing perceptions of mobility within a virtual system of services. In particular, Tzanelli (2020) highlights that digital travel is an act of irrealism:

not only does it open up new possibilities of performing travel as an imaginative/imagined form of movement, it also pluralizes the ways such travels are relayed to others. It is not limited to a "simulation" in consumerist ideological contexts, but also involves pluralizing representations of existing landscapes, heritages, and cultures of actual sites and increasingly tourismified destinations.

(p. 236)

Irrealism is a philosophical term originally conceived by Goodman (1978). It is not the antonym of realism but instead broadly means that the world consists of autonomous alternate versions. While these versions may concern specific modes of unreality in a very unique or unusual fashion, at the same moment they may also combine realistic scenarios in full compliance with human senses and universal physical laws. Their understanding can help us appreciate the multi-sided making of the world and the possibilities for constructing alternative worlds by transcending the conventional views of reality.

The irrealist capacity of digital travel put forward by Tzanelli reexamines understandings of world-making as a force that shapes tourism around the world. It revises Hollinshead's concept of *worldmaking*, which described that tourism realities were the preferred one-sided and exclusionary version staged by destination authorities (Hollinshead et al., 2009). On the contrary, digital travel encourages the human mind to produce several world versions, which may swerve from those designed by technical experts. While virtual tourists still come in contact with the objectified world made by digital experts and destination authorities, they have the liberty to making new worlds, both alone and in unison with other virtual travelers, with whom they can form a digital community. This enables fertile blends of world-making that combine different elements oscillating between fantasy and reality, subjectivity and objectivity, or internal (made in one's mind) and external (made by others) perceptions of the world. Such irrealist constructions of the world allow for more freedom and plurality of travelers' agency and representation of their various views. In this manner, they contribute to the hybridization of tourism in multidimensional niche products and its strengthening as a means to understand the world. In this regard, Tzanelli (2020) characterizes virtual travelers as popular culture worldmakers, being both romantic in a personal search for meaning and mundane "in constant dialogue with the technocraft of touristified digital business" (p. 239). This notion reconceptualizes Urry's tourist gaze to a hybrid irrealist perspective of popular culture. It also poses questions about how virtual 'popular culture worldmaking' and its tourist gaze are experienced, negotiated, and crystalized.

Thus, modern tourism is a complex form of mobilities, challenged by advanced digital consumption and production of popular culture, offering an alternative to the physical movement from home to the tourist destination. Today, consuming places at home has evolved as an alternative to being a tourist in the traditional sense. Consequently, one can become travelers in our everyday life, doing our routine activities and escaping at the same time, being in our ordinary world and entering other worlds concurrently, converging reality and imagination, hence making virtual travel part of our everyday life and popular culture, what is described with the term 'everyday tourism'. The notion of everyday tourism is eradicating the lines between home and away, the physical act of traveling and traveling through mind. One can argue that everyday tourism as an expressive practice of popular culture is a democratic form of tourism, accessible for the vast majority.

Nevertheless, from a sustainability perspective, everyday tourism raises questions about how everyday tourism mobilities change the consumer/tourist experience and behavior. Furthermore, it will lead to new groups of winners and losers in the tourism system, especially from a local business and community perspective, as well as the ecological footprint of popular culture induced tourism. A number of critical points can be made, using Foucault's understanding of power, critically applied in a tourism setting (Cheong & Miller, 2000). Power is omnipresent in the tourism system, and it includes the whole network of direct and indirect tourism stakeholders. Furthermore, power is not fixed to specific stakeholders in the tourism system; it works at many different levels and directions. The Foucauldian targets and agents may change over time and in different contexts. With that said, the tourist gaze is defined as the primary mechanism of power in the system and, lastly, power in the tourism system is both repressive and productive. A closer look and deeper understanding of the different dimensions of power of 'the theater of the mind' and the complex web of stakeholders included in the production of the tourism experience is needed (e.g., Lindström, 2019; Lundberg & Lindström, 2020). It will provide crucial information about stakeholders in control of the gaze, where the business opportunities are and, most importantly in the wake of the permacrisis, provide insights into the (un)sustainability of popular culture tourism.

The Everyday Tourist in Permacrisis

Returning to the everyday tourist and their everyday-extraordinary travels in the wake of turmoil and multiple crises, permacrisis signals a long-term turbulence in social organization and planetary economic, political, or cultural institutions and structures (Tzanelli, 2021; Ziakas et al., 2021). It marks an epoch of transition where boundaries are blurred as to what and how people may live, what they do, where they can go, how they can travel, and who they actually are. It is more of an existential crisis interrogating the very grounds of our hypostasis as individuals and makeup of society at large. The anxiety and fear about tomorrow put into question our current way of life and the extent to which it leads to sustainable futures. It is within this tempestuous context that expressions of irrealism take place and gain prominence as a way to adjust in the new conditions or new normal and find solutions to diachronic problems. Meta-tourism along with phygital forms of touristic interactivity that blend physical and digital elements (integrating technologies such as virtual reality, augmented reality, and artificial intelligence) to enhance the tourist experience play a significant part in enabling imagination to become a living medium of traveling in conjunction with the organic corporeality of tourists and destinations (Mieli, 2022). It is proffered here that meta-tourism is an emergent genre of popular culture that constitutes a new type of tourist: the everyday tourist.

The notion of everyday tourist is an analytical construct and ideational formation. It is neither merely a material category nor just a paradigmatic travel enactment. Instead, it is the archetype of a consumer who uses multimedia to travel across the world and experience destinations as much digitally as physically. The digital modes of traveling allow for an ontological transcendence of long-established dualisms that defined tourism: home–destination, ordinary–extraordinary, and mobile–sedentary. Specifically, the everyday tourist can be concurrently at home and away, traveling to extraordinary places of imagination as part of their ordinary life. In this fashion, the everyday meets the extraordinary extending the boundaries of reality, albeit confined within the spatio-temporal limitations of being home. Different blends of popular culture can be created to enable the confluence of imagination and reality (Ziakas et al., 2022). Consequently, meta-tourism takes on a different character and dimensions. Tourism traditionally has been viewed as an activity beyond ordinary life, a special time outside normal time and away from everyday routine, an activity exotic or extraordinary. On the contrary, the everyday tourist makes traveling part of their ordinary life, or in other words, it is the living embodiment of idealistic imaginaries taking place within their routine and mundane realities with the potential to reverse them when they move from the individual to collective level. This may occur when individual imagination of meta-travel experience is shared, discussed, negotiated, and evaluated via enduring participation in online fandom communities. Internet facilitates a shift from the individual to the collective through its 24/7 access to the audience's popular culture object of fascination and interaction with likeminded people. The audience's emotional ties to these 'fictional' places, characters, and storyline can shape their affectional feelings, loyalty, longing, and even shape their identity. Furthermore, it can provide an audience with a 'safe haven' or 'happy place' as a contrast from the strains and rejections of everyday life and contribute to an overall sense of well-being.

Perhaps this has never been more evident than in the global lockdowns in the wake of the COVID-19 pandemic in 2020–2021. In many countries across the world, citizens' mobility was highly restricted, allowing us only a few hours outside our homes a day. For many of us, this led to an everyday life of doomscrolling the news and checking in on real-time updates of rising COVID-19 cases and deaths, dealing with this new reality of life, which resulted in an all-time high need for bodily and mental escapism. As legal restrictions followed, prohibiting us from leaving our homes, and hefty penalties were enforced as well as physical traveling was banned, many of us turned to escapism by means of streaming platforms like Netflix, Amazon Prime, and HBO and social media platforms like YouTube, TikTok, and Instagram. This is where audiences went to escape our darkest fears and boredoms that made up their new dreary 'everyday life', to visit our favorite places and people, old and new. They visited a fashionable pre-pandemic New York to see our *Sex and the City* friends Carrie, Samantha, Charlotte, and Miranda. They also explored new fascinating places like a tiger zoo in Oklahoma and its owner Joe Exotic aka The Tiger King. This form of escapism allowed audiences to dream about old and new places. In crises that followed, such as the 2022–2023 cost-of-living crisis, other restrictions, such as money and time, have limited our physical mobility and challenged our mental health, and once again audiences have turned to escapism as an everyday tourist, traveling the theater of the mind by means of different hybrid forms of popular culture expressions.

Concluding Remarks

The increasing digitalization of tourism in the tumultuous conditions of permacrisis reconstitutes the traditional notions that defined tourism to date. A mobilities perspective allows for exploring and appreciating the transformation of tourism within its social organizing context of dynamic relationships, identities, and cultures. It brings to the fore the need to better understand how virtual connections are embodied, enacted, and elevated via online communities that convert and refine virtual travel from individual imagination to a social experience of world-making in terms of co-constructing alternate social realities. In this regard, the role of popular culture in bringing the everyday to tourism deserves much more attention. Especially, the kaleidoscopic spectrum of popular culture expressive practices enables the hybridization of tourism with different fandoms and activities (Lexhagen et al., 2022), thus engendering new forms and patterns of tourism like the archetype of everyday tourist that have been put forward in this chapter.

At a practical level, by understanding the everyday tourist, established practices on how tourism has been marketed and managed to date can be challenged. Time becomes less dominant as virtual travel can happen anytime. Importantly, the place is not a fixed destination anymore and can take on multiple irreal dimensions or world versions that are means to a perennial ontological end: how realities are built up and understood. Attributes of world-making and popular culture fandom visions should be incorporated into tourism products and online travel communities to cross-fertilize the daily realities with alternate world versions. In other words, tourism becomes a journey about the everyday as much inside as beyond the confines of the everyday.

The embodiment of everyday tourism provides possibilities for more freedom, pluralism, and emancipation of tourist agency to reconstruct social conditions or social worlds. It brings to the fore the ongoing hybridization of tourism with alternative modes of travel and permutations of popular culture. It transcends conventional dichotomies that have long constrained the ontology and breadth of tourism as a social phenomenon. In this sense, the everyday tourist may become an active agent of co-constructing alternate phygital orders and social worlds in a continuous process of world-making. In a similar fashion, the tourist gaze of the everyday can be described as a reflexive and dialectical practice between the idealistic and the mundane, which epitomizes different ways of looking at the world, reshaping what is seen, and how it is seen. It blurs the borderlines between reality and imagination bestowing an irrealist world-making theater of the mind that offers a temporary shelter from the pains of permacrisis. An imagined theater of everyday life about visions and embodiments of the ideal, extraordinary, or utopian performed by a diverse cast of actors ranging from individual tourists and online communities to media and destination authorities that altogether compose the hybrid phenomenon of everyday tourism.

The recognition of everyday tourism as an expression of popular culture can afford the critical study of this phenomenon. Fruitful prospects have been highlighted herein, but one should also be careful of its limitations, caveats, or possible negative effects. Here a major criticism is the extent to which virtuality via meta-tourism overcomes the risk of self-isolation within an individual's theater of the mind and translates to meaningful or equitable relationships and co-constructions of alternate worlds, then what these worlds are, where power lies, and whose interests are served in these realities. The economic power of high-tech apparatuses and affordability to everyone is a principal factor that needs to be considered in popular culture meta-tourism world-making. Future studies need to examine the economic and political tapestry of interests intertwined with meta-tourist configurations. Equally, the social context and nature of exchanges that reshape the tourist gaze have to be explored from a critical mobilities perspective. It might be unfortunate if what one has at the end is a return to where one was in the first place, reproducing or even magnifying existing inequalities, changing the world so that it remains the same. The critical study of everyday tourist journeys can give an insight into where fans are going to, how they move in-between, and eventually what kind of worlds they co-create.

References

Cheong, S. M., & Miller, M. L. (2000). Power and tourism: A Foucauldian observation. *Annals of Tourism Research, 27*(2), 371–390. https://doi.org/10.1016/S0160-7383(99)00065-1

Cohen, S. A., Duncan, T., & Thulemark, M. (2015). Lifestyle mobilities: The crossroads of travel, leisure and migration. *Mobilities, 10*(1), 155–172. https://doi.org/10.1080/17450101.2013.826481

Edensor, T. (2007). Mundane mobilities, performances and spaces of tourism. *Social & Cultural Geography, 8*(2), 199–215. https://doi.org/10.1080/14649360701360089

Fedorak, S.A. (2018). What is popular culture? In C. Lundberg & V. Ziakas (Eds.), *The Routledge Handbook of Popular Culture and Tourism* (pp. 9–18). Routledge.

Forbes. (2023). Permacrisis: Resilience emerges as key focus for digital transformation in 2023 (forbes.com).

Freeman, M., & Gambarato, R. R. (Eds.). (2018). *The Routledge Companion to Transmedia Studies*. Routledge.

Goodman, N. (1978). *Ways of Worldmaking*. The Harvester Press.

Haldrup, M. & Larsen, J. (2010). *Tourism, Performance and the Everyday: Consuming the Orient*. Routledge.

Hall, C. M. (2005). Reconsidering the geography of tourism and contemporary mobility. *Geographical Research, 43*(2), 125–139. https://doi.org/10.1111/j.1745-5871.2005.00308.x

Hollinshead, K., Ateljevic, I., & Ali, N. (2009). Worldmaking agency–worldmaking authority: The sovereign constitutive role of tourism. *Tourism Geographies, 11*(4), 427–443. https://doi.org/10.1080/14616680903262562

Innovasjon Norge. (2022). Tettere samarbeid mellom film og reiseliv skal trekke flere turister til Norge. www.innovasjonnorge.no/nyhetsartikkel/tettere-samarbeid-mellom-film-og-reiseliv-skal-trekke-flere-turister-til-norge

Jansson, A. (2020). The transmedia tourist: A theory of how digitalization reinforces the de-differentiation of tourism and social life. *Tourist Studies*, *20*(4), 391–408. https://doi.org/10.1177/1468797620937905

Lexhagen, M., Ziakas, V., & Lundberg, C. (2022). Popular culture tourism: Conceptual foundations and state of play. *Journal of Travel Research*, *62*(7), 1391–1410. https://doi.org/10.1177/00472875221140903

Lindström, K. (2019). Destination development in the wake of popular culture tourism: Proposing a comprehensive analytic framework. In C. Lundberg & V. Ziakas (Eds.), *The Routledge Handbook of Popular Culture and Tourism* (pp. 477–487). Routledge.

Lundberg, C., & Lindström, K. N. (2020). Sustainable management of popular culture tourism destinations: A critical evaluation of the Twilight Saga servicescapes. *Sustainability*, *12*(12), 5177. https://doi.org/10.3390/su12125177

McKinsey & Company. (2023). Tourism in the metaverse: Can travel go virtual? Tourism in the metaverse: Can travel go virtual? Report.

Mieli, M. (2022). *Smartphoned Tourists in the Phygital Tourist Experience*. Doctoral Thesis (compilation), Department of Service Studies, Lund University.

Morgan, N., & Pritchard, A. (1998). *Tourism Promotion and Power: Creating Images, Creating Identities*. John Wiley & Sons.

Reijnders, S., Boross, B., & Balan, V. (2022). Beyond the tourist experience: Analyzing the imagination of place and travel in everyday life. *Tourism Culture & Communication*, *22*(1), 31–44. https://doi.org/10.3727/109830421X16262461231783

Sheller, M. (2004). Demobilizing and remobilizing Caribbean paradise. In M. Sheller & J. Urry (Eds.), *Tourism Mobilities: Places to Play, Places in Play* (pp. 25–33). Routledge.

Sheller, M., & Urry, J. (2006). The new mobilities paradigm. *Environment and planning A*, *38*(2), 207–226. https://doi.org/10.1068/a37268

Sheller, M., & Urry, J. (2016). Mobilizing the new mobilities paradigm. *Applied Mobilities*, *1*(1), 10–25. https://doi.org/10.1080/23800127.2016.1151216

Tzanelli, R. (2020). Virtual pilgrimage: An irrealist approach. *Tourism, Culture and Communication*, *20*(4), 235–240. https://doi.org/10.3727/109830420X15991011535517

Tzanelli, R. (2021). *Cultural (Im)mobilities and the Virocene: Mutating the Crisis*. Edward Edgar Publishing.

Urry, J. (2000). *Sociology beyond Societies: Mobilities for the Twenty-first Century*. Routledge.

Urry, J. (2007). *Mobilities*. Polity.

Verma, N. (2013). *Theater of the Mind: Imaginations, Aesthetics, and American Radio Drama*. The University of Chicago Press.

Verma, S., Warrier, L., Bolia, B., & Mehta, S. (2022). Past, present, and future of virtual tourism-a literature review. *International Journal of Information Management Data Insights*, *2*(2), 100085. https://doi.org/10.1016/j.jjimei.2022.100085

Ziakas, V., Antchak, V., & Getz, D. (2021). *Crisis Management and Recovery for Events: Impacts and Strategies*. Goodfellow Publishers.

Ziakas, V., Tzanelli, R., & Lundberg, C. (2022). Interscopic fan travelscape: Hybridizing tourism through sport and art. *Tourist Studies*, *22*(3), 290–307.

7 From SDCC to Globalised/Glocalised Comic-Cons

Towards the Experience Economy of Co-existential "Event Fans"

Matt Hills

Introduction

Comics and media conventions aimed at fan participants constitute a major example of worlds of imagination. Such cons typically combine, in one convention centre or exhibition hall, an array of diegetic fictional worlds and their transmedia products, if not mixing a range of imaginary universes then at least abutting them together. The most studied of cons has been the mega-event San Diego Comic-Con (SDCC), run by Comic Con International (CCI) (see, for example, Hanna, 2020; Smith, 2022; Gilbert, 2017; 2018; Kohnen, 2020). Boosted by its proximity to Hollywood, SDCC has been analysed as a destination for fan pilgrimage (Geraghty, 2014). More generally, work has started to consider cons as tourist destinations in places outside the United States (Tang et al., 2023), while SDCC's international reach has itself been emphasised (Salkowitz, 2021, p. 152). In short, Comic-Cons have become a focus of study in their own right (Woo et al., 2020: Kohnen et al., 2023).

At the same time, work on media tourism has engaged with fan studies (and vice versa) to think about not only how film tourism or contents tourism can operate (Waysdorf, 2021; Yamamura & Seaton, 2020) but also how people can become fans of places (Williams, 2018, p. 104; Hills, 2021). In this chapter, I want to expand these approaches by centring on *fans of events* such as Comic-Con (Kohnen, 2020, p. 92; Edmunds, 2021, p. 176). Rather than such fannishness operating in addition to fandoms of text, brand and place, I will argue that these are *co-existential*: such fandom needs to be viewed as multi-stranded, with blended experiences of texts, brands, places and events being inter-connected.

There has been a tendency for theorists to separate out "brand fans" (Smith et al., 2017) from previous approaches to media fandom (Jenkins, 1992), or for work to see event fandom as something that is separable from text-focused fandoms (Kohnen, 2020, p. 96). But these lines between brand fans, text fans and place or event fans are somewhat artificial, I want to argue, and downplay the fact that for someone to be a fan of a convention, they must also have some level of investment in at least some of the texts and brands that are being celebrated at such an event. Even someone cosplaying a character from a media text that they know very little about at a con (see Mishou, 2021, pp. 5–6) will, over time, become more knowledgeable about and more invested in that text/franchise. I therefore want

DOI: 10.4324/9781003320586-10

to consider "event fandom" as co-existential, that is, it operates through a blend of fannish elements (text/brand/place), which cannot be meaningfully separated, particularly as fan experience connects and bridges these strands. Experiencing conventions is essentially intertwined with experiences of brands and textual materials (even if only via paratexts), and experiencing texts is increasingly also not separable from brands, transmedia, platformised spaces and a variety of places. I will go on to show that this sense of co-existential fandom calls for not only a "logistical turn" (Jansson, 2021, p. 18; Williams, 2020, p. 69; Kohnen, 2020) in terms of understanding how fans of events anticipate and work in advance to map out and maximise their enjoyment, but also a focus on third generation "experience economy" analyses (Alexiou, 2020, pp. 200–202), which can help to theorise how event experiences are imaginatively co-produced between fellow fans.

As a concept, co-existential fandom may also be helpful beyond conventions in terms of avoiding the silo-ing of discussions of brand fandom, place fandom or event fandom, as if these are special topics that can be contrasted with text-based fandom as supposedly constituting a normative core to fan studies. It also has the potential to be applied to other forms of media tourism, where place and text fandoms can be blended in a range of ways through cycles of fan experiences. Although I introduce the term through scholarship on SDCC, examining the transcultural iteration of fan conventions in China, India and South Africa enables me to consider how co-existential fandom may also be experienced in globalising and glocalising frames.

Although the "experience economy" as a concept has been used in work on SDCC, it has typically constituted an aside (Woo et al., 2020, p. 13; Hanna, 2020, p. 18 and 20) or has been used to think about experiential marketing "activations" outside the spaces of the con (Kohnen, 2021) rather than being fully interrogated to theorise the fan convention itself. Considering Comic-Cons more substantively through the "experience economy", I will argue that the cost of such events for their attendees means that a neoliberal return on investment is sought – this explains the need for "anticipatory labour" (Williams, 2020, p. 69) as well as the intense valuing of event experiences (commemorated through swag).

Co-existential event fandom typically celebrates "neoliberal mastery", thus acting as a form of the "late escapism" that has been theorised by Greg Sharzer (2022, p. 168). Such escapism is shaped by neoliberal forces which themselves structure global media and tourism through norms of marketisation. Adopting an experience economy approach to Comic-Cons highlights the neoliberalist struc-turing of fan experiences in these kinds of cons (indeed, it could be suggested that the concept of the "experience economy" is enmeshed in corporate philosophies). Where more ethnographic/anthropological approaches have illuminated fannish sociality and agency at major cons, I want to address the commercial structuring of types of fan convention. That said, I will ultimately argue that the experience economy approach still needs to be supplemented by ethnographic understandings.

I will begin by focusing on previous scholarship on SDCC and its information-seeking fans before moving on to address emergent work on the globalising (or glocalising) of Comic-Cons in China, India and South Africa alongside material on the rather differently oriented event Comiket in Japan. I have selected these last case studies because each has been the subject of pioneering academic analysis, enabling such work to be brought into dialogue with the proliferation of SDCC analyses. One of the key issues with fan conventions is that their take-up around the world indicates a certain conventionality of iterated event structures and invited forms of participation. As well as being platform-like in terms of serving industry and fandom (Kohnen et al., 2023), cons are themselves transculturally mobile patternings of activity, routine and expectation. The ways in which they grant ritualised access to the "media world" for non-media people (Gilbert, 2017, p. 355) tend to form a template for cons and their transcultural reinventions. "Non-media people" refers to the culturally subordinated category of ordinary people who do not appear in mass media texts, or do not have insider access to media production cultures via working in the industry; non-media folk are therefore excluded from the valued domains of media celebrity, presence and/or production, but conventions allow them to get closer to these desired "media worlds" (Couldry, 2003, p. 88). Whether occurring in the United States or on multiple other continents, cons thus involve the imagining, anticipating and production of memorable experiences for their consumers/fans, or what might be thought of as the production of imaginative experiences (Sager, 2014, p. 159).

Comic Con International: The (Re)production of Experience

Lincoln Geraghty (2014) has argued that Comic-Con represents a confluence of economic and cultural capital (p. 103). However, Erin Hanna's (2020) discussion of SDCC disaggregates these capitals, arguing that while "fans consuming Hollywood promotion at Comic-Con are largely operating in… frameworks of cultural capital, when Hollywood appears at the convention, it is seeking economic gains" (p. 20). Regardless of whether we see forms of capital as linked or as asymmetrically mapped onto industry–fan power relations (Gilbert, 2017, p. 355), Comic-Con attendees will clearly expect to spend sizeable amounts of money as consumers of the event, as well as through the "subconsumption" of flights and hotels (Stanfill, 2019, pp. 87–88), that is, related consumption supporting con attendance. SDCC's international reach can thus be viewed as an appeal to fans who recognise themselves within the con's interpellation of fan consumers (Gilbert, 2018, p. 320). Whether these fans are U.S. tourists travelling to San Diego, or fan-tourists travelling from further afield, they will tend to buy into SDCC as an elevated experience, making it important to focus on the experience economy of Comic-Con.

Coined by Pine and Gilmore (1999; 2011), the experience economy is about the staging of experiences for consumers; it assumes that goods and services can be sold at higher prices if they are perceived as having experiential worth, that is, if they form part of a memorable experience that consumers will remember fondly and nostalgically. Tourism has been one site for applications of this approach (Mehmetoglu & Engen, 2011). Abby Waysdorf (2021) has noted that "corporate places of fandom" (p. 131), such as theme parks, festivals or licensed pop-up events, can additionally "draw on the concept of the 'experience economy'" (p. 117) as "a

way of fostering engagement with a text as a brand through unique 'experiences'…
Ultimately, …these experiences reinforce particular producer-friendly readings of
the text-as-brand" (p. 112).

Despite resonating with the likes of Comic-Con, experience economy
approaches have remained under-developed in the con literature, perhaps due to
a suspicion that the "experience economy" concept fails to consider how staged
experiences translate into consumer/fan perceptions (Wood, 2020, p. 121), or
fails to address the role of fans and cosplayers in co-producing a con's atmosphere (Pine & Gilmore 2011, p. xx; Anderson, 2014, p. 22). That said, fans' sense
of belonging to an event could potentially be preconfigured by con organisers'
structuring choices, so this may not always constitute a clear binary between
experience economy and ethnographic approaches (Tang et al. 2023, p. 354).
But where people can be analysed as fans of an event, in combination with other
text/brand and place-related fandoms, then they are, in effect, concomitantly *fans
of an experience* that can be anticipated, planned for and commemorated. As
Melanie E.S. Kohnen (2020) has argued of con-bloggers, who share their tips for
getting the most out of a con:

> Con-bloggers are fans of comics, films, and TV shows, but first and foremost,
> they are fans of Comic-Con; even more specifically, *they are fans of an experi-
> ence situated in a specific time (four days in July) and space (the San Diego
> Convention Center…).*
>
> (p. 92, my italics)

The industry of "con-bloggers" surrounding SDCC is akin to the "anticipatory
labour" analysed in Rebecca Williams' study of theme park fandom and its participatory cultures. Both con-bloggers and theme park fans are concerned with
"practices of planning and preparing [whereby] … visitors/fans… draw on a
wide range of resources – both official and fan-created – to maximise their visits"
(Williams, 2020, p. 69). Indeed, the con-bloggers' "investment in mastery of
SDCC's schedule and spaces reveals most clearly that they are fans of an experience rooted in a specific space and time" (Kohnen, 2020, p. 96). This doesn't
simply challenge "scholarly and industrial definitions of fandom as coalescing
around specific texts" (ibid.), however; it also suggests that fandom can co-exist
in a series of interlinked ways, with this *co-existential fandom* simultaneously
involving affective relationships to media, brands, texts, places and – through all
of these – experiences. After all, it wouldn't make sense for individuals to become
fans purely of SDCC without any comics/media/film fandoms also forming part
of their self-narratives. Co-existential fandom is arguably the mode of fannish
activity encouraged by the mainstreaming of fan identity; what Sandvoss et al.
(2017) term "fanization" (pp. 22–23) involves fandom becoming articulated with
brands and branding. But co-existential fandom stresses that within neoliberal
brand culture, industrially policed "brandom" (Guschwan, 2012) and fandom not
only cannot be clearly separated out, but actually derive their cultural energies
from this entanglement. Neoliberalised fan practice is often a blend of brand

fandom, media fandom and place fandom where experience acts as a fusing component.

The con-bloggers studied by Kohnen (2020), and the theme park fans analysed by Williams (2020) both highlight the significance of what Andre Jansson (2021) has called a "logistical turn" (p. 18) to cultures of media tourism. By this, Jansson (2021) means that the desired experiences of tourism become increasingly "matters of travel planning, navigation, coordination, and evaluation" in advance and within everyday life – practices "that are… rarely associated with 'media tourism'" (p. 17). At the same time, tourist experiences are rendered more difficult to sequester away from everyday life, becoming permeable to intrusions from work or family obligations. The result "brings a new flavour to de-differentiation" (p. 17) as Jansson (2021) puts it, with ordinary life and extraordinary fan-tourist experiences inter-penetrating. For Kohnen's (2020) con-bloggers, logistics are about navigating the convention site itself, while Williams' (2020) theme park fans are also concerned with making the most of their visit. What these forms of anticipatory labour have in common is that they both emphasise how we can view convention-goers as planning the logistics of their day-by-day con attendance carefully in advance – which panels they will prioritise attending, whose autographs they most want, what exclusive merchandise they especially wish to get hold of and so on. Yet Jansson's account of this "logistical turn" indicates just how dependent fan tourism within an experience economy, or even a consumerist "experience society" (Miles, 2021), remains upon the fan's own co-production of this experience. Comic-Con is not merely staged as memorable for its attendees (though there is an element of this, as producers strive to create moments that will attain social media buzz (Jenkins, 2012, p. 26)). For fans to successfully get the most out of Comic-Con, they need to play a part in terms of knowledgeably planning their personal itineraries and hence properly navigating the event. The image of fans navigating through the pathways and schedules of Comic-Con may remind readers of Michel de Certeau's (1984) binary of powerful "strategy", controlling spaces and weaker "tactics", moving temporarily across those spaces, as well as its classic application in fan studies (Jenkins, 1992). In this instance, however, both commercial spaces and fans' logistical navigations are legible as neoliberal practices, making them less an oppositional binary and more part of a shared logic of experiential and imaginative production.

Bearing that resonance in mind, the very idea of precisely planning out one's desired convention experience resembles the concept of return on investment (ROI) that crops up in business analyses – for instance, see Kohnen (2021, pp. 170–172) on the "ROI" that brands hope for from experiential marketing "activations" created outside SDCC. Yet the notion of tourist fans seeking to maximise their visits sounds very much like an internalised or personalised fannish version of return on investment. Kohnen (2020) captures a sense of this in her discussion of the SDCC con-bloggers:

Almost everything at SDCC requires waiting in line; consequently, the dis-cussion of when to line up for which room or piece of merchandise is central to con-blogging. …Lining up costs time; as con-blogger James Riley puts it, "[t]ime is the currency you spend at Comic-Con to get things done."

(San Diego Comic-Con Unofficial Blog…) (p. 95)

Fans are focused on how to get the best return for their investment of hours (and money; they are acting as good neoliberals in terms of efficiency and economic rationale). Arguably, Greg Sharzer (2022) would view this as a "new form of escapism, for which the context is neoliberal mastery" (p. 168), a kind of "late escapism" that is fully aware of neoliberalist realities and pressures (p. 170). And a related appreciation for the staging of promotional events and the creation of branded worlds can arguably also be perceived in the plans and memories of Comic-Con's event fans. Con-bloggers actually aim to produce their own entrepreneurial "mastery", albeit of SDCC'S routines (Kohnen, 2020, p. 96). They help to engineer a sense of Comic-Con's "event community", that is, rather than only identifying as comics fans or *Star Wars* fans etc., fans also think of themselves as part of a "SDCC fandom" (Edmunds, 2021). This fits with what has been termed the "third generation" experience economy approach, as this emphasises how consumer-to-consumer interactions can produce memorable commercial experiences in addition to producers' stagings and producer-consumer co-productions (Alexiou, 2020, p. 200). Con-bloggers, in effect, seek to mentor those who lack their insider knowledge; by reliving past cons, they enable first-timers to imaginatively pre-live the convention. If there is a fan-to-fan or tourist-to-tourist (T2T) world of imagination, then it is one that shares convention highs by projecting them into other fans' anticipated futures. That the experience economy might make sense of con-going fan practices could be expected, for as Nicolle Lamerichs (2018) has argued:

> Although fans may want to rationalize this affect within fan communities, *fandom appears to be grounded in an aesthetic moment that is constantly relived…* It is important to note that affect is not something that simply arises but that requires preparation. Creating the right circumstances… is a central part of the aesthetic experience.
>
> (p. 209, my italics)

If fandom concerns reliving (and for related work on fans who return to an annual *Prisoner* convention, see Waysdorf (2021, pp. 61–86)), then it necessarily involves planning for and anticipating this transformed repetition – skilfully pre-living the moment of aesthetic/affective return in order to seek its likelihood. This suggests that "the circuit of the imagination" characterising media tourism (Reijnders, 2011, p. 110) may not only circle between place, text, appropriation and physical quest but also between moments of fans' performative consumption as these are lived, imaginatively appropriated/anticipated and then potentially lived out again, albeit differentially.

But if the high-profile SDCC can perhaps be understood experientially in these ways, then what of the globalisation of conventions? I will now turn to look at emergent work on cons in China, India and South Africa, as well as significant work translated into English on the event community of Comiket in Tokyo. To what extent can a third-generation experience economy approach make sense of these globalising/glocalising proliferations of cons?

Comic-Cons in China, India and South Africa: The (Re)production of Conventions

As noted, Comic-Cons are not just a major industry in North America, dominated by corporate players such as Reed Exhibitions and Informa (Salkowitz, 2021, p. 149). They also have a significant globalised presence. For example, Tang et al. (2023) have recently analysed cons hosted in Guangzhou in China; they drew up a list of major cons hosted in the city before selecting the "A-3" con on the basis of "popularity/brand awareness/number of tourists" (p. 350). Based on survey data (ibid.), these authors explore the tourist-to-tourist (T2T) interactions of cosplayers at the convention; they filtered participants to ensure they were surveying self-defined "tourists" who had travelled for the event, although all respondents were from China (Tang et al., 2023, p. 354). Hypothesising that T2T interaction can boost "emotional solidarity" and so positively enhance these fans' experiences of the con, it is demonstrated that such a correlation does indeed appear to be meaningful (Tang et al., 2023, p. 353). This echoes ethnographic work at SDCC, which also found that cosplayers contributed more widely to an environmental sense of the con's identity, vibrancy and success (Anderson, 2014, p. 22) – both findings suggest that fan-to-fan interaction can play a notable role in the experience of cons, not just in terms of con-bloggers' guidance and mentoring in advance, but also in situ. Rather than cons being fully "staged" experiences, in line with initial experience economy work (Pine and Gilmore, 1999), they seem to display similar structures of fan engagement and sociality transnationally. This suggests that it is important to supplement an "experience economy" approach to these cons with a more broadly anthropological or ethnographic approach. Such events are, in effect, staged between cosplayers and fan attendees more widely, with their corporate spaces being transformed by fan-to-fan interaction and emotional solidarity. And as such, it seems plausible to suggest that large-scale commercial fan conventions represent a template, or set of "media world" conventions, that can themselves become transculturally mobile.

With this in mind, it is striking that Sailaja Krishnamurti's (2023) discussion of cons in India argues that they indicate a powerfully globalising presence rather than a "glocalised" adaptation of conventions' iterated event structures and staging practices. Comic-Cons are relatively new in this cultural context, the first one having been held in Delhi in 2011 (Kamath & Lothspeich, 2023, p. 14). Krishnamurti (2023) notes that the "largest such events in the region are hosted by Comic Con India, a Delhi-based organization that organizes fan events in several Indian cities" (p. 191), and her analyses are based on attending three Comic-Cons, Delhi, Mumbai and Bengaluru, between 2015 and 2017 (p. 193). Though Krishnamurti (2023) acknowledges that cosplayers at these events can " 'play' with aspects of a character's gender, racial or sexual expression, sometimes incorporating elements of... the player's own body or persona" (p. 193), she emphasises how globally popular transnational flows of media frame Indian participants' views on supposedly authentic fan performance:

> I observed that Indian texts were less visible at these events and were not popular among cosplayers. The comic cons that I attended were dominated by the conspicuous consumption of and participation in American, Japanese, and Korean cultural texts and fan culture practices. ...Ellen Kirkpatrick uses the term "embodied translation" to describe how cosplayers translate fantastical representations... into real-world, tangible, material forms (2015, 4.9). ... Embodied translation, however, can also produce challenges. Indian cosplayers' impersonations of characters from Japanese, Korean and American texts prompt complex questions about the politics of racial mimicry in transnational cosplay culture.
>
> (2023, p. 193)

Here, the process of "embodied translation" is not merely between the imagined and the material; the "circuit of the imagination" highlighted by Reijnders' (2011) work is, in an even more convoluted sense, between global popular culture (Hollywood films, Japanese anime, etc.) and Indian con participants (p .110). As Kamath and Lothspeich (2023) summarise: "Indian cosplayers desire to situate themselves as 'authentic' fans within global cultural discourse" (p. 15). While it would certainly be possible to view such fans' mimetic cosplay as creatively transformational – embodying white characters, for instance, in ways that render that identity as marked rather than unmarked – Krishnamurti's reading is one which views local culture as nevertheless being overwritten by global media franchises. The experience economy that is implicitly represented is one where fan experiences are dominated by the stagings of official producers, with Indian Comic-Cons appearing to replay the same desired and transnational "media worlds" that dominate SDCC. The con's ritualised template of non-media people being granted symbolic proximity to media worlds (Gilbert, 2017, p. 355; Kohnen, 2020, p. 96) therefore seems to carry across these cultural and national borders without significant retooling.

By contrast, Keyan Tomaselli's analysis of his attendance at the 2022 Comic-Con Africa, hosted in Johannesburg, makes the case for understanding this event very much in relation to its national cultural setting, and as a potential glocalisation of the media world/ordinary world split. Firstly, Tomaselli (2022) stresses how the spaces of Comic-Con Africa displayed a sense of freedom:

> That positioning seemed to me to characterize the multiracial, multiethnic and multicolored sales environment constituted by Comic Con that would not have been permitted during apartheid (1948–1990). We were all at once in place but simultaneously 'out of place', both insiders and outsiders living through the body taking on roleplay enabled by the liminal environment and multiverses encouraged by Comic Con.
>
> (p. 35)

But more than this expression of post-apartheid generational change, Tomaselli (2022) goes on to argue that the con also enabled a distinctive cultural space and time, a kind of event community, where disciplining discourses of national identity could (at least temporarily) be suspended:

Entirely absent were officially imposed identities, prescribed ethnicities and ethnic determinations that position us socially in the so-called "new" South Africa. Imposed identities legislate… who we are and how we should think and behave, and they also delimit one's life's options. At Comic Con… starkly different self-imagined individuals… intermingled, talked and made new friends. In contrast, regressive conceptions of nation and nationalism were absent, replaced by much larger mythical universes.

(p. 36)

It should be noted that Tomaselli (2022), attending the convention with his son (p. 35), describes the event as more of an outsider rather than a fan-insider, and so his relative detachment from the franchise characters and scenarios being mimetically cosplayed might lead him to downplay how there could also be (differentially) disciplining discourses connected to global popular culture as well national identities. But that said, his reading of Comic-Con Africa is one which counterposes discursive national impositions within South Africa to a sense of the glocalised liberations of the con itself. In this case, Comic-Con templates carry over into a specific context, again demonstrating the transcultural mobility of the con industry, but the imaginative worlds that fans can symbolically enter into take on newfound and localised specificities.

I want to conclude by looking at the example of Comiket, or Comic Market, in Tokyo, Japan. Although this is a huge mega-event, it is very different in character to SDCC. Rather than acting as an intermediary between media industries and fans, it is focused on fans' self-published comics or *doujinshi* zines. Comiket is hosted at Tokyo Big Sight, or the Tokyo International Exhibition Centre, and has been studied by European scholars of cons on their own aca-tourist visits (see Lamerichs, 2018, p. 193) as well as by Japanese scholars whose work has been translated into English (Aida, 2016). Considering the emphasis of third generation experience economy analysis, where fan-to-fan interaction would be viewed as vital to the staging of an event experience, it is interesting to draw out a somewhat counter-intuitive finding from Aida's (2016) nuanced ethnographic study of Comiket (p. 59). *Contra* established views celebrating the way in which zine creators and buyers are brought closer together at Comiket, with buyers getting a chance to talk to artists (p. 64), Aida (2016) argues that non-communication in lines and between creators and fans is more common. She interprets this as a choice made by participants to avoid the precise details of their zine fandom being opened up to challenge, with fine-grained fan distinctions potentially causing displeasure and tension between fans, or between fans and artists. It might be assumed that these attendees are fans of specific texts rather than Comiket event fans, but Aida's analysis challenges any such binary, again implying that we need to take a co-existentialist view of such fandom. Because by choosing to self-silence, and so removing the possibility of being challenged for their tastes by unknown others within the event, these Comiket attendees can preserve a notion of belonging and togetherness among the community of Comiket:

> The space turns these 450,000 individuals into a single unity called Comic Market participants (*Komiketto sankasha*). If Comic Market participants can just close their eyes to individual differences in goals and tastes, then it becomes possible to recognize everyone that they see as the same – as Comic Market participants. …The term 'normals' (*ippanjin*) is used in contrast… people who do not understand the meaning of the Comic Market.
>
> (Aida, 2016, p. 66)

Aida (2016) is careful to note that her analysis cannot make a fully "generalized statement" (p. 64) about all attendees, and so it is possible that some fans may be more willing to assert vocal differences of fan taste over and above aligning themselves with the relatively silenced identity of "Comic Market participants", but this is not observed in her circle of creators. The result instead, due to a shared commitment to an insider "us" versus an outsider "them", is that "[i]ronically, given the event's consistent ideology of open participation, relations among participants (*sankasha*) at the Comic Market are characterized by non-participation (*fukanyo*) and self-containment (*jikokanketsu*)" (ibid.).

Rather than bringing zine creators and fans closer together in terms of social interactions, they remain somewhat divided (as do fans from their fellow fans) so as to maintain the imagined community of united *Komiketto sankasha*. Through this desire to avoid unpleasantness and disagreement, fan-to-fan interaction remains vital to securing a particular, and evidently valued, experience, but it is a fan-to-fan interaction that is unexpectedly characterised by "self-containment" rather than an excited sharing of fan passions. Yet this surprising difference does still resonate with concepts of event fandom established in relation to the likes of SDCC (Edmunds, 2021) — that is, there is still a strong sense of "event community" generated at Comiket according to Aida's (2016) work. Her discussion also fits into the precepts of third generation experience economy work (Alexiou, 2020, p. 200), where consumer-to-consumer or fan-to-fan interaction has to be considered as co-producing experiences in the experience economy approach. In this case, a unifying Comiket experience is produced by and among fans, yet this is achieved through a desire not to expose varied/oppositional fan tastes.

Conclusion: Event Fandom as Co-existential

I have sought to build on work addressing film tourism and place-oriented fandom within fan studies (Waysdorf, 2021; Williams, 2020) by focusing on event fans of SDCC and the international Comic-Con industry (Woo et al., 2020). However, rather than seeing such fandom as a challenge to text-focused concepts of media fandom (Kohnen, 2020), I have argued that we need to consider event fandom not as an additive type of fan activity – something to be added to potentially silo-ed ana-lyses of "brand fandom" or "place fandom" – but rather as a co-existential blending of text, brand, place and convention fandoms, with the experiential working to combine or fuse these aspects. Thinking of event fandom as co-existential means that rather than adding an additional and discrete fan object into the equation, that is, one might be a fan of SDCC *rather than* a fan of a specific media franchise, or a fan of an event *rather than* a text, event fandom is instead conceptualised as an inevitable blend of place, brand and text-oriented fan affects. Following this, I have also stressed the value of a "third generation" experience economy approach (Alexiou, 2020).

In terms of my first section on SDCC, it means acknowledging con-bloggers and their fan-to-fan advice (Kohnen, 2020) as well as how Comic-Cons are com-mercially staged. And it means acknowledging the contributions that fan sociality can make to the atmosphere of a major convention – as discussed by Tang et al. (2023) in the Chinese context – while also continuing to highlight the neoliberal structuring of fan experience within an experience economy stance.

My focus on globalised fan cons in the second section of the chapter allowed me to develop this point. Although prior work on major cons has touched on experi-ence economy theories (Pine and Gilmore, 1999, 2011; Gilbert 2017; Woo et al., 2020, p. 13), these have not been deployed centrally in analysis to date. Focusing on the neoliberalised fan experiences that are maximised through forward planning (Kohnen, 2020; Williams, 2020), and commemorated via subconsumption (Stanfill, 2019) as characteristic of major commercial conventions has enabled this chapter to consider how cons can themselves display transcultural mobility. Such global-isation of conventions, I have suggested, relies on iterating and reproducing the fascination of "media worlds" and their ritualised separation from "non-media people" (Tang et al., 2023; Krishnamurti, 2023) in different cultural contexts.

This fascination has already been viewed as central to U.S. mega-cons such as SDCC (Gilbert, 2017, p. 355). But transcultural iterations of this media world versus non-media world separation (and ritualised proximity) can also become glocalised through the meanings and event communities that form in relation to specific national histories (Tomaselli, 2022). A twinned focus on the performance of co-existentially fannish "event communities" (Aida, 2016; Edmunds, 2021) and the experience economies that underpin them remains to be further explored across a wider range of transnational and transcultural comics/media conventions, going beyond the starting points set out here.

Bibliography

Aida, M. (2016). The Contemporary Comic Market: A Study of Subculture (translated by Patrick W. Galbraith). *Journal of Fandom Studies*, 4(1), pp. 55–70.

Alexiou, M.V. (2020). Experience Economy and Co-creation in a Cultural Heritage Festival: Consumers' Views. *Journal of Heritage Tourism*, 15(2), pp. 200–216.

Anderson, K. (2014). Actualized Fantasy at Comic-Con and the Confessions of a "Sad Cosplayer", in Ben Bolling and Matthew J. Smith (eds), *It Happens at Comic-Con: Ethnographic Essays on a Pop Culture Phenomenon*. McFarland pp. 16–28.

Couldry, N. (2003). *Media Rituals.* Routledge.

De Certeau, M. (1984). *The Practice of Everyday Life.* University of California Press.

Edmunds, T. K. (2021). 'Comics and Comic Cons: Finding the Sense of Community', in Benjamin Woo and Jeremy Stoll (eds), *The Comics World: Comic Books, Graphic Novels and their Publics.* University Press of Mississippi, pp. 167–180.

Geraghty, L. (2014). *Cult Collectors: Nostalgia, Fandom and Collecting Popular Culture.* Routledge.

Gilbert, A. (2017). 'Live from Hall H: Fan/Producer Symbiosis at San Diego Comic-Con', in Jonathan Gray, Cornel Sandvoss and C. Lee Harrington (eds), *Fandom: Identities and Communities in a Mediated World,* second edition. New York University Press, pp. 354–368.

Gilbert, A. (2018). 'Conspicuous convention: Industry interpellation and fan consumption at San Diego Comic-Con', in Melissa A. Click and Suzanne Scott (eds), *Routledge Companion to Media Fandom.* Routledge, pp. 319–329.

Guschwan, M. (2012). Fandom, brandom and the limits of participatory culture. *Journal of Consumer Culture* 12(1), pp. 19–40.

Hanna, E. (2020). *Only At Comic-Con: Hollywood, Fans, and the Limits of Exclusivity.* Rutgers University Press.

Hills, M. (2021). 'The National Theatre, London, as a Theatrical/Architectural Object of Fan Imagination', in Nicky van Es, Stijn Reijnders, Leonieke Bolderman and Abby Waysdorf (eds), *Locating Imagination in Popular Culture: Place, Tourism and Belonging.* Routledge. pp. 297–311.

Jansson, A. (2021). 'Invited Contribution – The Janus Face of Transmedia Tourism: towards a Logistical Turn in Media and Tourism Studies', in Maria Månsson, Annæ Buchmann, Cecilia Cassinger and Lena Eskilsson (eds), *The Routledge Companion to Media and Tourism.* Routledge, pp. 13–19.

Jenkins, H. (1992). *Textual Poachers.* Routledge.

Jenkins, H. (2012). Superpowered Fans: The Many Worlds of San Diego's Comic-Con. *Boom: A Journal of California,* 2(2), pp. 22–36.

Kamath, H. M., & Lothspeich, P. (2023). 'Introduction', in Harshita Mruthinti Kamath and Pamela Lothspeich (eds), *Mimetic Desires: Impersonation and Guising Across South Asia.* University of Hawai'i Press, pp. 1–22.

Kohnen, M. E. S. (2020). Time, Space, Strategy: Fan Blogging and the Economy of Knowledge at San Diego Comic-Con. *Popular Communication,* 18(2), pp. 91–107.

Kohnen, M. E. S. (2021). The Experience Economy of TV Promotion at San Diego Comic-Con. *International Journal of Cultural Studies,* 24(1), pp. 157–176.

Kohnen, M. E.S., Parker, F., & Woo, B. (2023). From Comic-Con to Amazon: Fan Conventions and Digital Platforms. *New Media & Society,* pp. 1–22.

Krishnamurti, S. (2023). 'Cosplay, Fandom and the Fashioning of Identities at Comic Con India', in Harshita Mruthinti Kamath and Pamela Lothspeich (eds), *Mimetic Desires: Impersonation and Guising Across South Asia.* University of Hawai'i Press, pp. 191–211.

Lamerichs, N. (2018). *Productive Fandom: Intermediality and Affective Reception in Fan Cultures.* University of Amsterdam Press.

Mehmetoglu, M., & Engen, M. (2011). Pine and Gilmore's Concept of Experience Economy and Its Dimensions: An Empirical Examination in Tourism. *Journal of Quality Assurance in Hospitality & Tourism,* 12(4), pp. 237–255.

Miles, S. (2021). *The Experience Society: Consumer Capitalism Rebooted.* Pluto Press.

Mishou, A. L. (2021). *Cosplayers: Gender and Identity.* Routledge.

Pine II, B. J., & Gilmore, J.H. (1999). *The Experience Economy: Work is Theatre and Every Business a Stage.* Harvard Business Review Press.

Pine II, B. J., & Gilmore, J.H. (2011). *The Experience Economy,* updated edition. Harvard Business Review Press.

Reijnders, S. (2011). *Places of the Imagination: Media, Tourism, Culture.* Ashgate.

Sager, C. (2014). 'Tense Proximities Between CCI's Comic Book Consumers, Fans and Creators', in Ben Bolling and Matthew J. Smith (eds), *It Happens at Comic-Con: Ethnographic Essays on a Pop Culture Phenomenon.* McFarland, pp. 153–168.

Salkowitz, R. (2021). 'The Tribes of Comic-Con: Continuity and Change in the Twenty-First Century Fan Culture', in Benjamin Woo and Jeremy Stoll (eds), *The Comics World: Comic Books, Graphic Novels and their Publics* University Press of Mississippi, pp. 147–164.

Sandvoss, C., Gray, J., & Harrington, C. L. (2017). 'Introduction: Why Still Study Fans?', in Jonathan Gray, Cornel Sandvoss and C. Lee Harrington (eds), *Fandom: Identities and Communities in a Mediated World,* second edition. New York University Press, pp. 1–26.

Sharzer, G. (2022). *Late Escapism and Contemporary Neoliberalism: Alienation, Work and Utopia.* Routledge.

Smith, M. J. (2022). 'Pilgrimage to Hall H: Fan Agency at Comic-Con', in Vanessa Ossa, Jan-Noël Thon and Lukas R.A. Wilde (eds), *Comics and Agency.* De Gruyter, pp. 189–199.

Smith, A., Stavros, C., & Westberg, K. (2017). *Brand Fans.* Palgrave Macmillan.

Stanfill, M. (2019). *Exploiting Fandom: How the Media Industry Seeks to Manipulate Fans.* University of Iowa Press.

Tang, J., Song, B., & Wang, Y. (2023). Fandom in Comic-Con: Cosplay Tourists' Interaction and Emotional Solidarity. *Journal of Hospitality and Tourism Management,* 54, pp. 346–356.

Tomaselli, K.G. (2022). Comic Con Africa (2022): Cosplaying Identities. *New Techno-Humanities,* 2, pp. 34–40.

Waysdorf, A. (2021). *Fan Sites: Film Tourism and Contemporary Fandom.* University of Iowa Press.

Williams, R. (2018). 'Fan Tourism and Pilgrimage', in Melissa A. Click and Suzanne Scott (eds), *Routledge Companion to Media Fandom.* Routledge, pp. 98–106.

Williams, R. (2020). *Theme Park Fandom: Spatial Transmedia, Materiality and Participatory Cultures.* University of Amsterdam Press.

Woo, B., Johnson, B., Beaty, B., & Campbell, M. (2020). Theorizing comic cons. *Journal of Fandom Studies,* 8(1), pp. 9–31.

Wood, R.C. (2020). 'Tourism and the Experience Economy: A Critique?' in Saurabh Kumar Dixit (ed), *The Routledge Handbook of Tourism Experience Management and Marketing.* Routledge, pp. 119–127.

Yamamura, T., & Seaton, P. (eds) (2020). *Contents Tourism and Pop Culture Fandom: Transnational Tourist Experiences.* Channel View.

8 Multi-Vocality Induced by *Laid-Back Camp* among Chinese Audiences

Qian Jin

Introduction

In recent years, China has seen a rise in "camping fever", with the year 2020 being dubbed the "First Year of Camping"[1] in China on the internet. Unlike well-established outdoor pursuits abroad, camping has grown in popularity in China over the past three years propelled by social media. The data shows that 56.9% of opportunity for users to participate in camping comes from content sharing by people on social media, and 86% of users get camping information from social media-based platforms.[2] At the same time, the pandemic has made people crave an outdoor lifestyle, and television programs that portray the "slow life" further fuel people's desire to step out of the city and get closer to nature.

While "camping fever" is growing rapidly in China, a Japanese trans-media work, *Laid-Back Camp*, is often mentioned by camping enthusiasts. *Laid-Back Camp*, known as *Yuru Camp* in Japan, is a slice-of-life manga series created by Japanese manga artist Afro. It depicts camping excursions, outdoor activities such as cooking outside, and the daily lives of female high school students traveling to various campsites. The plot revolves around the characters: solo camper Rin Shim who enjoys peaceful outdoor activities, enthusiastic but inexperienced camper Nadeshiko Kagamihara, and their friends from the school's outdoor activity club. The girls embark on numerous camping trips and learn camping skills throughout the series, which is set in and around Yamanashi Prefecture where they live. They become closer when sharing heartwarming moments and appreciating the beauty and tranquility of nature.

Although *Laid-Back Camp* was first serialized in manga form in 2015, the story has also subsequently appeared in a variety of other media, as shown in Table 8.1, making *Laid-Back Camp* a truly cross-media work. With its uplifting narrative and superb visual portrayal, *Laid-Back Camp* has attracted a number of manga and anime fans and expanded its audience through trans-media production. Inspired by *Laid-Back Camp*, the vicinity of Yamanashi Prefecture, where the anime was shot and narratively situated, has grown to be a popular travel destination for anime enthusiasts.[3] Furthermore, this work has changed stereotypes about camping and become an access point for many people to learn about camping. This has allowed

DOI: 10.4324/9781003320586-11

Table 8.1 The main media form about *Laid-Back Camp*

Title	Media	Period
Laid-Back Camp	Manga	July 2015–
Hanamori, Introduction to Camping for Beginners	Program	Released on November 24, 2017
Laid-Back Camp	TV anime	Season 1: January 4 to March 22, 2018
		Season 2: January 7 to April 1, 2021
Room Camp	Short anime	January 6 to March 23, 2020
Laid-Back Camp	Live action	Season 1: January 10 to March 27, 2020
		Season 2: April 2 to June 18, 2021
Laid-Back Camp	Video game	Lake Motosu version: March 4, 2021
Virtual Camp		Fumoto Campgrounds version: April 8, 2021
Laid-Back Camp Movie	Theatrical anime	Released on July 1, 2022
Laid-Back Camp	Live action	Season 1: January 10 to March 27, 2020
		Season 2: April 2to June 18, 2021
Laid-Back Camp: Winter Starry Sky Camp for Everyone to View	Voice drama	Released on June 16, 2023

Note: Hanamori is the voice actress who voiced Nadeshiko in the anime.

for the development of a new wave of camping travel, especially among young people.

Laid-Back Camp is popular not only in Japan but also in China with the official release of the TV anime version by Chinese online video portals and in the context of the "camping fever". Against this backdrop, Chinese audiences have created various concepts of travel behaviors centered on *Laid-Back Camp*. To better immerse themselves in the world of *Laid-Back Camp*, many fans have made pilgrimages to the original sites where the stages that feature in the work were filmed. Unlike many works that create imaginary locations or anonymize stage prototypes, many of the locations featured in *Laid-Back Camp* exist, helping *Laid-Back Camp*'s fans identify the settings.

Even though many audiences indicated a desire to visit the locations where the work was performed in this research, most of them were unable to do so because of physical mobility restrictions imposed due to the pandemic. But they still experienced the world of *Laid-Back Camp* through camping in other non-specific locations. At the same time, as social media usage has increased in recent years, user-generated content and original texts have proliferated, providing audiences with a more diverse and creative understanding of media texts. Tourists can share their tourism experiences as well as their understanding of *Laid-Back Camp*, and the content created by tourists also influences the travel practices of other audiences.

This chapter focuses on the emergence of multi-vocality induced by *Laid-Back Camp* among the Chinese. Multi-vocality is a concept that refers to a method to depict a multidimensional reality through the symbiosis and interaction of numerous viewpoints (Bakhtin, 1984 [1963]). This chapter applies this concept to explore how audiences traverse barriers between representation and practice, fiction and reality, and geography through the consumption and production of media texts. It conducts a qualitative investigation based on the multimodal approach (Atkinson & Kennedy, 2016), choosing a visual and verbal approach from both online and offline, combined with the analysis of *Laid-Back Camp*, to examine the cultural practices of Chinese audiences around *Laid-Back Camp*.

From media tourism to contents tourism

At a time when the media landscape is changing rapidly, media representations continue to shape people's geographical imaginations, which makes traveling to the locations featured in media texts seem to be one of the most alluring types of tourism. Depending on the media forms of the text, it was previously referred to as "literary tourism" (Squire, 1996), "TV tourism" (Evans, 1997), "movie-induced tourism" (Riley et al., 1998), "film-induced tourism" (Beeton, 2005), "anime pilgrimage" (Yamamura, 2009), etc. Reijnders (2011) describes traveling to actual places related to film, television, music, books, or other media as "media tourism" and notes that the media texts may be fictional or real. The destinations of media tourism may be places that appear in media texts, places related to the creators of the texts, or even it can be a place imagined by tourists (cf. Norris, 2013). In short, in media tourism, the tourist enters the media world through the tourism experience and establishes a connection with imagination and reality.

However, it seems that the tourism triggered by *Laid-Back Camp* cannot be simply termed "media tourism" because the imaginary places (Reijnders, 2011, p. 17) it constructs are not only generated by one specific media but are constantly produced and in a state of flux in the intertextual relationship of multiple media texts. When people mention "Laid-Back Camp", they not only refer to animation, which is the most widely circulated media form, but also creations in other forms of media such as TV live action, games, and more. Meanwhile, the term "Laid-Back Camp" places greater emphasis on the story it tells about camping, and even a lifestyle centered around "leisurely camping" more than the media works.

Media tourism, which became a vital producer for the content industry, in Japan, began with the emergence of the "anime pilgrimage" phenomenon (Yamamura, 2008). The research about anime pilgrimage focused on the travelers' behavioral characteristics and the interrelationship between travelers and the local community (Yamamura, 2009; Okamoto, 2011), forming a media research trend on media tourism with Japanese characteristics. On the other hand, the Japanese government actively advocates the integration of media content with the tourism industry, and against the background of the "tourism-oriented country" (2007), the "Cool Japan" (2010) strategy was proposed to promote inbound tourism by promoting Japanese popular culture.

In 2005, the Japanese Ministry of Land, Infrastructure and Transport, the Ministry of Economy, Trade and Industry, and the Agency for Cultural Affairs jointly issued a policy document, which first proposed the keyword "contents tourism"[4], pointing out to the use of 'narrative' and 'thematic' elements of media content related to region and society and transform it into tourism resources to promote regional development. Since then, Japanese researchers have begun to use the term contents tourism to describe the phenomenon of tourism induced by media content, including anime pilgrimage, and contents tourism research has gradually been incorporated into the international academic discourse on media-induced tourism research.

Beeton et al. (2013) compared contents tourism with film-induced tourism, pointing out that contents tourism cuts across media forms to focus on the narratives. In 2017, Seaton et al. redefined contents tourism based on the context of international research on media-induced tourism, stating that contents tourism is travel behavior motivated fully or partially by narratives, characters, locations, and other creative elements of popular culture forms, including film, television dramas, manga, anime, novels, and computer games (Seaton et al., 2017, p. 3), which explored its nature of shaping travel motivation through "mediatization" from the perspective of popular culture.

Although the concept of contents tourism was first proposed by the Japanese government, and initially focused on domestic tourism in Japan, particularly the examples related to anime tourism, researchers have gone beyond the policy implications contained in contents tourism and used the term to refer to the cultural phenomenon of tourism triggered by media content in the context of trans-media production and cross-border consumption of media content. The phenomenon of tourism triggered by media texts, contents tourism, and media tourism, which are closely related to each other, seems to have different emphases.

Compared to media tourism, contents tourism is more concerned with trans-media content than with media. Although media tourism also recognizes the multimedia character of many contemporary examples (Reijnders, 2011, p. 5), which corresponds to the aspect of trans-media content in contents tourism, it mostly emphasizes the shaping of a specific place represented by a particular media text. On the other hand, contents tourism focuses more on the dynamism of the expanding narrative world. In the current situation where media texts are being used in multiple ways and audiences constantly engage in re-creation based on the original text, the narrative world created by media texts is always in expansion.

However, contents tourism not only focuses on the continuous proliferation of the texts themselves but also regards travel practices in physical places as a driving force for expanding the narrative world (Yamamura, 2020, p. 12). Based on this view, Yamamura redefined contents tourism as a dynamic series of travel practices and experiences motivated by content, through which contents tourists constantly access the expanding "narrative world" and attempt to embody it through "contensization" (Yamamura, 2020, p. 13).

The travel practices of contents tourists can be seen as embodying the interpretation of media texts through their physical experiences. However, there is often a gap between the intentions of the content creators and the interpretations of the contents tourists who consume it, and different interpretations may arise due to the different cultural backgrounds of contents tourists. Therefore, different travel behaviors in contents tourism reflect tourists' multi-vocality interpretations of media texts.

Multi-vocality is the concept proposed by Russian linguist Bakhtin (1984 [1963]) and refers to a method of writing novels that depict a multidimensional reality through the symbiosis and interaction of numerous viewpoints. Hall (2003 [1973]) imported the concept of multi-vocality into media studies, proposing the 'encoding/decoding' model to explain the opaqueness of media message transmission. The encoding/decoding model suggests that the media audiences do not always interpret the message in the same way as the message producers do, and there are multiple ways that are dominant, negotiated, and oppositional for media audiences to interpret media texts. While Hall points out the polyphonic nature of audiences' interpretations, this multi-vocality of audience interpretation remains structured. Furthermore, he does not point out how audiences produce multi-vocality.

In today's world where media is changing drastically, media audiences are no longer merely recipients of media messages, they can also act as producers. They produce content about their own interpretations of a media message and post it on social media. It can be claimed that when audiences create content, they not only present the multi-vocality of the original text but also induce multi-vocality through a secondary text, shaping a new intertextual relationship.

The various tourist behaviors presented in contents tourism are precisely related to the multi-vocality of text interpretation. The interpretations of the text are not just a single voice, so there will be various travel behaviors when the audiences become contents tourists. Content tourists' behaviors are directed at physical attempts to interpret the media text and expand the narrative world. Furthermore, this form of tourism, which is centered around popular culture, often includes other cultural elements, showcasing the cultural multi-vocality of travel practices. This research will apply the concept of multi-vocality to analyze the cultural practices inspired by *Laid-Back Camp* among the Chinese, to delineate how contents tourism allows multiple perspectives to coexist and the contemporary significance of contents tourism.

Methods

The multimodal approach, combined with online ethnography, participant observation, and semi-structured interviews, is employed in this research to explore the multi-vocality induced by *Laid-Back Camp*. Combining these research methods allows for a more comprehensive explanation of the interrelationships between media texts, the behavior of tourists who are also the audiences of media texts, and the tourist-created media content to explore how tourists consume *Laid-Back Camp*.

Online ethnography is used to gather the visual and textual data posted on social media to effectively discern which destinations tourists have visited, who were motivated by *Laid-Back Camp*, and how they demonstrate the relationship between their multi-vocal travel behaviors and *Laid-Back Camp*. Tourists' visuals are not merely a travel practice but also light up the tourist experience (Scarles, 2009).

When tourists share visual representations on social media, they share tourism experiences (Gretzel, 2017). By looking up "Laid-Back Camp" ("*yaoyeluying*" in Chinese) and "pilgrimage" on RED,[5] 35 user-generated posts posted during the period from July 2021 to September 2022 were collected. Contents are videos that are longer than three minutes to ensure that the samples are informative and relevant.

Given that the tourism experiences presented from the online ethnography show a connection to camping activities, participant observation was conducted by attending offline camping activities to identify correlations and differences between the camping activities represented by *Laid-Back Camp* and the reality of camping. In two camping events held in August and September 2022 in Hokkaido, Japan, some of the participants were Chinese who had watched *Laid-Back Camp*.

In addition, to explore the relationship between travel behaviors and interpretations of media texts from the tourists' perspective, semi-structured interviews were conducted with some audiences who had traveled motivated by *Laid-Back Camp*. In the preparatory investigation stage, the text of *Laid-Back Camp* was read in detail, and information such as prototype locations and props that appeared in the work was collected. The results from online ethnography and participant observation also informed the creation of the guide for the semi-structured interviews. Based on this information, an interview outline was organized. During the investigation stage, from April to October 2022, a total of 11 interviewees were conducted. The snowball sampling method was adopted to identify interviewees who had watched *Laid-Back Camp* and visited some tourist destinations. All interviewees were Chinese, including nine who were interviewed online and two who were interviewed offline (Table 8.2).

Interviews were conducted with specific research questions revolving around media works. For example, what prompted individuals to watch *Laid-Back Camp*? What was the earliest media form of *Laid-Back Camp* exposed to? How did the media work become attractive? Other research questions focused on tourist behaviors. For example, what were the motivations for visiting the locations associated with

Table 8.2 Interviewees

Code	Gender	Age	Identity	Tourism style
A	Female	20s	Audience/Content creator	Pilgrimage with Camping
B	Female	30s	Audience/Content creator	Pilgrimage with Camping
C	Female	20s	Audience/Content creator	Camping
D	Female	30s	Audience	Camping
E	Female	20s	Audience	Pilgrimage
F	Male	30s	Audience/Content creator	Camping
G	Male	20s	Audience/Content creator	Camping
H	Male	20s	Audience	Pilgrimage
I	Male	30s	Audience/Content creator	Camping
J	Female	20s	Audience	Pilgrimage and Camping
K	Male	30s	Audience	Camping

the work? What kinds of activities were performed? How were these activities influenced by *Laid-Back Camp*? The interviews led to an understanding of how tourists consumed media texts through their multi-vocal travel practices.

Multi-vocality influenced by *Laid-Back Camp*

Visiting physical places related to the work

Fans located the sites depicted in *Laid-Back Camp* and visited them to relive the work. Although manga is not a realistic style of media representation, as an outdoor enthusiast, the author incorporated personal experiences into the creation of the work, referencing scenes from real-world settings, and even engaging in one-to-one replication within the manga. This approach to representation also carries over to other media forms such as TV anime. In fact, the facilities at the campgrounds, surrounding attractions, and even the camping-related props are all authentic, and the campgrounds that serve as key locations in the plot correlate exactly with their real-world counterparts.

In *Laid-Back Camp*, the characters enjoy camping in various ways while appreciating the picturesque surroundings at campgrounds. Yamanashi Prefecture, where the characters live, is in the central region of Japan, and Mt. Fuji, the cultural symbol of Japan, is located on the boundary of Yamanashi Prefecture. The Fuji Five Lakes formed by the eruption of Mt. Fuji in this area is famous for outdoor activities. Campgrounds, which are the most important places as *Laid-Back Camp* depicts the camping lives of girls, are also the main sites for pilgrimage.

As an illustration, in the first episode of both the manga and TV anime, Rin meets Nadeshiko at the Kouan Campground, located on the shore of Lake Motosu, one of the Fuji Five Lakes. As a result, Kouan Campground has become an important location in the story, attracting numerous fans to visit on pilgrimage. Interviewee B, who once visited the Kouan Campground, said:

> After watching the first episode of the anime, I had the idea of going to Kouan Campground for pilgrimage. Everything that was displayed in the campsite, including the blue lake, the tents next to it, and the reception building, were all there at Laid-Back Camp. And at night, you can even enjoy the same scenery behind the 1000-yen bill of Mt. Fuji, while grilling by the campfire and pretending that you are Rin.

In addition to the Kouan Campground, as of the ninth volume of manga, that is, two seasons of TV anime, a total of 13 campgrounds have appeared, all of which have prototypes. The scenery of the campgrounds and the surrounding tourist attractions are described in detail in *Laid-Back Camp*, and Chinese visitors are particularly fond of the campgrounds that offer a glimpse of Mt. Fuji. Interviewee E said:

> Although I don't have camping experience, I can imagine that it would be a wonderful way to explore Mt. Fuji after seeing the characters enjoying different views of the mountain in different campgrounds.

It can be said that audiences enter the world of the story by visiting the places depicted in the work, and at the same time, they are also emotionally connected with the characters through the activity of "camping". Interviewee A said:

> As a loyal fan of Laid-Back Camp, I went on a pilgrimage to the model campgrounds after the TV anime aired. But I found that even while the scenery of the campgrounds was beautiful and just as they portrayed in the work, if I didn't set up camp there, I couldn't really experience the joy that character felt. So, I started trying camping. It seems as though I now understand why Nadeshiko fell in love with camping so quickly when I camped in these campgrounds. Now I am also a camping enthusiast, and even if I visit other campgrounds that don't appear in *Laid-Back Camp*, I also feel like I am in the world of the story it tells.

In contents tourism, the most common and basic tourist behavior appears to be visiting physical locations related to the work. However, it is worth noting that because the story of *Laid-Back Camp* is centered around the activity of camping, in most scenes, especially in the campgrounds that serve as tourists' pilgrimage destinations, the characters often engage in camping-related activities. As a result, the way tourists sanctify the campsite is not merely by "visiting", but they also yearn to approach the narrative world through physically engaging in camping-related actions. At the same time, the physical tourist destinations of contents tourism not only include the places that appear in the work but also the places that tourists associate with the work through imagination. Therefore, even though under the influence of the pandemic, most Chinese audiences were unable to make pilgrimages to the locations in Japan, they connected other campgrounds to the narrative world through keywords such as "Camp" in the title, and still presented the relevant nature of contents tourism through their physical mobilities to visit some other campsites.

Re-perform the behaviors of the characters

When analyzing *Laid-Back Camp* videos, some user-created content with the title "Camping in an Anime Way" is particularly eye-catching. In these videos, people imitate *Laid-Back Camp*'s outdoor cuisine, character-specific camping behaviors, and even certain scenes from the anime by using the characters' camping equipment. Taking a photo with the same scene from the same angle as in the anime is considered one of the important behavioral characteristics of contents tourism (Okamoto, 2015, p. 24), but the content presented in these photographs is often fragmentary and does not have a specific meaning, merely copies of the characters' poses. However, the replication of the actions in *Laid-Back Camp* was originally a practice about camping. People can more profoundly connect with the narrative world by physicalizing the experiences of the characters than by simply taking a photo.

In a video, F uses a split-screen technique to place Rin's camping process from the anime and his camping process in the same frame and uses the same scenery, angle, special effects, and montage as the anime to make his actions as consistent as Rin's as possible. F said:

> Although my gender, age, and camping location are completely different from Rin's, and even if the scenery may not be as beautiful as it is in the anime, by replicating Rin's behavior and using the same camping equipment as her, I feel like I experience her joy. Moreover, filming this video is already full of fun, and many people say they want to watch the Laid-Back Camp after seeing my video.

In addition to videos that faithfully reproduce the work, some people imitate characters' daily camping itineraries and compare them with their camping experience. In the video, G uses cross-cutting to describe Rin's actions as 'imaginary camping' and his own behaviors as 'real camping'. Although 'imaginary camping' is more ideal than 'real camping', G believes that

> It's not that the anime lied, but that camping itself has many variables, which is what makes camping interesting. I feel like I am living a day in the character's life in a parallel universe, at least, we see the moon the same way.

Furthermore, it is impossible to deny the significant part that camping gear played in these performative actions. As mentioned earlier, all the camping gear from *Laid-Back Camp* has an actual counterpart. For example, in F and G's videos, they both use the Mont-bell Moonlight 2 Tent that Rin used, which allows them to stay consistent with Rin's actions when setting up the tent. Some of the camping gear used in the work is also very practical, which has attracted audiences to use these same items in their camping adventures. J, a fan of *Laid-Back Camp*, showed off her new Coleman lantern – the same one that Nadeshiko uses – at a camping outing. In the work, when Nadeshiko first saw the lantern at an outdoor store, she fell in love with it, and using the money she earned from her part-time job, she finally bought the lantern. J said:

> I also bought this lantern with the money I earned from working. Although the process wasn't as difficult as Nadeshiko, when I lighted this lantern on a camping night and saw the flickering flame, I deeply felt the beauty of camping life, much as how Nadeshiko cherishes it.

Although these audiences simply visited a representational location such as a campground, rather than the actual places featured in *Laid-Back Camp*, they entered the narrative world by taking on the experiences of the characters. And from the perspectives of 'tourism motivated by content' and 'expanding narrative world through practice and experience', going to non-specific locations and imitating the behaviors of characters in the work is indeed a bodily practice related to contents tourism. Moreover, when people create their behaviors as new content and upload them to the internet for more audiences to watch and communicate, the narrative world constructed jointly by the works, places, and tourist practices will attract more potential contents tourists.

Produce vertical texts based on social media

As discussed earlier, people not only mimic the actions of characters in the work but also capture the practice through visual content. The "mediatisation of travel experiences" (Beeton et al., 2013, p. 150) continuously expands the narrative world created by the original work, and "mediatisation" itself has become an important component of the "contentization" process. The expansion of the narrative world in contents tourism is always based on the original work. In the content created by tourists, the original works always exist in the form of visual reproduction or textual description. However, while social media has deeply permeated into people's lives, the user-generated content that was originally created as a derivative work is consumed and adapted again, and enters a spiral structure, the discourse power of the original work is diluted in such "simulation" (Baudrillard, 1994[1981]).

A type of content draws attention when analyzing user-generated content related to *Laid-Back Camp*. Although all of this content is tagged with "Laid-Back Camp (yaoyeluying)" and even includes this word in their titles, there is no visual information related to the work. Moreover, there is no mention of the work in the text description, which only uses 'yaoyeluying' to describe their camping excursions. In fact, 'yaoyeluying' does not fully conform to Chinese grammar structures, and the Chinese word '*yaoye*' more actually means 'sway'. It is reasonable to believe that it functions as a proper noun. Interviewee I said:

> Before knowing about *Laid-Back Camp*, I always thought that 'yaoyeluying' is comparable to 'glamping' and is just a phrase to describe a kind of camping style.

This question is also raised during offline camping activities. Camping enthusiast K stated that

> The camping culture in Japan is well-established and has a relatively mature market. Chinese consumers favor Japanese camping brands. Perhaps it is the depiction of a Japanese camping style in *Laid-Back Camp* that inspires those who have only recently seen Chinese camping culture to use 'yaoyeluying' to refer to Japanese-style camping.

In such a dissemination process, the popular culture works, which were originally the driving force behind contents tourism, seem to have been 'invisible'. However, the "aura" (Benjamin, 2009[1935]) of the original work has not faded due to replication. Interviewee D said:

> In fact, I initially didn't watch *Laid-Back Camp*. I first became interested in camping after watching some videos on social media titled 'camping in an anime

style'. It wasn't until I started camping myself that I watched the *Laid-Back Camp* for the first time and realized that the videos I had previously watched were imitations. I also started to imagine trying solo camping like Rin.

If a popular culture work and its related trans-media works were regarded as the 'original' that constitutes people's tourism motivation in contents tourism, and if the travel practices of tourists around the 'original' were regarded as 'rituals' in the past, now, the mediatization of travel experiences may also be seen as 'original' in the context of social media. It is worth noting that tourist-generated content is also presented in trans-media. A travel experience may be described in various forms such as text, photos, and videos, and the same text may be posted on different social media platforms. Therefore, the narrative world constructed by tourist-generated content itself is constantly changing. Maybe the tourist-generated content could make the "imagined audience" (Marwick & Boyd, 2010) contents tourist, triggering new tourist rituals and expanding the narrative world in intertextuality.

Conclusion

This chapter discusses a series of Chinese multi-vocal travel practices inspired by a trans-media work based on the Japanese manga *Laid-Back Camp* and its adaptations. It manifests itself in a complex way in transnational, trans-cultural, and trans-media travel practices in multi-voiced contents tourism, allowing multiple perspectives to coexist, ensuring cultural diversity.

The travel induced by *Laid-Back Camp* is a transnational phenomenon. The activities related to *Laid-Back Camp* are not just localized practices but have enabled new forms of tourism through transnational consumption of the text. On the one hand, a Japanese popular culture work that has gained popularity among Chinese audiences has made the locations in Japan where the work is set sacred to the audience, encouraging transnational tourism. On the other hand, *Laid-Back Camp* transcends national boundaries and has an impact on China's emerging camping culture, helping to create a locally imagined sacred place. The sacred place of contents tourism is often exclusive to a constitutive element of narrative, which is the one and only place, but the sacred place created by *Laid-Back Camp* is anonymous, mostly plural, and even transnational. The travel practices of Chinese tourists induced by *Laid-Back Camp* complement the transnational case of contents tourism and exemplify the fact that contents tourism provides a multi-voiced travel environment for people with different nationalities and cultural backgrounds, creating a field of cultural exchange and mutual understanding.

At the same time, the tourism induced by *Laid-Back Camp* incorporates real-world places and events into the narrative world, creating tension among popular culture, tourism culture, and even more. Travel behaviors induced by *Laid-Back Camp* seem to have a distinct feature, which is that contents tourism and camping as a physical activity are always entwined. It is because *Laid-Back Camp* itself is a story about camping trips, and 'camping' plays a significant role in constructing the narrative world. While the hallmark activity of contents tourism is the reenactment of a scene from the media contents, in the case of *Laid-Back Camp*, reenactment shifts from a symbolic to functional travel behavior.

Tourists sanctify the campsite, a functional location, through camping rituals, connecting the narrative world with physical locations. Visiting places related to the work is originally a physical practice related to media culture, or rather, to popular culture, but in the case of *Laid-Back Camp*, it is also a physical practice related to camping culture. Japanese popular culture represented by anime and camping culture converge through *Laid-Back Camp*, triggering a cross-cultural phenomenon, showing the cultural multi-vocality of contents tourism.

Finally, and most importantly, as a trans-media text, it not only expands the narrative world at the level of the creators' power but also multiplies exponential proliferation through the secondary creative works of audiences, creating a distinct cultural phenomenon from the past. If the trans-media adaptations such as anime, live action, and movies that are based on the original work constantly expand the narrative world of *Laid-Back Camp* at the creator level horizontally, then tourist-generated content expands the narrative world vertically in the context of social media.

Tourist-generated content presents multi-vocal interpretations of the media text and becomes an original text that induces polyphonic interpretations in other audiences. It's worth considering how the original text produced by professional creators will exist when tourists generate more narrative-related interpretations. But such vertical intertextuality intervened by travel behavior discredits the text's hegemony and enables the text to spread in polysemy. People thus become the active audience and gain a mysterious sense of power by controlling the texts. Contents tourism breaks away from the structured text interpretations, presenting a tourist-driven multi-vocality.

Travel practices of Chinese tourists motivated by *Laid-Back Camp* present a form of multi-vocal contents tourism and refine the notion of multi-vocality. In such transnational, trans-cultural, and trans-media multi-vocality practices, Chinese tourists approach *Laid-Back Camp* in their own way, expanding the narrative world and enriching the meaning of contents tourism. The multi-vocality induced by *Laid-Back Camp* among the Chinese presents complex intertextuality in contents tourism. This chapter researches a transnational tourism phenomenon, which has energized media-induced tourism research in Asia. It not only focuses on the impact of media culture on tourism but also hints at the cross-cultural possibilities of media-induced tourism by exploring how media culture forms intertextuality with other cultures.

At the same time, this chapter strengthens the meaning of tourist-created content and complements the trans-medial aspect of contents tourism. One of the vital responsibilities of the field of tourism is to ensure respect for others and diversity through cultural exchange. The transnational, trans-cultural, and trans-media multi-vocality contents tourism provides a field for creators, audiences, and tourists with different cultural backgrounds to communicate and promote multi-vocal cultural exchange. Today, there are more diversified phenomena induced by media cultures that are appearing, and there will also be more transnational, trans-cultural, and trans-media practices creating the dynamism between fiction and reality that are worthwhile to be found.

Notes

1 The phrase often appears on social media in China and is also referred to as the "First Year of Glamping". Mafengwo. *2022 Camping Quality Research Report*, viewed 30 August 2023, www.mafengwo.cn/gonglve/zt-1029.html.
2 Data obtained from YiGuan Analysis. *China Camping Market Insights 2022*, viewed 30 August 2023, www.analysys.cn/article/detail/20020563
3 Japan Anime Tourism Association selects 88 influential anime worth preimage every year. *Laid-Back Camp* has been consistently on the list since 2019. Japan Anime Tourism Association, *88 anime sacred places in Japan* (2023 edition). Viewed 30 August 2023, https://animetourism88.com/ja/88AnimeSpot
4 'contents' is used in plural not only to match the Japanese pronunciation, but also to better represent the variety of contents in any given work that may create fan affinity or induce tourism. See Beeton, S., Yamamura, T., & Seaton, P. The mediatisation of culture: Japanese contents tourism and pop culture. In *mediating the tourist experience: from brochures to virtual encounters*. Farnham: Ashgate, 2013: 140.
5 RED, also known as 'Xiaohongshu' in China, is a social media platform that allows users to upload various texts, photos, and videos; it has been compared to Instagram in China. It particularly attracts youngsters to share their daily lives, including product recommendations, travel activities, and interests in popular trends. In recent years, as an important venue for camping promotion, RED has contributed significantly to China's "camping fever" interest.

References

Atkinson, S. A., & Kennedy, H. From conflict to revolution: The secret aesthetic, narrative spatialisation and audience experience in immersive cinema design. *Participation: Journal of Audience and Reception Studies, 13*(1), 2016: 252–279.

Bakhtin, M. *Problems of Dostoevsky's Poetics*, trans. Emerson, C. Minneapolis: University of Minnesota Press, 1984.

Baudrillard, J. *Simulacra and simulation*, trans. Glaser, S. F. Ann Arbor: University of Michigan Press, 1994.

Beeton, S. *Film-Induced Tourism*. Bristol: Channel View Publications, 2005.

Beeton, S., Yamamura, T., & Seaton, P. The mediatisation of culture: Japanese contents tourism and pop culture. In *Mediating the Tourist Experience: From Brochures to Virtual Encounters*. Farnham: Ashgate, 2013: 139–154.

Benjamin, W. *The origin of German tragic drama*, trans. Osborne, J. London: Verso books, 2009.

Evans, M. Plugging into TV tourism. *English Tourist Board, 8*(4), 1997: 302–332.

Gretzel, U. Influencer marketing in travel and tourism. In *Advances in Social Media for Travel, Tourism and Hospitality*. New York: Routledge, 2017:147–156.

Hall, S. Encoding/decoding. In *Culture, Media, Language: Working Papers in Cultural Studies, 1972–79*. London: Routledge, 2003:117–127.

Marwick, A. E., & Boyd, D. I tweet honestly, I tweet passionately: Twitter users, context collapse, and the imagined audience. *New Media & Society, 13*(1), 2011:114–133.

Norris, C. J. A Japanese media pilgrimage to a Tasmanian bakery. *Transformative Works and Cultures*, *14*, 2013: 1–16.

Okamoto, T. Tourism as a circuit of interaction: Considering traveling communication in informational society from *anime* pilgrimage. *Journal of Japanese Society for Artificial Intelligence*, *26*(3), 2011: 256–263.

Okamoto, T. Otaku tourism and the anime pilgrimage phenomenon in Japan. *Japan Forum*, *27*(1), 2015:12–36.

Reijnders, S. *Places of the Imagination: Media, Tourism, Culture*. Farnham: Ashgate Publishing, 2011.

Riley, R., Baker, D., &Van Doren, C. S. Movie induced tourism. *Annals of Tourism Research*, *25*(4), 1998: 919–935.

Scarles, C. Becoming tourist: Renegotiating the visual in the tourist experience. *Environment and Planning D: Society and Space*, *27*(3), 2009: 465–488.

Seaton, P. A., Yamamura, T., Sugawa-Shimada, A., & Jang, K. *Contents Tourism in Japan: Pilgrimages to 'Sacred Sites' of Popular Culture*. Amherst, MA: Cambria Press, 2017.

Squire, S. J. Literary tourism, and sustainable tourism: promoting 'Anne of Green Gables' in Prince Edward Island. *Journal of Sustainable Tourism*, *4*(3), 1996: 119–134.

Yamamura, T. Study of birth and development of 'sacred place for anime fans': Discussion of tourist promotions based on animated work "luckey star" focused on Washimiya, Saitama Prefecture. *The Journal of International Media, Communication, and Tourism Studies, 7*, 2008: 145–164.

Yamamura, T. Anime pilgrimage and local tourism promotion: An experience of Washimiya town, the sacred place for Anime "Lucky Star" Fans. *Journal of Tourism and Cultural Studies Hokkaido University, 14*, 2009:1–9.

Yamamura, T. Introduction: contents tourism beyond anime tourism. In *Contents Tourism and Pop Culture Fandom: Transnational Tourist Experiences*. Bristol: Channel View Publications, 2020:1–16.

9 Bollywood Tourism among the Hindustanis in the Netherlands

A Transnational Perspective[1]

Apoorva Nanjangud and Stijn Reijnders

Introduction

The Dutch Hindustani community is a prominent group in the Netherlands that has its roots in various parts of North India. It is a community belonging to the Indian diaspora that is generationally away from India given its colonial history of indentured labor migration from British India to the Dutch colony of Suriname in the late eighteenth and early nineteenth century. Consequently, after Suriname gained independence in 1975, there was a wave of migration of this community to Holland. This pattern of double migration led them to be connoted as the 'twice-migrants' (Verstappen & Rutten, 2007). Decades of spatial detachment later, they still actively maintain a keen interest in their distant first homeland India. This connection is reflected in, for example, their voracious consumption of Indian popular culture, practicing Indian performing arts or the manner in which they celebrate Indian religious festivals at home. Today, the Netherlands has the second largest Indian diasporic community in Europe (Longkumer, 2013).

According to Basu (2004), in this global world of movement, the notion of 'home' has become a powerful part of the contemporary debate to relocate identity. Many members of the diasporic communities therefore visit their 'homelands' to re-root their identities, soothe their nostalgia and find nourishment. This temporary diasporic visitation to homeland is acknowledged in academic circles as 'diaspora tourism' or 'roots tourism' (Timothy, 1997; Timothy & Teye, 2004). Sara Ahmed's seminal work discusses the situational feeling of 'home' that comes with these transnational journeys. Her work dwells on the possibility of having multiple homes, and an individual's movement between them, drawing a distinction between a space which almost feels like home but is not (Ahmed, 1999, p. 331). This adds valuable insights to how the Dutch Hindustanis may situationally feel at home in the Netherlands, India, and Suriname at the same time.

This chapter investigates the role and significance of Bollywood cinema for developing feelings of belonging among the Dutch Hindustanis and furthermore seeks to understand the extent to which Bollywood cinema motivates them to physically conduct a touristic visitation to places represented by Bollywood (Bhattacharya, 2018). The research question that guides this chapter is 'What are

DOI: 10.4324/9781003320586-12

the cinematic imaginations of India developed by the Dutch Hindustanis under the influence of Bollywood cinema, and to what extent do these imaginations influence their travel decisions to India?'

With this chapter, I aim to contribute to existing debates within the field of film tourism studies while also diversifying it. Existing research on film tourism has a strong Western focus and is often based on isolated, high-profile examples of box-office hits resulting in tourism. Besides a few studies (see Bandyopadhyay, 2008; Bhattacharya, 2018; Biswas & Croy, 2018; Laing & Frost, 2018), a non-Western film industry like Bollywood or a destination like India is seldom discussed in film tourism research. This is remarkable considering that Indian film industry is one of the largest producers of films annually and caters to one of the largest diasporic audiences globally. Thus, this chapter aims to depart from the monolithic definition of 'the' film tourist and signal a departure from the classical *3W's* tourists – *wealthy, white and western* – by exploring the potential of tourism across various groups by reflecting on a specialized section of film tourists, who are in fact diasporic film audiences.

This chapter utilizes insights from 17 in-depth semi-structured interviews conducted between June and October 2017, across different cities in the Netherlands. The participants were asked questions about their imagination of India, the role of Bollywood in their lives as well as its impact on their travel decisions to India. The collected data was transcribed verbatim and then subjected to thematic analysis.

The role of cinema in home and belonging

The phenomenon of film tourism is not limited to the actual act of tourism alone but is deeply rooted in the prior processes of consumption of the media narratives, fantasizing about the locations concerned and then ending with a reflection of the finished journey (Larsen & Urry, 2011; Reijnders, 2016). This process of imagination is triggered when confronted with visual or auditory cues. Being a combination of both robust visuals and songs, Bollywood readily tends to fuel the imaginative process. Appadurai's (1996) concept of *mediascapes* – that is, 'image-centered, narrative based accounts of strips of reality' (p. 35) – can be understood as the many media outlets – in this case cinema – that shape the 'imagined world' we inhabit, where indeed, narratives and images from the media have become pivotal for how people form an initial image about 'other' places and cultures. This is especially true for diasporic communities, like the Dutch Hindustanis, who are generationally away from their homeland and therefore substantially rely on the imaginations of a place as propagated by cinema and other media. As Appadurai (1990) writes, 'The further away these audiences are from the direct experiences of metropolitan life, the more likely they are to construct "imagined worlds"' (p. 299). For these diasporic communities, popular culture holds significance not only for providing a tourist gaze of foreign countries but also for guiding them construct an image of their own 'homeland' and possibly to inspire future travel plans.

The Indian diaspora is one of the biggest around the globe, and many of its members actively seek to maintain and condition their (multicultural) 'Indianness', which has led to the formation of a global Indian diasporic identity. Bollywood films and songs are tremendously popular within these communities (Bal & Sinha-Kerkhoff, 2003). Existing research shows that the cultural (and religious) expressions associated with India grant the Dutch Hindustanis the visibility and distinctiveness among the various communities in the Netherlands, a very multi-cultural country in itself (Bal, 2012). As Gowricharn (2009, 2016) notes, travels to India are popular among the Hindustani community, which are often a combin-ation of pilgrimage, tourism, the quest for tracking their roots, and shopping, while helping them retain their language, fashion, identity, gender patterns and social intercourse (Verstappen & Rutten, 2007).

As vastly researched, Bollywood cinema has gained considerable signifi-cance in the domain of cultural studies globally from multiple perspectives. From understanding Bollywood's changing linguistic norms (Ahmad, 2018), to studying the representation of whiteness in films (Acciari, 2017), the scope and study of Bollywood cinema from various socio-cultural angles has been rather robust. In addition, works by Dudrah (2002), Mishra (2002), Punathambekar (2005), Mohammad (2007), Bandyopadhyay (2008), Takhar et al. (2012), Marwah (2017), among others, illustrate an ever-growing presence of the diasporic viewpoints in the studying of popular Hindi cinema. However, what remains predominant in many of these studies is the perspective of the Indian diasporic audiences from the United States. Often there is lower emphasis on Persons of Indian Origin (PIO), that is, people who migrated under varying situations such as post-colonial circumstances, despite their large presence in the global Indian landscape. For example, few studies discuss diasporic groups such as the Indo-Fijians in Australia (e.g., Lal, 1990; Voigt-Graf, 2004), or the East African diaspora in the United Kingdom (Mattausch, 2011).

Several studies have addressed the popularity of Bollywood among Hindustani diasporic communities (Gowricharn, 2009; Gowricharn & Choenni, 2006a; 2006b). Gowricharn (2009), for instance, discusses how Bollywood constitutes a powerful 'source culture' and enhances transnational ties between second-generation Hindustanis and India, supporting the idea that Bollywood is a powerful transnational force which influences cultural bonding among the Hindustani dias-pora, its influence on their everyday lives and how they watch more Bollywood movies over Western cinema while living in Europe (Gowricharn, 2009). This insight also provides an important stepping-stone to this chapter as to how the Dutch Hindustani community reconnects with India through cinema, by not only providing entertainment, emotional and cultural anchorage, but also by being a medium that creates both a mental and a real travel route between India and the Netherlands. It therefore becomes important to understand their process of identity construction through cinema, and how these diasporic imaginaries generate film-induced tourism to their ancestral land. There is still a remarkable gap to address this through the lens of diasporic media studies and film tourism studies. This chapter aims to take a first step in bridging this gap by empirically exploring the connections between Bollywood consumption, image-building and the creation of travel itineraries among the Dutch Hindustani community.

Imagining India through Bollywood

Ryan (44) moved to the Netherlands at the age of one. His grandfather was Indian, but he and his parents were born in Suriname. At least two generations have spent their entire life physically detached from the Indian subcontinent. In such a family set-up, the stories of Bollywood cinema have been a big inspiration in the lives of many like Ryan to be mentally connected with India and its popular culture. It is this visuality and robustness of the medium that mirrors the Indian landscape and soothes their curiosity about what India must look like:

> Yeah, it (life) is always inspired by the (Bollywood) movies, I think... . The movies are our guide because we are living in Holland and... . through the movies we see India... .
>
> (Ryan, 44, Rotterdam)

While there are many new mediated ways of communication that facilitate the 'connection' with India, it is the 'infotainment' aspect of Bollywood cinema that makes it a desirable medium. For most participants, like Ryan, Bollywood facilitates a connection with India and is likewise used by Hindustani parents as a tool to keep their children in touch with the place they themselves originally came from. Consumption of Bollywood movies not only aids them in learning Hindi but also helps to pick up Indian cultural codes and rituals far away from their current cultural context in the Netherlands which is largely European. For Akash (23), who was born and raised in the Netherlands and has never set foot on Indian soil, Bollywood narratives of India have not only increased his knowledge of the Hindi language but has also shaped his popular imagination of India, adding to it a more positive connotation:

> If Bollywood wouldn't have been there, I wouldn't have known a lot of things... . It wouldn't seem so beautiful in my head, because in the news it doesn't seem as beautiful as it seems in the movies.
>
> (Akash, 23, Rotterdam)

This cultural transfiguration returns among most of the participants. Merlin (40), for example, was born and raised in Holland and has never visited India. She has a strong desire to go there, but she also feels scared to actually travel to India due to her unfamiliarity with the country. However, she credits Bollywood cinema for keeping her interested and informed about her roots. She interprets these movies as replacements for actually traveling there. This supports the idea of film spectators as potential film tourists (Corbin, 2014). Bollywood movies tell her about her distant homeland, its practices and places, and in some cases even provides her with the longing to be there herself:

> I don't know if I will go, because I find it a bit scary... . But when I see it in movies... . there are places I think 'Oh, those are nice, that looks nice!', maybe

I can go, and see how it is for myself.... [. . .] I have learnt about India that the traditions there are more sacred than they are here in the Netherlands. They are more meaningful over there than they are here.

Having these cinematic imaginations of India without having actually been there reflects the extent of Bollywood's influence in the lives of the Hindustani participants. The Dutch Hindustanis emotionally depend on the imaginaries propagated by Bollywood cinema, which is potentially also an important reason for their affinity with Bollywood cinema. However, many recent movies tend to be shot abroad for the most part, and therefore risk hampering the process of imagining India among the diasporic audiences. In similar light, Jaswina (38) says,

Nowadays movies are less interesting when they are set abroad and a lot of them are set abroad in the West. You know, like, Europe or America. I actually want to see India.... I think [. . .] my motivation is to see India. And all those movies from now.... they're only showing me (the west). Especially after my time over there, I really feel more connected to India in different ways than before I went to India [. . .] A way to keep the bond with India or to see India, is through the movies.

(Jaswina, 38, The Hague)

In response to these recent production trends, many participants like Jaswina show a renewed interest in the classics from the 1980s and 1990s, when popular Hindi cinema often portrayed India in a relatively social-realist style and was arguably more 'relatable' than the current crop of contemporary movies. As Jaswina puts it, a definite way of keeping the bond with India for the participants is to watch Bollywood movies and this may hold true for Indian diaspora in general. While the shooting of Bollywood films abroad entices the domestic Indian audiences to go explore places abroad (Nanjangud & Reijnders, 2022), to some extent it also seems to affect the connection between the diasporic audiences and India.

A particularly interesting fact is that many of the participants consider Bollywood cinema to be the *first* source for fueling their imagination, thereby making cinema akin to guidebooks, educating them about various rituals and festivities of Indian culture. This also imparts a strange sense of familiarity when they finally visit India, with expectations largely guided by the cinema:

I have grown up with movies since I was little.... . I watched every weekend only Bollywood movies, I went with my dad to the video store and got all the movies that I wanted, 6 or 7 and watched them really the whole weekend.... . And it appealed to me because it had singing, dancing, laughter.... . And I could identify myself with those actors, in the sense of 'Oh, I am also a part from India', my great-grandparents are from India, and when I saw the movies I was like 'Oh, there's a part of me!' and the language is also a part of me.... . And growing up I learnt the language from Bollywood movies as well, not only from my parents but also from the movies.... .

(Chan, 54, Arnhem)

As these quotes show, Bollywood creates idealized cinematic imaginaries of the Indian 'homeland'. But why does India constitute the 'homeland' for these participants? How does Suriname or the Netherlands fit in this picture? And what is the role of cinema in these different identity processes?

Cinematic identities: Homeland, hostland and the land in-between

Verstappen and Rutten (2007) state that the Dutch Hindustanis identify with India and the Netherlands and consider both countries to be their home. Gowricharn (2009) also indicates the growing absence of Suriname in the self-perception of Hindustanis. Even after so many years and miles of distance from India, the Dutch Hindustanis interviewed for this study still relate themselves first with India and, second, with Holland. The relation to Suriname was rarely acknowledged during the interviews and often this relationship with the 'intermediary country' was dismissed completely. As Gowricharn (2009) points out, there has been a constant process of 'ethnification' of Indian culture among the Suriname Hindustanis, which he defines as 'the modified reproduction of the ethnic community, taking the form of institutionalization and the establishment of cultural identity' (Gowricharn, 2009: 10). This has consequently resulted in an unequal representation of India and the Netherlands in the media consumption of the Hindustanis, with consequences for the underlying identification processes. After moving to the Netherlands at the age of one and living there ever since, Ryan currently feels more 'Indian' than 'European':

> I don't balance it (my identity), because I'm feeling more Indian than European. And that's why I always mention it, you know, and I don't act like a European also. I love my roti, you know, and I love my masala... . I'm more Indian than European.
>
> (Ryan, 44, Rotterdam)

Similarly, this feeling of situational identities is experienced at all age groups. Asha, who left Suriname at the age of 10, feels similarly:

> . ...When I'm at work... . I am Dutch. But when I introduce myself or they ask me where I come from, then I always say, my roots lie in India. I was born in Suriname but my grand ancestors came from India. Uh, but at festivals, I feel Indian. You know, I'm a Hindustani. Not an Indian, but a *Hindustani*.
>
> (Asha, 57, Rotterdam)

According to Bal (2012), the affinity toward Suriname, the intermediary country, is greatly dependent on the generation one belongs to. She notes that many from the first generation of migrants from Suriname in the Netherlands identify as Surinamese, but many from the younger generation of Hindustanis born in the Netherlands do not share the same feelings about Suriname. For example, Chietra

(37) explains that she finds it easier to connect with 'her Indian self' over 'her Surinamese self':

> My first time in India was [. . .] five years ago. I feel so connected with everyone. I love my Saree, and, and my bangles and 'payals'. And there it's so normal... . I feel more connected to India than with Suriname and that's so strange because my grandparents and my mummy and daddy are all from Suriname.
>
> (Chietra, 37, Almere)

It is duly noted that the connection that many participants felt with India was more of a cultural connection – something they felt proud to associate themselves with. How can this lack of identification with Suriname be explained? It may perhaps be attributed to the mediatized presence of India in global popular culture, making identification with it more relatable for the Hindustanis in the Netherlands. Chiming with the previous interviewee's thoughts on home and belonging, Lisa (24) points out the strong presence of Indian movies in global popular culture and how this provides her with something of her own in an overly European society in which it is sometimes hard to belong:

> You know, Bollywood kind of became my saving, like, this was mine and nobody really knows what's going on with that [. . .] I grew up with a lot of people of colour, so there's always this everything and this was just mine... . And also, to have a language in a Dutch speaking bubble it's very nice... .
>
> (Lisa, 24, Rotterdam)

This quote from Lisa shows how the clear presence of Bollywood in global popular culture provides diasporic audiences with an important tool for identity work: one can identify as Indian because there are global references to 'Indianness'.

For people like Lisa and many others, growing up and trying to find one's own voice in a multicultural, Western world can be quite challenging. Bollywood seems to provide a welcome 'tool' in this process. Lisa pointed out how she tried hip-hop multiple times but failed at it, because it wasn't hers. Her personality and interests always resonated with Bollywood; her cultural identity was Indian. In the process of imagining India through Bollywood, and relating to it and building one's own identity through it, it becomes the question if these cinematic associations result in concrete travel interests in Indian destinations.

Cinematic itineraries: Traveling in India 'as-seen-on-screen'

After having seen many Bollywood movies, and identifying with the country and culture as depicted in these storylines, many participants experience a growing desire to actually visit this 'homeland'. It becomes all the more interesting when diasporic audiences are involved, as a part of their family history is embedded in India. For example, Chan proclaims to be a vivid fan of Bollywood movies and explains that these cinema-viewing experiences eventually drove him to book his first trip to India at the age of 35:

> When I started watching Bollywood movies, I was becoming a part of India...
> . And I think it was 1999, when I decided, okay let's go, I am so curious about India, and I have seen a lot of movies and places, Juhu beach... . Goa... . So the first step I put in Delhi, I somehow felt emotions in my body and I started crying... . And since then [. . .] I go almost every year to India.
>
> (Chan, 54, Arnhem)

Similar to Chan, Joy (41) was also inspired by cinema to conduct not only his first travels but also the subsequent trips to India:

> The movies and the interest in India itself really took me. I went four times to India. Just because of Bollywood. Because otherwise I didn't know about India. I saw India through the movies. And yeah, it took my interest and I thought okay, I need to see this. I need to experience this... .
>
> (Joy, 41, Rijswijk)

However, one must remember that the 'Bollywood lens' provides only a sanitized and idealized version of India that does not necessarily meet the expectations of diasporic tourists who have developed and sustained Bollywood-inspired imaginaries of India. Asha, for example, shares her experience of being in India for the first time:

> I was really shocked. Because I saw so much poverty. I was really scared... . because I saw so many people. And many poor people that you didn't see in the movie. But what I saw there was such a big contrast.
>
> (Asha, 57, Rotterdam)

Such experiences are common among the Dutch Hindustanis as their affiliation to India is often limited to the cultural codes transmitted through the Bollywood cinema, while being away from the ground realities of what India has to offer. Seema, who visited India for the first time at a young age of 9, shares:

> I already had images that what I saw in the movies, that's India. And I was very excited, like, 'yay! Everything what I see in the movies that will become true. So beautiful, so nice'. So I was really happy. It was just a dream come true for me. And when I came there, I was feeling 'oh gosh. It's so dirty'. And when you're walking, 'oh my god'. I just walk into the *gobar* [cow excreta]. And everywhere you got the cows and everything and it smells dirty.

Going to India due to being inspired by Bollywood cinema is something that happened often, but where exactly do these 'roots travelers' go? Joy explained how he created a travel itinerary through India based on his knowledge of Bollywood movies, traveling from place to place. More participants used Bollywood associations in a similar direct way to create a travel plan for themselves which they followed through the trip. It not only included film studios or film tours, but also locations that felt familiar through the movies:

. …There are a lot of movies that have good places to show what India is. And it keeps you like, 'oh I want to go there' So, I think a lot of people from the diaspora, who go see India is eighty percent because of the movies… . I saw the movies and I took it [straight] with me.

(Chan, 54, Arnhem)

Chietra who agrees to the idea that Bollywood is often instrumental in influencing people's travel decisions toward India says,

Bollywood movies have an influence. The beautiful places which we can see in the movies and then you can imagine, 'Hey, when I'm going to travel, I want to see this. I want to see the Taj Mahal. Because you see it on your screen, but then you can see it in real life'.

(Chietra, 37, Almere)

Jaswina also recalled how cinema was a strong factor when planning her recent short trip to Mumbai. Her travel decisions were, to a large extent, driven by her interest in popular Bollywood actors:

Because we are just here for one day (we thought) 'okay, let's go to the place where Bollywood actors are living'. Not like, let's see a fort or something or go to a museum. So the only thing I have seen in Mumbai is Marine drive. And the area where the stars are living.

(Jaswina, 38, The Hague)

These findings are interesting as they seem to point at practices that go beyond the standard itinerary of the film tourist. These participants are not in search of one specific movie or one particular scene, as is commonly the case, but they use a collection of Bollywood movies and celebrity hotspots to develop a multi-sided tour through India, traveling from one movie to another and experiencing India through the lenses of Bollywood as a cohesive whole. I would like to refer to these practices by coining the concept of 'cinematic itineraries': travel itineraries that are composed of several sites associated with multiple movies or associated film stars.

What makes these cinematic itineraries through India so exciting for the participants? For some, going to see the spots which mark their fandom for Bollywood often equals a dream-like situation:

. …I think the one thing that really stood out to me was when we went to Marine Drive and I had just seen *Wake Up Sid*, and I was like 'Yes!' you know, amazing… . That was my fake Bollywood moment, and then also the fact that we went along *Mannat*, you know, and then the guide was just like 'Yeah, this is *Mannat*, this is where Shahrukh Khan lives'.

(Lisa, 24, Rotterdam)

From the interviews, it became clear how much influence films have on the manner in which the diaspora connects with and discovers locations beyond the usual Indian cities. Rajnie says,

I saw *Lootera* a few months ago, and I think it was shot in Dalhousie, so ever since I saw *Lootera* I wanted to go to Dalhousie… . In Raaz, they went to Shimla, and I still want to go to Shimla… .

(Rajnie, 35, Rotterdam)

Lisa expresses her desire to go back to India, and Mumbai in particular, to re-live her Bollywood moments:

I want to go back to Mumbai again, just to have that quintessential Bollywood experience… . I really want to be like that uber tourist, you know, and (see) everything that you've seen in movies so far, I would like to see that for real and have these experiences.

(Lisa, 24, Rotterdam)

Chietra (37) speaks about why she loves Bollywood movies and how they influence her association and experience of India:

. …In movies you see a lot of places where I went to. Let's say Mumbai. Then you see the central station and you see Juhu beach. And in one of my vlogs I said, 'Look! I am here just as in the Bollywood movies you see the Juhu beach, this is the Juhu beach!'

(Chietra, 37, Almere)

For these participants, going to India rarely meant going in search of their roots or their ancestral home, or finding long lost relatives and friends. The visit to India does elicit a strong emotional response, but the primary motivation has largely been to visit a generic and idealized India as imagined in Bollywood movies. This can be also corroborated by previous research that for the second-generation Dutch Hindustanis, the affective ties to India function as 'source cultures' rather than as 'Home' (Gowricharn, 2009). There are no specific spots in India called Bollywood, unlike its US-American counterpart. There are, however, a growing amount of locations that are identified as being quintessentially 'Bollywood' locations based on the indelible association film tourists draw between locations and Bollywood films. These 'places of the imagination' (Reijnders, 2011) have become a standard part of the diaspora's Bollywood tourism experience.

Conclusion

This chapter has revealed that Bollywood cinema has a considerable influence on the imagination of India among the Dutch Hindustanis. For participants who were born in Suriname and migrated to the Netherlands from the 1970s onwards, or who were born in the Netherlands itself, Bollywood cinema is a major reference point to Indian landscape and culture. For the older participants who did not visit India during their youth and only started going there in their middle age, for a long time, the images of India as propagated by Bollywood cinema have been a rather dominant source for defining the participants' relationship and imaginations toward India, which ensured a sense of familiarity when they arrived in India. This also holds true for the younger participants who have never been there but rely on the mediatized imaginaries created by Bollywood. This at times led to a dissonance between the imaginaries of India idealized by Bollywood and their real-life experiences in India. The younger participants not only relied on cinema for comfort, but also appropriated it to their present context by utilizing Bollywood elements in their current cultural expression. Many participants adopt cultural codes concerning the 'Indianness' from cinema and practice rituals, clothing and religious festivals (Verstappen & Rutten, 2007).

The mediatized presence of Bollywood culture often tends to weigh heavier in comparison with Suriname, making Indian identity more relatable, approachable and 'workable' as a tool for identity work. This also meant that Bollywood cinema was a source of cultural education, but also provided a 'pull' factor to make the participants travel to India to soothe their cultural affiliations and not really in search of 'Home'. However, the recent trend among Bollywood producers to shoot movies outside India slightly hindered the process of imagining India through cinema for the participants. This results in them falling back on the classic Bollywood films from 1970s, 1980s and 1990s, to keep them engaged with the materiality and authenticity of the homeland.

It became clear that through the repeated consumption of Bollywood cinema, the participants also became acquainted with different places and landscapes, stirring the desire to actually visit India. Many participants had a clear list of specific places they desired to visit. Particularly, Bollywood cinema turned out to be a predominant source of travel inspiration for most of the participants, as they designed their India trips based on associations with famous scenes from Bollywood's film history. These trips were not based on one movie or one particular scene, but on a string of movies that together formed a 'Bollywood filter' of India. In addition to film locations, these visitations included homes of famous Bollywood celebrities as well as film studio tours. Through this chapter, the concept of 'cinematic itineraries' is introduced to suggest these multi-sided and comprehensive film tourist practices based on the Bollywood lens, where not one film but a collective imaginary of the story-world of Bollywood, inspires travels for these diasporic audiences.

Acknowledgments

Parts of this chapter have been published before in *European Journal of Cultural Studies*. We thank the editors for their permission to reuse this material. This study is part of the project Film Tourism, which has kindly received funding from the European Research Council (ERC) under the European Union's Horizon 2020 research and innovation program (grant agreement no. 681663). Note on the division of work: the first author has done the fieldwork and first analysis, while the second author has contributed to the research design, concept building and analysis.

Note

1 A version of this chapter has been previously published in *European Journal of Cultural Studies*: https://doi.org/10.1177/1367549420951577

References

Acciari, M. (2017). Bollywood's variation on the firanginess theme: Song-and-dance sequences as heterotopic offbeats. *South Asian Popular Culture, 15*(2–3), pp. 173–187.

Ahmad, R. (2018). My name is Khan. …from the epiglottis: Changing linguistic norms in Bollywood songs. *South Asian Popular Culture, 16*(1), pp. 51–69.

Ahmed, S. (1999). Home and away: Narratives of migration and estrangement. *International Journal of Cultural Studies, 2*(3), pp. 329–347.

Appadurai, A. (1990). Disjuncture and difference in the global cultural economy. *Theory, Culture & Society, 7*(2–3), pp. 295–310.

Appadurai, A. (1996). *Modernity at Large: Cultural Dimensions of Globalization.* University of Minnesota Press.

Bal, E.W. (2012). Country report: Indian migration to the Netherlands (CARIM-India; No. RR2012/7). San Domenico di Fiesole: Robert Schuman Centre for Advanced Studies.

Bal, E.W., & Sinha-Kerkhoff, K. (2003). Eternal call of the Ganga: Reconnecting with people of Indian origin in Surinam. *Economic and Political Weekly, 38*(38), pp. 4008–4021.

Bandyopadhyay, R. (2008). Nostalgia, identity and tourism: Bollywood in the Indian diaspora. *Journal of Tourism and Cultural Change, 6*(2), pp. 79–100.

Basu, P. (2004). My own island home: The Orkney homecoming. *Journal of Material Culture, 9*(1), pp. 27–42.

Bhattacharya, T. (2018). Impact of Hindi films (Bollywood) on the Indian diaspora in Honolulu' Hawaii. In: Kim S. and Reijnders S. (eds), *Film Tourism in Asia.* Springer, pp. 239–250.

Biswas, J., & Croy, G. (2018). Film-tourism in India: An emergent phenomenon. In: Kim, S. and Reijnders, S. (eds), *Film-Tourism in Asia.* Springer, pp. 33–48.

Corbin, A. (2014). Travelling through cinema space: The film spectator as tourist. *Continuum, 28*(3), pp. 314–329.

Dudrah, R. (2002). Vilayati Bollywood: Popular Hindi cinema-going and Diasporic South Asian Identity in Birmingham (UK). *Javnost – The Public, 9*(1), pp. 19–36.

Gowricharn, R. (2009). Changing forms of transnationalism. *Ethnic and Racial Studies, 32*(9), pp. 1619–1638.

Gowricharn, R. (2016). Shopping in Mumbai: Transnational sociability from the Netherlands. *Global Networks, 17*, pp. 349–365.

Gowricharn, R., & Choenni, C. (2006a). *Indians in the Netherlands the Encyclopedia of Indian Diaspora*. Editions Didier Millet.

Gowricharn, R., & Choenni, C. (2006b). Indians in the Netherlands. In: Lal, B.V. and Reeves, P. (eds), *The Encyclopedia of the Indian Diaspora*. Didier Millet, pp. 346–349.

Laing, J., & Frost, W. (2018). Imagining tourism and mobilities in modern India through film. In: Kim, S. and Reijnders, S. (eds), *Film-Tourism in Asia*. Springer, pp. 21–32.

Lal, V. (1990). The Fiji Indians: Marooned at home. In: Clarke, C., Peach, C. and Vertovec, S. (eds), *South Asians Overseas: Migration and Ethnicity*. Cambridge University Press, pp. 113–130.

Larsen, J., & Urry, J. (2011). Gazing and performing. *Environment and Planning D: Society and Space, 29*(6), pp. 1110–1125.

Longkumer, L. (2013). A cultural narrative on the twice migrated Hindustanis of the Netherlands. Migration Policy Centre, CARIM-India Research Report, 2013/23. San Domenico di Fiesole: Robert Schuman Centre for Advanced Studies.

Marwah, S. (2017). Kehte Hain Humko Pyar Se Indiawaale: Shaping a contemporary diasporic Indianness in and through the Bollywood Song. *South Asian Popular Culture, 15*(2–3), pp. 189–202.

Mattausch, J. (2011). A chance diaspora: British Gujarati Hindus. In: Oonk, G. (ed.), *Global Indian Diasporas: Exploring Trajectories of Migration and Theory*. Amsterdam University Press, pp. 149–166.

Mishra, V. (2002). *Temples of Desire*. Routledge.

Mohammad, R. (2007). Phir Bhi Dil Hai Hindustani (yet the heart remains Indian): Bollywood, the 'homeland' nation-state, and the diaspora. *Environment and Planning D: Society and Space, 25*(6), pp. 1015–1040.

Punathambekar, A. (2005). Bollywood in the Indian-American diaspora: Mediating a transitive logic of cultural citizenship. *International Journal of Cultural Studies, 8*(2), pp. 151–173.

Reijnders, S. (2011). *Places of the Imagination: Media, Tourism, Culture*. Ashgate Publishing.

Reijnders, S. (2016). Stories that move. Fiction, imagination, tourism. *European Journal of Cultural Studies, 19*(6), pp. 672–689.

Takhar, A., Maclaran, P., & Stevens, L. (2012). Bollywood cinema's global reach consuming the diasporic consciousness. *Journal of Macromarketing, 32*(3), pp. 266–279.

Timothy, D. (1997). Tourism and personal heritage. *Annals of Tourism Research, 24*(3), pp. 751–754.

Timothy, D., & Teye, V. (2004). American children of the African diaspora: Journeys to the motherland. In: Coles, T. and Timothy, D. (eds), *Tourism, Diaspora and Space*. Routledge, pp. 111–123.

Verstappen, S., & Rutten, M. (2007). Bollywood and the Indian diaspora: Reception of Indian cinema among Hindustani youth in the Netherlands. In: Oonk, G. (ed.), *Global Indian Diasporas: Exploring Trajectories of Migration and Theory*. Amsterdam University Press, pp. 211–233.

Voigt-Graf, C. (2004). Twice migrants' relationship to their ancestral homeland: The case of IndoFijians and India. *The Journal of Pacific Studies, 27*(2), pp. 177–203.

Part 3

Making Place in a Mediatized World

Heritage, Community and Social Change

10 Whose Homestead Is It?

Little Houses on the Prairie and the Cultural Politics of White Colonial Settlement in the United States

Nancy Reagin

Introduction

Strung across the prairies of the United States, there are small towns whose claim to fame – and draw for visitors – is that they were once home to Laura Ingalls Wilder (1872–1957), author of the *Little House* series, which began publication in 1932. Wilder's wandering family lived in homes and farms in Wisconsin, Minnesota, Kansas, Missouri, Iowa, and South Dakota. Some of their original dwellings still stand.

Although it is largely unknown outside the United States, the *Little House* series was widely loved in the United States during the twentieth century: almost 60 million copies were sold by 2000. For generations, many American children read it in school. Originally a literary fandom, after 1970 the *Little House* canon spun off a long-running television series and several films. Over time, it thus developed into a transmedial world and fandom, anchored and made tangible by the Wilder memorial "homesites" and museums scattered across the Midwestern United States, which form focal points for fan pilgrimages. At the homesites, visitors can experience the historic locations where Wilder actually lived, but which were also settings for the novels and television shows based on her life. They can see and examine objects that she or family members made or used, which also played important roles in her stories.

This chapter explores the ways in which the Wilder homesites and memorials function as both *lieux de mémoire* ("sites of memory") and as *lieux d'imagination* ("places of the imagination"). My approach draws on both Nora's work on *lieux de mémoire* (1996), and Reijnders' influential *Places of the Imagination* (2011). Nora defined *lieux de mémoire* as "entities" (including concepts, physical locations, or objects) that have "become a symbolic element of the memorial heritage of any community" (Nora, 1996). Museums, monuments, or specific locations can be seen as historically significant in popular collective memory and serve as physical points of reference that help to define national identity and collective memory. Reijnders (2011) applies Nora's insights to media tourism, arguing that "material reference points like objects or places.... . for certain groups in society serve as material-symbolic references to a common imaginary world"

DOI: 10.4324/9781003320586-14

(2011, p. 8); these places of the imagination thus become "symbolic anchor[s] for the collective imagination of a society" (2011, p, 14). The Wilder homesites serve as focus points for both collective memory (for white Americans) and collective imagination for her fans.

Visitors to the Wilder homesites (and their staff) see these locations as sites of collective heritage, which make tangible a key chapter in American history and national identity. But for many visitors, they are simultaneously places of imagination, where fans can connect to the storyworld's characters, and extend how they imagine important scenes. This chapter explores how homesites anchor fans' participatory culture, offering sensory immersion and experiential engagement. But because the homesites and storyworld also function as sites of collective memory and national history, they have become implicated in political conflicts, as American culture increasingly polarizes. This chapter examines the impact of America's culture wars on the Wilder fandom, and how media tourism sites can be drawn into debates about public history, if they simultaneously function as *lieux de mémoire*.

Methodology

During the summer of 2019, I visited six Wilder memorial sites and homes in upstate New York, Wisconsin, Minnesota, South Dakota, Kansas, and Missouri. I interviewed eight homesite staff members, asking them about the profile of those who visited their site, their interactions with visitors, and their motives for devoting time and effort to the homesites (most were also *Little House* fans). These interviews were supplemented by participatory observation during tours of each homesite, along with short interviews with a dozen visitors to the sites who were chosen at random. I asked visitors about their motives for coming to the homesites, and how they evaluated and experienced these locations. Short interviews were supplemented by longer, in-depth interviews with 18 Wilder fans; these people were active in the fandom, and many ran fan organizations or infrastructure. Most had longstanding connections to the homesites. In many cases, they had conducted research on Wilder or her period; some had published scholarly or popular books on Wilder, produced podcasts, or made public presentations (e.g., at libraries).

I also conducted research at two of the Wilder memorial homesites' archives: the Laura Ingalls Wilder Historic Home and Museum, in Mansfield, Missouri; and the Laura Ingalls Wilder Memorial Society in De Smet, South Dakota. These archives had records about both the author and the fandom going back to the 1930s: correspondence between fans and Wilder, or fans and the homesites, and decades' worth of homesite and fan newsletters. Finally, during 2020 I conducted "virtual ethnography" by following the discussions of *Little House* fans on a variety of social media sites (Facebook, message boards, and Instagram); I did not participate in online conversations with other fans, however (see Williams, 2020, p. 21). Throughout this research, my engagement with fans was guided by the approaches described in *The Routledge Companion to Media Studies* (2017).[1]

Little Houses and the U.S. Master Historical Narrative

The *Little House* fandom traces its roots to the 1930s, when white settler colonialism was widely celebrated in American popular culture and was central to white Americans' national identity. Wilder's stories aligned well with the broader narrative of American history promoted by Frederick Jackson Turner, an influential U.S. historian. In an 1893 essay, Turner argued that white settlers' westward expansion underlay American national identity and the success of American democracy.[2] Wilder's books were written for the popular cultural "marketplaces of remembrance" of white settler–Indian relations that dominated early twentieth-century American culture.[3] In Wilder's novels (as in Turner's history), white settlers replace "Indian" characters or their communities, and then build farms, small towns, and flourishing participatory democracies in a "wilderness."

In Wilder's stories, white women on the frontier take center stage: their housekeeping skills, childrearing, and social lives are at the heart of the stories. *Little House* is thus a thoroughly domestic frontier series.[4] Reflecting the author's conservative values, the series celebrated the virtues of allegedly self-sufficient settler families, downplaying the federal government's support for settler colonization.

Wilder's fans saw her stories as authentic accounts of the frontier experience because Wilder persuasively linked her own story to the "master narrative" of U.S. history that most white Americans shared for much of the twentieth century. She presented her story representative of all frontierswoman, a pioneer who was present for the "opening" of the frontier (Fraser, 2017, p. 395). The homesite exhibits carry forward the author's vision of her place in American history, connecting Wilder's family history with this widely accepted narrative of pioneers who "opened the West." For example, an exhibit I saw on "prairie heroes" at the De Smet homestead farmed by the family asserts:

> the pioneer men, women and children who settled the American West were heroes. Their struggle is a story of adventure, daring and courage. Laura Ingalls Wilder, her Pa, Charles, her Ma, Caroline, and sisters, Mary, Carrie and Grace were part of this great westward movement.

Similarly, the Pepin, Wisconsin replica log cabin (where Wilder was born) includes a memorial plaque affirming that "the pioneer experiences she has shared in her books are a precious heritage," at least for those descended from white settlers.[5]

Thus seen a quintessentially American eyewitness account, the *Little House* novels were popular from the start with American librarians and teachers, who shared the books with generations of American children during the twentieth century. Many U.S. teachers built both reading and social science curriculums around Wilder's novels. Librarians' support was equally crucial, as evidenced in the American Library Association's creation of a lifetime achievement award in children's literature in 1954: the Laura Ingalls Wilder Award (Fraser, 2017, p. 479).

A second branch of the *Little House* story world and fandom grew from the successful television series *Little House on the Prairie*, which premiered in 1974;

it ran for eight seasons and remains a syndication today. Three films, spun off from the television series *Little House* world, were also made during the 1980s. The television show was loosely based on the novels and reflected the conservative values and ideas of Michael Landon, the show's executive producer. Landon, who played Laura's father, had previously starred in *Bonanza,* a globally popular Western television series. *Little House* was said to be Ronald Reagan's favorite television show, harnessing nostalgia to support modern conservatism (Fellman, 2008).

The *Little House* world is thus transmedial and has been further built by the fans themselves over the last 70 years. Both individuals and groups of fans have carried out extensive research on the Ingalls and Wilders, creating a participatory knowledge culture that greatly expanded the information we have about Wilder's life and family.[6] The Wilder homesites physically anchor the entire story world and fandom. Over decades, the homesites provided the fandom's infrastructure, publishing newsletters in which fans could exchange news and publish their research and inspiring thousands of fan pilgrimages.

Walking in Laura's Footsteps: The Homesites' History

Soon after Wilder's death in 1957, residents and town leaders in both Mansfield, Missouri and De Smet, South Dakota (where she and her family had spent many years) formed organizations to memorialize Wilder. The Mansfield group, the Wilder Home Association, began to advertise the fact that Laura's home was still open; a hostess showed visitors around this homesite.[7]

In De Smet, leading citizens who were fans of Wilder's stories formed the Laura Ingalls Wilder Memorial Society. The De Smet Society built its homesite gradually over the next 15 years, refurbishing two homes where Wilder's family had lived, and the schoolhouse that she attended as a child. Nearby, the farm originally homesteaded by Wilder's parents was turned into an attraction for visitors. The number of annual visitors to De Smet and Mansfield grew in step with the attractions available in each town for Wilder enthusiasts. In De Smet (population 1,100 today) the number of annual visitors grew from several hundred in the 1960s, to roughly 25,000 per year by 2019.[8] The Mansfield homesite (population 1,296) attracted almost 28,000 visitors in 2018.[9]

People in other small towns where Wilder had lived soon followed suit, building replica log cabins, restoring original homes, and creating museums dedicated to Wilder during the 1960s and 1970s.[10] Volunteers at the Walnut Grove, Minnesota location developed a Wilder museum with a regional history focus, along with a successful annual summer pageant based on the family's lives in Walnut Grove, which they began to offer in 1978.[11] Tour guides organized homesite tours for groups starting in the 1970s.[12]

Besides the schoolchildren who come on field trips, nearly all visitors to the homesites are women (sometimes with family members). Their signatures in guest books (including addresses), my interviews with staff, and my conversations with homesite visitors made it clear that many come from rural conservative communities. Whether they are descended from frontier settler families or not, almost every visitor I spoke to shared a strong identification with Laura and her family. Women make up the overwhelming majority of the fans who visit the homesites or are otherwise active in the fandom. One Mansfield homesite staff member, Vicki Johnston, estimates that about 85% of their non-school field trip visitors are women with friends or children, sometimes bringing their husbands with them. Groups of middle-aged female friends often travel to the homesites together on vacation, as well.[13]

If the homesites' visitors reflect the Wilder fan community as a whole, then it is an overwhelmingly white fandom, as well. While my own experience may not be completely representative, during my visits to the Wilder homesites I saw hundreds of other visitors and interviewed a dozen Wilder fans at the homesites, and conducted 18 more interviews with active fans. All of them appeared to be white, and all but one were women.[14]

This fandom is focused on the material culture and handicrafts of white female settlers, along with research and debate about their lives. Some of the homesites are open air museums, offering heritage interpretations of nineteenth-century white settler culture. Particularly iconic objects of prairie settlers' culture – a one-room schoolhouse and a covered wagon – are found at several Wilder homesites. The Ingalls' homestead site, just outside De Smet, South Dakota, is a living history farm, with guides who demonstrate for visitors a variety of skills, artifacts, and building types associated with nineteenth-century homesteading. In the town of De Smet, costumed guides offer visitors tours of two homes where the Ingalls actually lived.

The guides and many visitors and fans I spoke with saw the homesites as embodying and making tangible not only Wilder's family history, but also as sites that symbolized regional and national heritage, *lieux de mémoire*. Both the exhibition texts and guides' scripts connected Wilder's experience to those of frontier pioneers, for example in the hope expressed by the Walnut Grove volunteer staff in 2003 that "visitors will take with them a sense of history and a deeper appreciation of the joys and hardships that challenged our ancestors when settling on the prairie."[15] At the De Smet homesite, I spoke with a woman from North Dakota who had brought her daughter. She had read the books as a child but had not come to De Smet because she was a Wilder fan. She had brought her daughter to see the exhibits, she said, because "this is our heritage, this is our history, too." Regional school districts send thousands of children on field trips to Mansfield and De Smet each year for similar reasons.

But for *Little House* fans, the homesites offer more than the usual attractions of living history sites. They are also places of imagination, making a storyworld tangible. They showcase the actual artifacts that feature prominently in the stories: Pa's fiddle (mentioned in most of the books), Laura's Spode china, and the china box that Laura received in *On the Banks of Plum Creek*. These are storied possessions, which link the stories to a material reality.

While many *Little House* fans do engage in research and writing (i.e., storytelling, text-based fan engagement), the experience of the homesites means that *Little House* is also a place-based fandom, offering fans sensory immersion and physical experiences which become central to how they imagine Laura's life.[16] Like Laura, they can wade in Plum Creek, near Walnut Grove, or ride in a covered wagon in De Smet. In the actual one-room schoolhouse attended by the Ingalls daughters in De Smet, visitors can reenact scenes from one of the books. They can practice crafts (e.g., quilting) discussed in the novels. Their engagement with *Little House* and the homesites in some ways resembles the experience of theme park fans studied by Rebecca Williams, who noted that the fandom of "the theme park is…. resolutely rooted in specific places and, therefore, physical experiences of an extended storyworld."[17]

But at the same time, the homesites are not only places linked to a storyworld, but also (similar to a religious pilgrimage) the historic locations where Wilder actually lived with her family; the objects on display were actually touched, made, or used by her or family members. As with religious pilgrimages, visits to the homesites can "generate feelings of an extraordinary, authentic encounter" with Laura and her world (Toy, 2017, p. 253; Brooker, 2007). The physical, experiential nature of the homesites also allows them to expand on how they imagine Laura's life. But unlike the media tourists interviewed by Reijnders (2011), whose relationship to their stories "was ultimately characterized by the suspension of disbelief" (2011, p. 105) that nonetheless "played itself out within clear boundaries" (2011, p. 105) Wilder fans are not required to suspend disbelief: the locations and objects are authentic relics.

Julie M., a deeply engaged fan, helps to offer such sensory, physical experiences for visitors at the homesites involving actual Wilder artifacts, which (as she recalls) gave visitors an emotional, intuitive sense of connection with Laura, shaping how they imagined her life. A teacher and Wilder fan, Julie volunteered to teach quilting to the hundreds of fans who attend the annual "Laura Days" celebration held at a homesite not far from her home, showing them how to make the same quilt patterns that Laura had made. Volunteers decorated its log cabin to include objects described in the novels, recreating Laura's childhood home. At another homesite, Julie once played hymns sung at Pa's funeral on Laura's own pump organ for an audience of fans for six hours:

> To hear it and to know that Laura played [this very organ] and so did Rose [Laura's daughter], well, people cried…. it's an important part of people's lives when they come to the homesites and can participate in these things and remember that for the rest of their lives. …[it makes real] the simple life [of the settlers].[18]

Indigenous Resurgence, New Western Historiography, and Culture Wars

Until the 1980s, the frontier history that Wilder presented as the backdrop for her childhood enjoyed broad assent from white Americans across the political spectrum. Staff and visitors saw (and still see) the homesites as embodying an important chapter in American history, one which was generally viewed positively. But this consensus understanding of American history and identity began to fragment after 1990.

The historiographical and political framework that Wilder's fandom operated within was changing by the 1960s. Indigenous resistance – expressed by Indigenous political organizations, tribal governments, writers, and artists – grew steadily after 1970. Tribal governments were increasingly successful at regaining lands and fostering economic development, and Indigenous American cultures achieved considerable resurgence after 1980.

Indigenous activists and a new school of academic historians (the New Western historiography, which included some Indigenous scholars) challenged accepted narratives of colonial settlement. This historical scholarship became the basis for new approaches to public history education, articulated in school textbooks and museums. These new narratives contradicted earlier popular understandings of American history, in which concepts like "pioneers," "westward expansion," and "the frontier" had long played a central role. American academic historians began to approach the story of the West quite differently, as a complex multicultural tale of encounters between varied national, social, and ethnic groups. This historiography readily acknowledged cross-racial violence, conquest, and settler colonialism, along with the costs of economic exploitation of natural resources; their research was gradually absorbed into public discussions.

The research findings of the New Western historians entered public school curricula and museum exhibitions during the 1980s and 1990s (Peers, 2007). Americans on the political left pioneered multicultural pedagogical approaches, with their often bleak assessments of white settlers' impact on other cultures and the environment offered by the new historiography. But other Americans, including the Christian Right and Ronald Reagan, vehemently rejected the New Western historians' findings. Politicians, school boards, and academic historians became engaged in what became known as the "culture wars," still ongoing in the United States today (Hartman, 2019).

Wilder's Stories and the New West Histories

In this changed and increasingly polarized political context, Wilder's stories now carried different meanings for many readers; some openly rejected the idea that the appropriation of Indigenous lands had been a necessary part of "building America." Except for *Little House on the Prairie*, Wilder's books either erased local Indigenous cultures entirely even when they had been there, historically, or depicted them briefly, but usually positively. Indian characters (without tribal identification) appear as threats, however, in *Little House on the Prairie*, the second novel in the series, which became increasingly controversial among Wilder's readers after 1980.

Laura was almost three when the family squatted land on the Osage Diminished Reserve in what is now southeastern Kansas, land which was not open to white settlers, a fact not discussed in the novel (Fraser, 2017). The book describes Indian characters who entered the family's cabin without invitation as "naked wild men," who force Ma to cook for them and stole Pa's tobacco when they leave.[19] On another occasion, two Indian characters enter the cabin and attempt to take the furs that Pa has stockpiled but ultimately leave them behind. With no explanation as to why the Osage might have felt aggrieved toward white settlers squatting on their lands and using their resources, the book simply depicts the Indian intruders as frightening thieves and predators: "Their eyes were black and still and glittering, like snake's eyes.... Their faces were bold and fierce and terrible."[20]

The book's characters express a range of attitudes toward Indigenous cultures. But even the character with the most positive attitude, Laura's Pa, also believes that "an Indian ought to have enough sense to know when he is licked," saying that "when white settlers come into a country, the Indians have to move on."[21] The book presents the white settlers as having a right to expect that the land that they have developed as farmland would become theirs, when in fact the Osage legally owned the land (Fraser, 2017).

Little House on the Prairie's racist descriptions of the Indigenous characters' appearance and behavior attracted little comment or notice among white readers during the decades when the homesites were being established. But after 1990, some parents began to object to Wilder's place in their children's curriculum, throwing fans on the defensive.

Little Houses in the Classroom and Library

Critiques of Wilder's depiction of "Indians" arose during the 1990s first in public schools and children's libraries. In a few school districts, Indigenous school employees and parents began to question the book's place in elementary school curriculums, where exposure to it was mandatory for all children. Their attempts to remove the books from curriculums failed but provoked extensive public discussions about the books' place in elementary schools.

One Lakota parent, Angela Cavender-Wilson, mounted a sustained protest in Minnesota against the mandatory use of Wilder's book by her daughter's class in 1999. Cavender-Wilson argued to her local school board that "The underlying message in this work is that the white settlers are the heroes and the Indians are the villains." Cavender-Wilson's petition failed.[22] Over the long run, however, Wilder's books assumed a lower profile in the school where Cavender-Wilson's daughter attended, and in other American schools. While I found no examples of Indigenous Americans engaging directly with the fandom, many remained critical of its canon. In reading textbooks given to elementary school students, the use of Wilder's books declined; instead, some districts used historical novels told from the point of view of nineteenth-century Indigenous children, such as Chippewa author Louise Erdich's *Birchbark House* stories.[23]

Children's librarians, central to Wilder's original success, also began to distance themselves from her work. Some members of the American Library Association argued that the organization should change the name of the lifetime achievement medal for authors of children's literature, named for Wilder in 1954. In 2017, the librarians' association, which administers the award, asked its members whether Wilder's name should be removed from the medal. Two-thirds of the membership voted to strip Wilder's name off the award, which in 2018 was renamed the Children's Literature Legacy Award.[24]

Wilder Homesites, the New West Histories, and Culture Wars

The Wilder fandom watched the mounting public criticism of Wilder with very mixed emotions. Homesite newsletters reported the attacks on Wilder's portrayal of "Indian" characters. The *Little House* stories still had fans across the political spectrum, but as Wilder scholar and fan Pamela Smith Hill noted, "Laura Ingalls Wilder's reputation is crumbling" among liberal readers.[25]

Staff at some homesites and fan organizations began to take into account their contemporaries' changing views of Western and Indigenous history. The staff of the Laura Ingalls Wilder Museum in Pepin, Wisconsin, made substantial changes to their displays on local history, writing Indigenous history back into Wilder's first novel (which depicted the Wisconsin woods as empty, before whites arrived). The displays now trace the history of regional Indigenous cultures between 10,000 BCE and 1650 CE, discussing changes over this period in local climate and Indigenous cultures' food sources, technologies, arts, architecture, and political systems. In 2007, they also updated the "Laura Contest" held at the town's annual Laura Days celebration. Winners had been previously selected based on their ability to create costumes and look like the book's character, advantaging those who could resemble white settler children. Thereafter, the contest was based a verbal quiz and questions about Wilder's novels, rewarding analysis rather than physical appearance.[26]

In Mansfield, Missouri, the homesite's leadership convened a series of meetings in 2003 to create a new master plan. It included a fundraising for a new museum complex and larger set of displays on the Ingalls and Wilders, opened in 2016. It also noted that

> the impressions of an ancient road still in use until at least 1894 are clearly visible on Rocky Ridge Farm. Worn down by many centuries of Native American use prior to the arrival of settlers, it was also used by the settlers, and was even an alternate route taken by one of the last contingents on The [Cherokees'] Trail of Tears. The Master Plan honors the location of the former road by leaving it undisturbed and viewable to visitors from newly created walking trails.[27]

In 2021, the Laura Ingalls Wilder Legacy and Research Association (LIWLRA; an important fan organization) went even further, responding to the national discussion within the United States about racism, inclusion, and equity by announcing a new commitment to diversity in its work. The LIWLRA acknowledged that Wilder's books

> have racist depictions... .the norms that were accepted in the 1930s about the 1870's and 1880's should not have been tolerated then and are not tolerated

today by our organization. We share the pain that is palpable across the Little House community.[28]

These homesites and fan organizations thus mirrored changes happening in American public history sites and education as a whole.

Some homesites and fans, however, have yet to make changes that incorporate more recent understandings of Indigenous history and white settlement: not all were willing to walk back their commitment to Wilder's depiction of white settler colonialism. Some Wilder sites (e.g., the De Smet homestead site) in 2019 still presented a traditional narrative of "heroic" white settlers, omitting Indigenous history. And even as public schools reduced or eliminated the use of Wilder's stories in their curriculums, many Christian conservatives responded to broader changes in public education by placing their children in conservative private schools, or by homeschooling their children. In both private schools and homeschooling, Wilder's stories were still cherished.

Today, homeschoolers come from across the political spectrum; American parents homeschool their children for varied reasons. But Christian conservatives led this movement in the beginning and remain an important segment of homeschooling parents today. Their political and religious values are often reflected in speeches given at homeschool convention circuits, for example, in a presentation given by Phil Robertson, star of *Duck Dynasty*, who told Christian homeschooling parents in 2019 that, "based on my observation of the last 60 years of no God, no morals in the school system, I highly recommend taking responsibility for your child's education. The Department of Education has become the department of indoctrination".[29] For Christian conservative homeschoolers, the *Little House* stories, which describe communities in which most settlers (including Laura's family) are active in church life, and champion familial self-sufficiency and limited government, align well with a conservative Christian worldview, and fit easily into a conservative Christian homeschooling curriculum. Homesite staff told me that their sites were popular field trips for homeschooling families.

During 2019, I surveyed sample homeschooling curricula for homeschoolers on a variety of platforms including Pinterest, a popular site with homeschoolers, and teaching materials offered to homeschooling parents. I also examined the websites of companies and nonprofits that create learning materials for homeschoolers, and websites of homeschooling conventions. I found a variety of deeply conservative religious curricula online for American history homeschooling that included *Little House* books and visits to homesites, based on an almost Turnerian view of white settlement.

So like public history institutions, many of the fans who manage the homesites and Wilder fan organizations have responded to progressive and Indigenous critiques of the book. At the same time, however, some Christian conservatives have "doubled down" on using Wilder's books to support older white American narratives regarding colonial settlement and the displacement of Indigenous peoples and cultures, because they see them as important sites of (white) collective memory. Many conservatives would have agreed with the fan who saw the removal of Wilder's name from the children's literary award as historical erasure, imposing current values on past cultures. This fan wrote to a newsletter that

> Conservative opinions are excluded more and more in education and literary world [sic]. This decision represents the opinions of only one portion of literary leaders... . The latest craze is to change the name of anything that is tied to racism... . [It is] an attempt to erase [our] history.[30]

Conclusion

The history of the Wilder homesites illustrates how fans and media tourism sites can contradict, reject, or accept political changes in their own cultures. In this case, changing public history narratives about white settlers and Indigenous cultures fragmented white Americans' understanding of their history. As a result, some *Little House* homesite staff and volunteers, fan organizations and fan scholars – like the public history sector that they overlap with – are working through the process of including Indigenous Americans in their depictions of the historical world where the stories are set. At the same time, conservative fans amplified their support for older, Turnerian understandings of Western history. Some use the *Little House* books and homesites to support a conservative Christian approach to their children's education. The homesites and books can be (and are) repurposed in opposing ways.

Neither group seems interested in walking away. Across the political spectrum, fans continue to visit the homesites, since these locations make tangible a cherished, imagined world, allowing sensory immersion in Laura's life. And both liberal and conservative fans still perceive the stories and homesites as important *lieux de mémoire*. Liberal fans value the stories because (as one said) they brilliantly "encompass the central contradiction and conflict of the history of the American West, which is the conquest of native peoples," while conservatives see the stories as affirming their own values and history.[31] But all of them see the homesites and Wilder's world as central to their collective memory, and to the origins of American identity.

Notes

1 See Busse's guidelines in Melissa Click and Suzanne Stott, *The Routledge Companion to Media Studies* (New York: Routledge, 2017), 9–17. Some of the people I interviewed allowed me to use their names, while others requested to remain anonymous.
2 Turner's "frontier thesis" became the founding paradigm for the history of the American West. See Frederick Jackson Turner, "The Significance of the Frontier in American History," http://xroads.virginia.edu/~hyper/turner/ (accessed 29 August 2019).
3 See Boyd Cothran, *Remembering the Modoc War: Redemptive Violence and the Making of American Innocence*, (Chapel Hill: University of North Carolina Press, 2014), 21.
4 For a discussion of how gender functions in the stories, see Ann Romines, *Constructing the Little House: Gender, Culture, and Laura Ingalls Wilder* (Amherst, NY: University of Massachusetts Press, 1997).

5 I saw the plaque when I visited the homesite on 12 June 2019.

6 For a discussion of 'virtual ethnography' in fandoms, see Williams, *Theme Park Fandom*, 21

7 "The Story of the Laura Ingalls Wilder Home Association," *Rocky Ridge Review* (Summer 2007), 2.

8 Aubrey Sherwood letter to Mrs. Addison Wilson, dated 17 July 1965, in "Sherwood" folder, De Smet Wilder Memorial Society Archive; and "Society Celebrates 50 Years, *Lore* Fall 2007 33, no 2 (Fall 2007): 1. See also Anderson, "In the Beginning," *Lore* 36 no. 1 (Spring 2010), 1.

9 The De Smet homesite staff and Vicki Johnston, who oversees the Mansfield Wilder Museum, shared estimates regarding homesite annual attendance.

10 These later homesites are in Pepin, Wisconsin (Wilder's birthplace); Walnut Grove, Minnesota; Burr Oak, Iowa; Independence, Kansas; Spring Valley, Minnesota, and Malone, New York.

11 See the history of the Walnut Grove pageant in the 2007 production program, "Laura Ingalls Wilder Pageant, 'Fragments of A Dream'."

12 I spoke with several fans who were completing this homesite circuit in 2019. For the organized group tours of homesites, see the *Little House Site Tours* offerings at http://lhsi tetours.homestead.com/ (accessed 2 December 2019).

13 Interview with Vicki Johnston on June 20, 2019. This matched my own observations of visitors.

14 Interviews with DeSmet homesite staff (14 June 2019) and Sarah Uthoff (5 August 2019), Barbara Walker (22 November 2019) and Kitty Latane (26 November 2019), and also my own observation of hundreds of homesite visitors in the summer of 2019.

15 "Laura Ingalls Wilder Pageant: Fragments of a Dream," 2003 Walnut Grove homesite pageant program.

16 For fans' sensory immersion in place-based fandoms, see Williams, *Theme Park Frandom*, 25–26.

17 Ibid., 6.

18 Interview with Julie M., 19 November 2019.

19 Laura Ingalls Wilder, *Little House on the Prairie* (New York: Harper C1ollins, 1971), 134–143. The depiction of Indian characters in Wilder's book has generated considerable scholarly analysis. Examples include Frances Kaye "Little Squatter on the Osage Diminished Reserve," *Great Plains Quarterly* 23, no. 2 (May 2000): 123–40; and Sharon Smulders, "The Only Good Indian," *Children's Literature Association Quarterly* 27, no. 4 (2003): 191–202.

20 Wilder, *Little House on the Prairie*, 134–139.

21 Ibid., 236.

22 See the account of Cavender-Wilson's presentation and the school board discussion in Waziyatawin, "Burning Down the House," 69–70. See also the De Smet Archive, "Racism" folder, containing correspondence related to the controversy. The De Smet archives offered several examples of Indigenous Americans who were critical of Wilder's novels, but I found no evidence that any participated in the fandom.

23 I am indebted to Dr. Laura McLemore, who discussed changes in basal readers and the *Birchbark House* series with me.

24 See the ALSC Awards Program Review Task Force's report (dated 15 May 2018) to the Board, ALSC Board AC 2018 DOC #25 here: www.ala.org/alsc/sites/ala.org.alsc/ files/content/awardsgrants/bookmedia/wildermedal/DOC%2025%20Award%20Prog ram%20Review%20Task%20Force%20Wilder%20Award%20Recommendation.pdf (accessed 5 December 2019).

25 Interview with Pamela Smith Hill, 20 December 2019.
26 Interview with Pepin museum staff, 9 June 2019. The description of the museum's exhibits comes from my visit to the Laura Ingalls Wilder Museum in Pepin in June 2019.
27 See description of planning meetings in the Summer 2003 edition of the *Rocky Ridge Review*, and the description of the final Master Plan in the Winter 2009 issue of the *Rocky Ridge Review*.
28 See the LIWLRA's announcement at www.liwlra.org/about-us/ (accessed 1 March 2021).
29 See https://greathomeschoolconventions.com/speakers/phil-robertson (accessed 2 December 2, 2019).
30 See the discussion on the name change here: www.slj.com/?detailStory=alsc-changes-name-wilder-award-childrens-literature-legacy-award (accessed 10 December 2019).
31 Interview with Pamela Smith Hill, 20 December 2019.

References

Brooker, W. (2007). Everywhere and nowhere: Vancouver, fan pilgrimage and the urban imaginary. *International Journal of Cultural Studies*, 10(4), 423–444.
Fellman, A. C. (2008). *Little House, Long Shadow: Laura Ingalls Wilder's Impact on American Culture*. Columbia: University of Missouri Press.
Fraser, C. (2017). *Prairie Fires: The American Dreams of Laura Ingalls Wilder*. New York: Metropolitan Books.
Hartman, A. (2019). *A War for the Soul of America: A History of the Culture Wars*. Chicago, IL: University of Chicago Press.
Nora, P. (1996). *Realms of Memory: Rethinking the French Past*. New York: Columbia University Press.
Peers, L. (2007). *Playing Ourselves: Interpreting Native Histories at Historic Reconstructions*. Lanham, MD: Alta Mira Press.
Reijnders, S. (2011). *Places of the Imagination: Media, Tourism, Culture*. London: Routledge.
Toy, J. C. (2017). Constructing the fannish place: Ritual and sacred space in a Sherlock fan pilgrimage. *Journal of Fandom Studies*, 5(3), 251–266.
Williams, R. (2020). *Theme Park Fandom: Spatial Transmedia, Materiality and Participatory Cultures*. Amsterdam: Amsterdam University Press.

11 Popular Music Heritage in Ekaterinburg

From Seeking Authorisation and Nostalgia for Soviet Rock to Participatory Place-Making

Alexandra Kolesnik and Alisa Maksimova

Ekaterinburg (located in the Urals region, Russia) is one of the few Russian cities where popular music heritage has become present in the urban space in the last 15 years. During "perestroika" in the 1980s in Ekaterinburg (formerly Sverdlovsk) there existed the Sverdlovsk Rock Club – an officially authorised community of rock musicians. Currently, ex-club members are engaged in multiple initiatives to commemorate the rock club as an important component of urban cultural heritage. At the same time, active fan communities in Ekaterinburg have been creating their own memorials dedicated to Western popular musicians such as the Beatles or Michael Jackson. These memorials have become remarkable urban places attractive to locals and tourists alike, despite the fact there are no immediate links between the musicians and local history. The chapter compares and explores processes of place-making through memorialisation of Western music and heritagisation of late Soviet rock music as a part of post-Soviet nostalgia.

Popular music increasingly turns out to be an important part of cultural heritage in different cities, regions, and countries. As Brandellero and Janssen (2014) note, "while official definitions of popular music heritage may be missing," there is "a variety of heritage practices of preservation, exhibition, education and remembrance" (2014, p. 236). The very emergence of popular musical heritage marks the broader processes of revision of the concept of cultural heritage and the problems of its commodification (Lowenthal, 1989) as well as the growing role of popular culture in the development of transcultural tourism (Ziakas and Lundberg, 2018) through its imaginary worlds and accessibility of media (Reijnders, 2016). Popular musical heritage often becomes a landmark of the city or a vibrant urban attraction (Liverpool and Manchester are notable examples) (Leonard and Strachan, 2010), contributing to the creation of new public places for citizens and tourists. At the same time, popular musical heritage often turns out to be in conflict with the approaches of official heritage institutions (Roberts and Cohen, 2014).

While studies of popular music heritage have mostly focused on Anglo-American cases and large-scale cultural initiatives, examining other locations and practices can be fruitful to understand the variety of its forms (Bennett, 2022). In Russia, popular music heritage is relatively scarce and understudied. Nevertheless,

DOI: 10.4324/9781003320586-15

there are various attempts at construction and authorisation of such heritage, which, as we claim, is a part of both more global processes for the development of cultural tourism and post-Soviet nostalgia. Our research aims to investigate its different forms and corresponding cultural processes in one of the largest Russian cities, Ekaterinburg. We will focus on the cultural practices for the creation and maintenance of popular music heritage, features of its "embedding" in the urban space and the types of the agents that form this heritage.

Popular Music Heritage between Fan Communities and Cultural Economy

The study of popular music and related cultural practices as a heritage is a new area in the field of heritage studies. "Music heritage" often works as an umbrella concept that helps to identify and describe the different agents, memory and identity that arise from interest in popular music history.

Many studies have focused on strategic labelling and use of popular music heritage to stimulate urban development. Since the 1980s, in many cities of UK, Europe and the United States, turning to cultural history in general and to musical history in particular has become an important economic and symbolic resource for their revival within "culture-led regeneration" projects (Vickery, 2007). The projects were aimed at renewing cities and stimulating the local economy through the development of the cultural sector – creation of public spaces, opening of new museums, development of tourist routes, and organisation of thematic events (festivals, fairs, concerts). In the 1990s and 2000s, many British post-industrial cities – Liverpool, Manchester, Birmingham, Coventry, Sheffield – implemented urban renovation projects with an emphasis on their musical past, contributing to the growth of transcultural tourism in the Merseyside region. In Liverpool, the city centre has been rebuilt to create new and more visible sites of music history (Cohen, 2017).

On the other hand, significant attention is paid to a variety of fan activities, mostly bottom-up, in creating and maintaining musical memorial sites and commemorative practices. The focus is made on do-it-yourself (DIY) memorials to deceased musicians, places of remembrance for the fan communities (clubs, pubs, hangouts), fan archives and collections, and commemorations and associated traditions (Cohen et al., 2014; Baker et al., 2018; Baker and Collins, 2015; Istvandity, 2021). Of particular interest is not only the diversity of fan practices and forms of memory consolidation, but also the issue of power in defining music heritage. One of the most discussed questions is how music heritage is authorised and is integrated into the work of official institutions (mainly museums) and historical politics at the local and national levels. Analysing music heritage in the UK, Les Roberts and Sarah Cohen distinguish three types: officially authorised, self-authorised and non-authorised heritage (Roberts and Cohen, 2014) In every type, a key role is assigned to fan communities, musicians and citizens; however, the role of institutions in the process of recognising the cultural significance of music heritage is different.

Researchers emphasise that the divide between official, or authorised and "heritage from below" (Muzaini and Minca, 2018), is often blurry and ambiguous. They note that many amateur and fan practices with time undergo professionalisation and institutionalisation that DIY initiatives often become involved in place marketing and tourism, and that commemorative acts largely reproduce the existing power structures (Brandellero and Janssen, 2014). These ambivalences need to be considered to analyse the processes of the inclusion of the musical past in the local cultural policies, agents involved in the construction of music heritage, as well as areas of tension and conflict related to its authorisation and commodification.

Ekaterinburg

Ekaterinburg (named Sverdlovsk in Soviet times; also romanised as Yekaterinburg) is a fourth largest city in Russia, located close to the Ural Mountains. Formerly, it was an important industrial area. The city is well known in Russia for its rich and peculiar culture. In recent years, in particular, this was due to the street art festival Stenograffia, museum and cultural centre named after the first Russian president Boris Yeltsin (Yeltsin Centre), music festival Ural Music Night and the Ural Industrial Biennial of contemporary art. It is also home to authors of famous literary and cinematic works such as Alexei Ivanov and Alexei Balabanov. Lipovetsky (2018) traces the origins of today's cultural scenes in a local "cultural revolution" of 1980s and 1990s. However, beside "progressive" cultural scenes concerned with contemporary art, theatre and music, there exist (sometimes leading to clashes) (Kormina, 2023) conservative cultural symbols, for example, Cathedral on the Blood on the site of the killing of tsar Nicholas II's family, or pseudo-folklore Ural tales by the early twentieth-century writer Pavel Bazhov (Wilmers, 2023).

Ekaterinburg culture is a contradictory entanglement formed under the influence of globalisation and neoliberal conditions as much as of nationalist and conservative Russian ideology. Improvised, albeit ambiguous, public spaces and practices are still possible in the in-between of rigid and controlled social structures (Müller and Trubina, 2020). Authorities, corporations, artists and cultural intermediaries together create an image of the city that citizens and visitors experience. As Trubina (2018) notes regarding local street culture, "producers of street art in Ekaterinburg...successfully sold the street artists' skills not only to the urban public but also to the powerful stakeholders, namely, the municipality and the state corporation" (2018, pp. 695–696).

The cultural events taking place in Ekaterinburg attract tourists, primarily from other parts of Russia and from larger cities. People visit the city for business purposes too, while seizing the opportunities to enjoy historic sites and contemporary culture, entertainment and nightlife. Although the share of foreign tourists is low, large international events held in the city, such as summits (Trubina, 2012) or sport events (for instance, 2018 FIFA World Cup), attracted a significant number of visitors from abroad. The case discussed in the chapter, therefore, does not deal with international tourism as much as with domestic tourism and the local citizens' experiences. Yet we see that initiatives concerned with music heritage share some global tendencies and potentially are likely to be relied on for building tourist routes and attractions.

Ekaterinburg is one of the few Russian cities where popular music heritage has become represented in the urban space in the last 15 years. Although the construction of music heritage there can be viewed as a unified element of developing urban culture, in what follows we will show the two different types of "heritagisation".

Methodology

The analysis is based on ethnographic data, gathered in 2016–2022. The main corpus of in-depth interviews was collected in July 2019 in Ekaterinburg during short-term fieldwork. Interviews were conducted with former and current musicians, activists, fans, museum workers and city officials (see Table 11.1 for an overview of interviewees). Several additional field trips were conducted to observe the memorials and places related to musical heritage and to document changes. Supplemental data include observations, video and photographic materials, as well as publications in local online media and informal social media communities.

The interviews were subsequently transcribed and coded according to the research questions of the study. As requested by some informants, we did not identify their names. In other cases, we give the names of informants with their permission. Our aim was to analyse how the music heritage is constructed by the active participants of the cultural scenes of the city and what meanings the informants assign to the places and stories connected to music heritage.

Table 11.1 List of informants

Mikhail Simakov	Interview 1	Ex-member of Sverdlovsk Rock Club (SRC), musician, radio producer
Alexander Pantykin	Interview 2	Ex-member of SRC, musician, active in commemorating SRC
Anonymous	Interview 3	Ekaterinburg History Museum, curator
Vladimir Vedernikov	Interview 4	Owner of a private museum of local rock culture, musician
Dmitry Karasyuk	Interview 5	Journalist, local historian, author of books on SRC and Russian and Soviet rock culture
Anonymous	Interview 6	Initiator of Tsoi memorial in Ekaterinburg
Anonymous	Interview 7	Initiator of Michael Jackson memorial in Ekaterinburg
Anonymous	Interview 8	Member of a local Michael Jackson fan club
Anonymous	Interview 9	Activist, architect, curator of exhibition on the building where SRC was located at the Yeltsin Centre
Anonymous	Interview 10	City administration official, department of culture
Anonymous	Interview 11	Leader of the Ural Beatles club, one of the initiators of the Beatles memorial
Anonymous	Interview 12	Member of the Ural Beatles club, radio presenter

Sverdlovsk Rock: Authorisation and Contested Memorial Sites

Studying different forms in which the musical past is present in Ekaterinburg urban space, we discover a noticeable interest in the history of the Sverdlovsk Rock Club (SRC). The rock club was opened in 1986 after the founding of similar organisations in Leningrad and Moscow. At the time, it was the authorities' way to oversee potentially controversial youth culture. While being a form of control and censorship, the official rock club became a way of legitimisation and helped popularise rock music. Among musicians and rock bands connected to the SRC that have become popular in Sverdlovsk and far beyond its borders, the following should be noted: *Nautilus Pompilius, Agatha Christie, Chaif, Egor Belkin and Friends, Nastya, Urfin Jus, April March, Sphinx, Top, Semantic Hallucinations*, etc. The SRC existed for five years.

Over the past decade, attempts to commemorate the SRC history have been made in Ekaterinburg – a private museum was opened, exhibitions are held, memorial plaques are installed and ideas for opening monuments are being discussed. It was the ex-members of the SRC who began to make attempts to memorialise the history of the rock club and Sverdlovsk rock.

Many of the SRC ex-members who did not move to Moscow or St. Petersburg now occupy prominent positions in the business sphere in Ekaterinburg: Mikhail Simakov (*April March*) heads a large radio station, businessman Evgeny Gorenburg (*Top*) organises one of the largest festivals in the region Ural Music Night, Nikolai Grakhov (SRC ex-director) runs the largest media holding in the Ural region and Alexander Pantykin (*Urfin Jus*) heads the Ural Composers' Union and runs his own theatre. In addition, almost all of the active ex-musicians of the SRC had the support of the former mayor of Ekaterinburg, Evgeny Roizman (in November 2022, he was declared a foreign agent in Russia; in 2023, he was fined for discrediting the Russian army), who in his youth was a big fan of the Sverdlovsk rock music. As a result, the consolidation of the memory of the SRC, provided by the ex-members of the club, is quite traditional and tends towards official authorisation that is acceptable to state institutions.

The tensions and conflicts that arise in this case remain at the level of the official city authorities (or rather, personal ties between business and authorities) and other state institutions (Orthodox church, universities, and schools). Citizens, as well as representatives of different generations of Sverdlovsk rock music fans, are excluded from this process and often do not know at all about new memorial signs and sites. Likewise, memory associated with the biographies of musicians, poets, film directors who collaborated with them, and the SRC itself does not become a part of the urban historical policy. The consolidation of this memory remains private initiatives, focused primarily on a narrow circle of friends and relatives.

The most common form of memorialisation is the installation of memorial plaques. All places chosen for this purpose are "auratic" and have a significant symbolic meaning for the history of the SRC or biographies of musicians. In 2015–2022, memorial plaques, shown, for example, in Figure 11.1, were installed for Vladimir Mulyavin (*Pesnyary*), poet Ilya Kormiltsev (author of lyrics for *Nautilus*

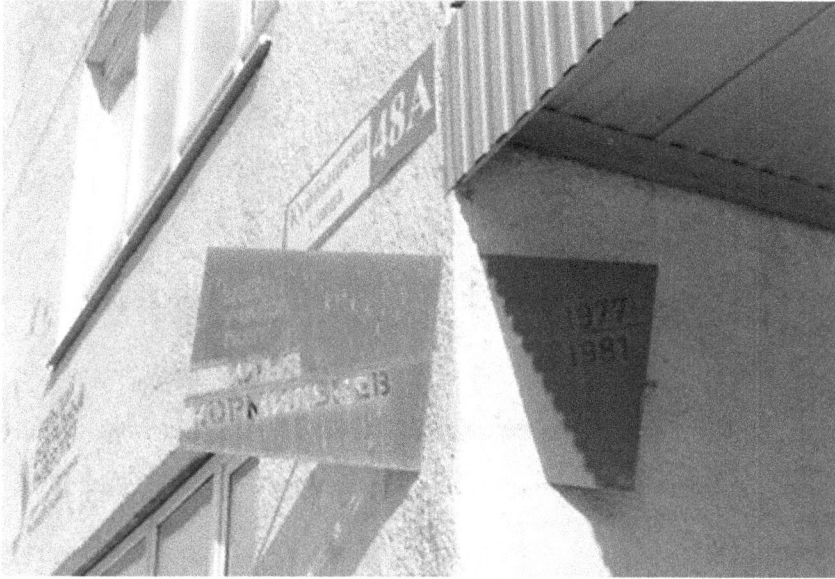

Figure 11.1 Memorial plaque to Ilya Kormiltsev, installed on the building of the Ural Federal University. The inscription on the plaque says "The poet Ilya Kormiltsev studied here, 1977–1981".

Photo by the authors, July 2019.

Pompilius), Alexei Balabanov (director, author of video clips and documentaries for many SRC bands), *Chaif,* and *Nautilus Pompilius*. These initiatives are aimed primarily at the relatives and friends of deceased musicians and the musical community itself. As a result, many citizens are not even aware of the existence of such signs. Mikhail Simakov recalls, "About 60 people gathered at the opening of the memorial plaque [to Ilya Kormiltsev]. We came, and there was almost no one from the younger generation, because it is already history for them" (Interview 1).

Opening of the Ural Rock History Museum in 2015 was indicative, albeit hardly noticeable both in the musical community and among the citizens. The museum is a private initiative of an ex-member of the SRC, Vladimir Vedernikov (*Sphinx*). Located in the district of Uralmash, it can be accessed only by appointment. The place is only partly connected with SRC history. There was no permanent "headquarters" of the club, but musicians often performed there before the formation of the SRC and later in the 1980s. The museum exhibits Vedernikov's private collection of predominantly Ural-made musical instruments (see Figure 11.2) of the late Soviet period, as well as DIY guitars and amplifiers used by some SRC rock bands. All presented exhibits are loosely related, as a result the museum does not attempt to present any specific narrative about Ural rock or the SRC. As Vedernikov notes, he aimed to "mark the [Uralmash] Palace of Culture as a place

Figure 11.2 Ural-made guitars from Vladimir Vedernikov's collection, Ural Rock History
        Museum.

Photo by the authors, July 2019.

connected to biography of some Ural rock musicians, while the building where the
SRC was located [in the 1980s], is closed to the public now" (Interview 4).

The most controversial is the building where the SRC was based in the second
half of the 1980s. Previously it was a church that was rebuilt in 1928–1930 in the
style of constructivism and it housed the Profintern[1] Club (later the Sverdlov Palace
of Culture, where in 1986 the SRC settled in a small room). At the moment, the
building belongs to the Russian Orthodox Church, but is not used, because of its
poor condition. The ideas of reconstructing the building and opening a museum of
Ural rock and SRC history, or erecting a monument to the SRC, have been repeatedly discussed, but the issue of renting it remains unresolved to this day. Dmitry
Karasyuk would prefer to install a monument to the SRC to share musical history:

> the idea of the monument is granite boulders ..., a guitar is stuck on top, from
> which cracks radiate along the boulders, and the inscription 'Rock and roll was
> here.' Moreover, on the stones there are plates with the names of groups and QR
> codes that lead to a website with information and songs.
>
> (Interview 5)

The building is contested not only in the context of musical history, but also as an
important constructivist landmark. Music community and urban activists are in
favour of erecting a monument to the Sverdlovsk Rock Club near this building,
both in order to draw attention to the musical and architectural past of the city
and turn it into a tourist attraction, and to raise funds for its reconstruction. Some

believe that "neither musicians nor fan community, but the city administration should initiate the installation of the monument" (Interview 2). Some, however, suggest that the problem partially lies within the community itself:

> We can't get a rock museum in Ekaterinburg, no matter what. I think it's because there are too many stakeholders. That is, too many truths. [Ex-members] still can't set things straight between themselves, and when it comes to the rock club, it's just hell [dispute].
>
> (Interview 10)

Nostalgic Exploration of the City through Soviet Rock Music History

Along with memorial plaques and plans to commemorate the SRC, there exists work with the musical past of the city that is more visible and in-demand. It is done at and with the help of institutions connected to public history: museums, cultural centres and libraries. The history of the SRC becomes one of the optics of learning about urban history and development of cultural tourism. Such initiatives offer a much more diverse set of ways to build a narrative about the history of Sverdlovsk rock music: with an emphasis on the late Soviet musical practices, on the specifics of the Sverdlovsk sound, musical places in the city and, in general, rock music as an important part of the cultural landscape of Sverdlovsk/Ekaterinburg. The local history projects are organised with the assistance of the former SRC members, but, importantly, imply a process of mediation of their stories through institutional expertise of public historians – museum curators, journalists, archivists and tour guides. Storytelling opportunities are limited in case of memorial plaques or collections of memorabilia like Ural Rock History Museum, but they are much more robust in the city museum exhibition and walking tour. The former is aimed at "insiders" and "music pilgrims," while the latter target citizens of different generations, who both want to learn more about the late Soviet rock music and the history of the city.

One of the most resonant museum projects was the exhibition "Sverdlovsk Rock Club," held in 2015 at the Ekaterinburg History Museum. The musical past was represented not only more conventionally through the personal belongings of the musicians, but, importantly, in the broader context of the urban history of Sverdlovsk in the 1980s. Attention was paid to the practices of recording, performing and listening to rock music. In this case, Sverdlovsk rock music history was exhibited in accordance with broader nostalgic trends in modern Russian culture. For the older generation of visitors, the exhibition has become a powerful nostalgic experience associated with memories of their youth: "A lot of visitors were recalling their youth, these grey-haired, covered in chains gentlemen and ladies. It all lives as good memories" (Interview 3). For the younger generation, it was the Soviet musical instruments and the practices associated with them, which had already become distant and exotic, that were in the spotlight.

Another prominent example is a walking tour "Sverdlovsk Rock: Legends of the Underground." The tour was launched for the 35th anniversary of SRC in May 2021. It was sponsored by one of the largest mobile communications companies in Russia, MTS. The tour is not conducted on a regular basis and needs to be booked in advance. The three-hour walk assumes a rich programme: the story about the SRC history, venues for concerts and rehearsals, places where music videos were shot, where hippies hang out in the 1970s and where vinyl records were bought in the 1960–1980s. The tour offers a broader and more detailed look at the history of the SRC and demonstrates that many places related to music have become part of the city's everyday life.

Even though both these projects are widely discussed in Ekaterinburg media and praised by cultural elites, one should not overestimate their popularity and attention they attracted. Primarily citizens who are already deeply interested in local history and the history of Soviet rock music participated in the walking tour or visited the museum exhibition.

The interest in the history of the Sverdlovsk Rock Club can also be understood as a part of post-Soviet nostalgia, which has become a significant phenomenon in modern Russian culture. The attributes of the late Soviet period not only formed a second-hand market but are also actively used in various fields, from fashion to themed cafes and restaurants. As Boele et al. note, in post-socialist contexts, nostalgia "packaged" in an abundance of goods and media images, "fulfil[s] the function of smoothing the traumatising effects of the political and socio-economic transitions" (Boele et al., 2019, p. 12) after the collapse of the USSR. In Russia, in many cases, nostalgia has become a real attraction for different audiences, promoting tourism and markets for cultural goods, where the musical past is featured prominently.

Celebrating Internationally Famous Musicians in Ekaterinburg

There is another type of urban heritage connected to popular music in Ekaterinburg. In contrast to the previous type, fans of musicians are usually the driving force behind these memorials, instead of it being the members of the musical community themselves. Two examples analysed in the study are devoted to internationally famous performers (*the Beatles* and Michael Jackson) and one to a Soviet one (Victor Tsoi, leader of a rock band called *Kino*). Neither the Beatles nor Michael Jackson gave concerts in Ekaterinburg, and no part of Tsoi's biography concerns the city.

The places associated with international pop music in Ekaterinburg can be classified as aesthetic cosmopolitan music heritage (Regev, 2013). Development of tourism means that "heritage can begin to take very similar forms in different places," which "is amplified with popular music heritage because popular music is already a highly globalised cultural form" (Strong et al., 2017, p. 85). As a result, popular music heritage emerges, which stands for global, rather than local, music cultures and the objects and aesthetic features of which could be found almost in any part of the world.

The Beatles memorial was opened in 2009 on Iset river embankment. The idea started in a local Beatles fan club after their trip to Moscow to attend a Paul McCartney concert in 2003. After much work of the members of the Ural Beatles Club, who organised a public crowdfunding campaign, and material and political support of a local businessman, finally a site was found, as well as resources for its development and the manufacture of the monument. One of the fundraising activities was a musical "marathon," an idea borrowed from Hamburg Beatles fans. The memorial itself, as shown in Figure 11.3, depicting just the contours of the band members, also reminds of Beatles-Platz in Hamburg.

Viktor Tsoi was a charismatic rock musician, frontman of the band *Kino*. By the late 1980s, the Soviet band gained national and even international fame. After his death in 1990 in a car accident, Tsoi walls started to form in post-Soviet space as non-auratic places of fan commemoration (Zaporozhets and Kolesnik, 2020). The most well-known Tsoi Wall exists in Moscow, but there are local ones in other cities. The Tsoi Wall in Ekaterinburg is of the later ones (2012) and is situated in an underpass in the very centre of the city (see Figure 11.4). The walls are covered with graffiti-like portraits of the musician and art devoted to his songs. The man who came up with the idea liked rock music, but was not a big Tsoi fan: "We just wanted to make a wall for the 50th anniversary of a famous rock musician, and we have a pretty mighty rock scene in our city. Why don't we have a wall in our city?" (Interview 6). The head of the company where he was working wanted to organise something interactive for the city that would get media coverage. The idea

Figure 11.3 The Beatles memorial.

Photo by the authors, July 2019.

Figure 11.4 Tsoi's underpass.

Photo by the authors, July 2019.

was approved by the city administration and received support from the citizens. Although over the course of a decade, several times the underpass was threatened to be painted over, every time it created public uproar, forcing the authorities to abandon those plans.

A statue of Michael Jackson is situated on a pedestrian Vainer street (see Figure 11.5), near a shopping mall called Greenwich. It was built in 2011 as a fan project with the agreement of shopping mall owners and sponsored by a businessman who wished to stay anonymous. In terms of the initial financial involvement of local businessmen, it is similar to both the Beatles memorial and Tsoi's underpass. Moreover, the strategy of using the aura and fame of musicians to get the attention of the media (and therefore attract new customers or gain a certain reputation) is evident in Tsoi and Jackson cases alike.

All three places devoted to famous musicians are located in the city centre, on popular pedestrian routes, are open to the public and visible in urban space. The Beatles memorial is usually mentioned in lists of local places of interest in the internet and guidebooks. Other two places, although rarer, also appear in online articles about sightseeing highlights of Ekaterinburg. Tourist route called "Red Line" passes by all three sights, although only the Beatles memorial has its own designated stop on the map.

Figure 11.5 Michael Jackson memorial.

Photo by the authors, July 2019.

Playful and Affective Practices: Paying Tribute, Performing, Taking Pictures

Unlike memorial plaques devoted to the SRC heritage, installation of which usually left the surrounding environment unchanged, creation of the international stars' memorials implied beautification. Before the Beatles memorial was opened, the site was neglected and covered with litter. The same was true for the "tube" – underpass later dedicated to Tsoi and repainted. Although traditionally street musicians hung out there, it was not a pleasant place, dirty and dark. The monuments are also viewed by their creators as incomplete. For example, the initiators of the Beatles monument have plans for its expansion and improvement, and Tsoi's underpass is updated to freshen the paintings and add new ones.

The historical authenticity does not matter here in the way it matters in the SRC case. For instance, the site of the Beatles memorial was not connected with any past events or popular music. At the same time, certain associations were involved, as the place reminded one of the activists of images of Liverpool, with red brick and a chimney in the background. Thus, imagination was as important in the monument creation as fact and history, if not more important. Vainer street where Jackson's statue is located, also has a statue of a popular Russian sitcom's character (who according to the sitcom plot worked in a shoe shop in Ekaterinburg) and a statue of lovers. It is a space of playful interaction, where recreation and consumption intertwine.

The primary intended audience of the monuments are the respective fan communities. The main goal of fans is to celebrate the musicians that they like, even when there is no direct connection of the performer to their city. Another important objective of the fans is sharing their love for the musicians with the wider public. As the musicians in question are well known, citizens notice the memorials, and they become places of interest for tourists. In comparison to more niche histories of the SRC members, these musicians are recognisable pop-stars, accessible to audiences with varying amounts of cultural capital, no matter if they like their music or just heard it.

The memorials are used for various purposes. Events are held commemorating important dates at both Jackson and the Beatles memorials. Even despite Michael Jackson's posthumous fall from grace following child sexual abuse allegations, his fans in Ekaterinburg are still very active. They stage flash mob dances and bring flowers and candles on the days of the musician's birth and death. Ordinary passers-by also engage in small rituals such as taking photos. The monuments are of human height, which creates photo opportunities: people pose as *the Beatles* on stage or mimic dance moves next to Jackson's statue. Tsoi underpass is a place where street musicians perform; it also lies on a commonly used route for people walking in the city centre. The Beatles memorial provides a conveniently located area, which different groups use for their distinctive activities: some do exercise, others (in fact, including Michael Jackson's fans) rehearse dance moves, stage photoshoots or record videos. Benches and a nice background make it attractive for people who just want to relax as well as tourists looking for pretty and peculiar corners of the city.

The brick wall around the Beatles memorial became a "canvas" where fans of various stars can manifest their love. Besides drawings and quotes from *the Beatles* songs, there are inscriptions devoted to other bands (for example, K-pop bands) and popular culture franchises (such as the *Doctor Who* series). There are also tributes to the recently deceased musicians: at the time of our study, one could see graffiti-style portraits of Chester Bennington from *Linkin Park* and *the Prodigy's* Keith Flint. Not only does this place enable wider practices of memorialisation concerned with popular music, even the memorial itself is already celebrated as heritage: in 2019, an exhibition was held in the Sverdlovsk Regional Museum of Local History in honour of the tenth anniversary of the Beatles memorial.

Conclusion

The case of Ekaterinburg sheds light on the cultural dynamics of the emerging popular music heritage and diverse practices of its construction. It reflects how global phenomena in media heritage, fan cultures and cultural consumption are locally enacted by idiosyncratic formations of power structures and stakeholders' agency.

Although the memorials analysed in the previous section are dedicated to musicians not affiliated with Ekaterinburg, they reflect the stories of local fan communities. They do not merely emerge as new *objects* in the urban environment, like plaques – they form new *places*, where people can gather, play music, dance, write or draw, take photographs, bring flowers, fan art, candles and air balloons. While the musicians do have special meaning to the fans, their status is not sacred, and that makes the memorials open to change and participation. As one of the creators of the Beatles memorial jokingly recalls: "A guy with dreadlocks said: and what if I put a small Bob Marley statue here? – Well, who's against it? That's what music is. Why not?" (Interview 11). While these places do function as ritual sites, they are also associated with spaces and practices of recreation, tourism and consumption.

The heritagisation of the SRC is rather aimed at achieving official authorisation, "solidifying" the memory of the rock musicians and Sverdlovsk rock music of the 1980s. The methods of construction of the musical heritage chosen by representatives of the SRC community turn out to be quite traditional, in line with the general urban historical policy–installation of monuments or memorial plaques, opening of museums. However, the nostalgic appeal to Sverdlovsk rock provides more opportunities to attract the attention of citizens and visitors, potentially contributing to the development of cultural tourism.

Although in every case there is a grassroots initiative, nothing was possible without, first, agreement of the city authorities, and second, financial support from local businessmen. To an extent, popular music heritage in Ekaterinburg depends on these and so far has been sanctioned by those who have political and economic power. Yet, the future of the music heritage is unclear, the existing memorials have an uncertain status. At the moment, objects belonging to the SRC heritage as well as international popular music memorials lend to an informal, "hip" image of Ekaterinburg. They serve as tame enclaves of, respectively, underground and globalised culture, against the backdrop of political repressions and spreading conservative ideas. Although there exists no consistent strategy of using the musical heritage for tourism and cultural development, it is called upon whenever it can play a convenient part in cultural projects.

Note

1 Profintern, or the Red International of Labour Unions, was an international body established in 1921 by the Communist International with the aim of coordinating communist activities within trade unions.

References

Baker, S., & Collins, J. (2015). Sustaining popular music's material culture in community archives and museums. *International Journal of Heritage Studies*, 21(10), 983–996.
Baker, S., Strong, C., Istvandity, L., & Cantillon, Z. (Eds.). (2018). *The Routledge companion to popular music history and heritage*. London: Routledge.

Bennett, A. (2022). *Popular music heritage:* P*laces, objects, images and texts.* London: Palgrave Macmillan, 2022: 22–24.

Boele, O., Noordenbos, B., & Robbe, K. (Eds.). (2019). *Post-Soviet nostalgia: Confronting the empire's legacies* (Vol. 76). New York: Routledge.

Brandellero, A., & Janssen, S. (2014). Popular music as cultural heritage: scoping out the field of practice. *International Journal of Heritage Studies*, 20(3), 224–240.

Cohen, S. (2017). *Decline, renewal and the city in popular music culture: Beyond the Beatles.* London: Routledge.

Cohen, S., Knifton, R., Leonard, M., & Roberts, L. (Eds.). (2014). *Sites of popular music heritage: Memories, histories, places.* London: Routledge.

Istvandity, L. (2021). How does music heritage get lost? Examining cultural heritage loss in community and authorised music archives. *International Journal of Heritage Studies*, 27(4), 331–343.

Kormina, J. (2023). Fervent Christians: Orthodox activists in Russia as publics and counterpublics. *Religion, State & Society*, 51(1), 11–29.

Leonard, M., & Strachan, R. (Eds.). (2010). *The beat goes on: Liverpool, popular music and the changing city.* Liverpool: Liverpool University Press.

Lipovetsky, M. (2018). The strange case of a regional cultural revolution: Sverdlovsk in the perestroika years. In *Russia's Regional Identities* (pp. 143–159). London: Routledge.

Lowenthal, D. (1989). Material preservation and its alternatives. *Perspecta*, 67–77.

Müller, M., & Trubina, E. (2020). Improvising urban spaces, inhabiting the in-between. *Environment and Planning D: Society and Space*, 38(4), 664–681.

Muzaini, H., & Minca, C. (Eds.). (2018). *After heritage: Critical perspectives on heritage from below.* Cheltenham: Edward Elgar Publishing, 1–21.

Regev, M. (2013). *Pop-rock music: Aesthetic cosmopolitanism in late modernity.* Cambridge: Polity.

Reijnders, S. (2016). *Places of the imagination: Media, tourism, culture.* London: Routledge.

Roberts, L., & Cohen, S. (2014). Unauthorising popular music heritage: outline of a critical framework. International journal of heritage studies, 20(3), 241–261.

Strong, C., Cannizzo, F., & Rogers, I. (2017). Aesthetic cosmopolitan, national and local popular music heritage in Melbourne's music laneways. International *Journal of Heritage Studies*, 23(2), 83–96.

Trubina, E. (2012). International events in the non-capital post-Soviet city: between place-making and recentralization. Region: Regional Studies of Russia, *Eastern Europe, and Central Asia*, 1(2), 231–253.

Trubina, E. (2018). Street art in non-capital urban centres: between exploiting commercial appeal and expressing social concerns. *Cultural Studies*, 32(5), 676–703.

Vickery, J. (2007). The Emergence of Culture-led Regeneration: A policy concept and its discontents. University of Warwick: Centre for Cultural Policy Studies, 2007.

Wilmers, L. (2023). The local dynamics of nation building: Identity politics and constructions of the Russian nation in Kazan and Ekaterinburg. *Nationalities Papers*, 51(2), 258–279.

Zaporozhets, O., & Kolesnik, A. (2020). Music geography in Russia: non-auratic places and institutionalization "in becoming". *Journal of Cultural Geography*, 37(1), 1–25.

Ziakas, V., & Lundberg, C. (2018). *The Routledge handbook of popular culture and tourism.* London: Routledge.

12 Negotiating Dark and Light Magic

Witch-Themed Tourism in Harz, Germany, in a Transcultural Context

Timo Thelen

Introduction

MEPHISTOPHELES
Hear'st thou voices higher ringing?
Far away, or nearer singing?
Yes, the mountain's side along,
Sweeps an infuriate glamouring song!
WITCHES (in chorus)
The witches ride to the Brocken's top

(Goethe, 1808/2005, XXI Walpurgis Night)

Magic and witchcraft have long been steady topics in literature and media, but the last decades have also seen them become increasingly popular in relation to film/media tourism. The *Harry Potter* franchise is especially noteworthy here as it attracts a global audience with events and theme parks in the United Kingdom, the United States, and also Japan (Lee, 2012; Waysdorf & Reijnders, 2016; Lovell, 2019; Schiavone & Reijnders, 2022) – even though the author J. K. Rowling sparked controversies in recent years. In contrast to film/media tourism that focuses on fictional stories, there are also sites with a history of witch trials, most prominently the so-called Witch City Salem in Massachusetts (Nugua, 2006; DeRosa, 2009). The tourism to sites of such tragic events – commonly called dark tourism – faces the difficulty of reconciling "dark" truths of the past with more entertaining and "light" elements like pop-cultural media content, which tend to be more appealing to visitors (Sharpley, 2009).

Another site well known for witch-themed tourism but not yet investigated in this regard, is the Harz region in Germany around Mount Brocken (also called Blocksberg). Witches have been the major tourism concept there for almost three decades (Behringer, 2000, p. 98). This region has a famous heritage of witches' gatherings around the Brocken, which was prominently featured in Goethe's classic drama *Faust* (1808). Every year on the night before the first of May, numerous municipalities of the Harz region celebrate witch-themed events of the Walpurgis night that oscillate between tradition and tourism; the largest party attracts close to 10,000 visitors. However, in the last decade, the local history of witch trials

DOI: 10.4324/9781003320586-16

also gained attention from news media (e.g., Streckel, 2015; Sielaff, 2017), which raised the question of how such a dark heritage can coexist with cheerful witch-themed tourism events.

This chapter thus examines the "dark" and "light" elements of witch-themed tourism in the Harz region. The research data is based on ethnographic fieldwork in the Harz region and qualitative interviews with local stakeholders, such as representatives of the tourism offices and local souvenir producers. As there are around 20 municipalities with Walpurgis night parties and tourism attractions related to witches in the Harz region, this study focuses on the two major sites: Schierke (ca. 550 inhabitants, today part of Wernigerode, ca. 32,000 inhabitants) and Thale (ca. 17,000 inhabitants), both located in the state of Saxony-Anhalt (former East Germany).

The first part of the following literature review will offer an overview of the local history of witch-themed tourism and how this tourism relates to media content and popular imaginations of witches. The second part will draw on the well-investigated case of Salem and offer a short theoretical discussion of dark tourism. The methodology section will provide further information on the research design and data collection. In the findings, three central aspects will be analyzed: (1) global and transcultural context: a comparison between witch-themed tourism in Salem and the Harz region, and how transcultural ideas and globalization influence them; (2) questions of power: how was witch-themed tourism affected by the history of East Germany, and how do the sites react to claims about their dark past of witch trials? (3) cultural practices: how is witch-themed tourism connected to local cultural heritage, and how was the COVID-19 pandemic handled? Lastly, the conclusion will summarize the main results and take a brief look at recent events.

Literature review

Witch-themed tourism and media of the Harz region

The image of the Harz region as a hotspot for witchcraft and witch gatherings dates back to the mid-seventeenth century (Leiste-Bruhn, 1995). The center is Mount Brocken (1,142m), where the witches are believed to meet on Walpurgis night (from 30 April to 1 May) to celebrate (Figure 12.1), in some legends with Satan himself or other devilish creatures (the term "witch sabbath" is also used; the word "Walpurgis" derives from the Christian missionary Saint Walpurga). According to the legends, the witches meet at different places near the Brocken and then fly to the summit at midnight. As scholars pointed out, there were many witch-gathering mountains in what is now Germany and its eastern neighbors, often called "Blocksberg" or similar terms. When the legends of these places were compiled, they were assumed to relate to the one in the Harz region, making this Blocksberg the supposedly genuine one (Köhler-Zülch, 1993, p. 56). In this sense, the transferring and false attribution of witch stories about any Blocksberg to the

Figure 12.1 Walpurgis night illustration in Johannes Praetorius's *Blockes-Berges Verrichtung* (1668).

one now known as the Brocken is akin to what is nowadays called "displacement" in media tourism studies – the filming of scenes in a different location than the one they are supposed to portray (Bolan et. al., 2011).

Around the early eighteenth century, the image of the witch mountain began attracting visitors to the Brocken and the Harz region. Poet Johann Wolfgang von Goethe (1749–1823) came here three times in the years 1777, 1783, and 1784. He visited various municipalities – including Schierke/Wernigerode and Thale – and climbed to the Brocken's summit three times. Many signs in the region inform tourists of his visits, and there is a mountain path following his hiking journey. Goethe's travels to the Harz region inspired the Walpurgis night episode of the drama *Faust* (1808), in which the titular protagonist and his devilish companion Mephistopheles participate in the bewitched festivities at the Brocken. As one of Germany's most canonical works of literature, *Faust* is regularly taught in high schools and still frequently performed in theaters, with the most famous staging being the one by Gustav Gründgens, which was filmed 1960. Many other German artists traveled to the Harz region and the Brocken in the nineteenth century, including Heinrich Heine (1797–1859) in 1824, whose travelogue *The Harz Journey* (1826) also refers to Goethe's *Faust* episode of the Walpurgis night and the imagination of the Brocken as a witches' gathering place.

During the mid-nineteenth century, the "witches' dance floor" (*Hexentanzplatz*) of Thale gained attention as a tourist spot. A witches' dance floor is a site where the witches are supposed to gather for a "pre-party" on the eve of Walpurgis night before flying to the Brocken's summit together for the "main event" at midnight. Similar to the Brocken/Blocksberg as a collective imaginative site encompassing several witch mountains, this witches' dance floor in Thale became the most famous one, probably accumulating the stories of other such places in the region (Köhler-Zülch, 1993). The witches' dance floor in Thale is a plateau (454 m, accessible by a gondola lift), which is thought to have originally been a pagan ritual site of the Celts until ca. 500 BC and later of Germanic tribes (Leiste-Bruhn, 1995). In 1901, the Walpurgis Hall was built there, a Jugendstil building with paintings by Hermann Hendrich of the Walpurgis night episode from Goethe's *Faust*. Around this time, the tourist infrastructure of the witches' dance floor was supplemented by a hotel (1895) and an open-air theater (1904), which was expanded in 2022 to include 1,900 seats in total. In 1899, a steam locomotive railway from Schierke to the Brocken's summit was installed, and it is still in service today with historic trains, which also contributes to the region's tourism appeal.

After a nearly 30-year-long hiatus during the East German era, witch-themed events and tourism resumed in the 1990s, and soon witches became omnipresent as local mascots and souvenirs (Köhler-Zülch, 1993). Mencej (2017, p. 26) reminds us, however, that commodified witch-themed tourism as well as other contemporary imaginations from the media have nothing to do with the "traditional witchcraft" of agricultural communities in central and eastern Europe's peripheries, which in some cases is still relevant today. The tourism in the Harz region ultimately feeds off of modern imaginations of witches, that is, predominately "light" interpretations of funny (usually old) women with broomsticks and magical pets, using their witchcraft for good purposes, which sometimes backfires with curious results. Stone (2006) argues that for older sites of dark tourism in particular, the dark atmosphere is transformed into something more entertaining, and historical accuracy is substituted with pop-cultural reinterpretations. In this regard, it is no surprise that, since World War II, the majority of witch-themed media content related to the Harz region has turned out to be more lighthearted than the Walpurgis night in *Faust*.

The famous German children's novel *The Little Witch* (1958) by Otfried Preußler (1923–2013) describes the coming-of-age story of an "only" 127-year-old witch and her struggle against the older witches, who ban her from participating in the Walpurgis night party on the Brocken's summit. The novel was adapted into a movie in 2018. The German media franchise of the friendly young witch *Bibi Blocksberg* (1980–ongoing) is named after the alternative term for the mountain. The image of the Harz region as a place of witchcraft and magic is further cultivated by the castle of Wernigerode serving as a filming location for "spooky" children/youth movies like *The Little Ghost* (2013) – which is also based on a novel (1966) by Preußler – or the *School of Magic Animals* movies (2021, 2022), about a school where the students have talking animals as companions. The Harz region and its witches, furthermore, appear in some international pop-cultural media: for example, the fantasy manga series *Berserk* (1989–ongoing) contains a witch named Schierke. In this way, the legacy of the Harz region's witch-themed heritage and tourism continues reaching younger generations, domestically and internationally. However, as the Harz region is very centrally located in Germany, foreign tourists – usually from the neighboring countries – are a minority, making up only 10% of the visitors to Wernigerode.

Witch-themed tourism in Salem and a critical consideration of dark tourism

The most well-known site of witch-themed media and tourism is Salem, Massachusetts; during the Salem witch trials of 1693–94, 20 people were executed as witches and hundreds of people were accused. These cruel events have been referenced countless times in literature and other media, ranging from horror narratives of the supposed "real" events to comedic mentions of the city's name as a metonym for any place of witchcraft. Since the 1970s, tourism related to the city's "dark" past and witchcraft has grown into a major industry (DeRosa, 2009; p. 26). Media supported this development with the seventh season of the popular comedy TV series *Bewitched* (1964–1972), set and filmed in Salem in 1970 (Nugua, 2006, p. 57). In 2005, a memorial bronze statue of the protagonist (or respectively, of Elisabeth Montgomery) was erected in the city's center. Another famous show set in Salem is the comedy witch movie *Hocus Pocus* (1993). Comparable to the Walpurgis night as an annual event, the time around Halloween is the peak tourism season in Salem with many "haunted happenings" (costume walks and gatherings) since 1982 (Nugua, 2006).

However, there is a crucial tension between two fractions in the local tourism industry about how to handle the city's past and modern image:

> On the one hand, educational and high art museums and sites battle to de-emphasize Salem's witch past, which gets marked as entertaining, tacky, and trivial. At the same time, however, the city is witchcrazy; tourists are obsessed with witches, the Wiccan community is thriving, and witchcraft-related attractions continue to draw the greatest crowds in Salem.
>
> (DeRosa, 2009, p. 156)

Likewise, the two largest museums in Salem have very different approaches, as DeRosa (2009) has studied. The Peabody Essex Museum focuses on global art and keeps the original materials related to the witch trials in storage, only partly presenting them on their web channels. The museum shop is one of the few in Salem that does not sell any witch souvenirs, although recently, the Peabody Essex Museum seems to be shifting in a more commercialized and entertaining direction (Berger, 2022). In comparison, the Salem Witch Museum uses puppet reenactments, theatrical lighting, and sound techniques, as well as narration to provide an uncanny experience for its visitors, but it does not display any real artifacts. The museum's sign is a stereotypical witch with a hat and broomstick, and the museum's shop mainly sells similar items. Likewise, most other museums in the city related to witchcraft or popular horror tropes like Dracula tend to focus more on entertainment than education, which is more attractive to the majority of tourists who visit Salem for its infamous reputation.

This phenomenon of a site focusing more on contemporary entertainment than its authentic past is common at many dark tourism sites (Stone, 2006). Dark tourism, that is, the visiting of sites related to death and tragedy, was prominently investigated, for example, for Holocaust and World War II memorial sites but became a global tourism trend in the last decades with a huge variety of sites and contexts. However, as the case of Salem exemplifies, dark tourism can best be regarded as a continuum: the "dark" end is historically accurate, educational, authentic, and thus often considered boring by tourists, while the "light" end is entertaining, thrilling, commercialized, staged, and reflects the conventional beliefs and expectations of most tourists (Sharply, 2009, p. 21). Thus, today's dark tourism commonly includes "light" aspects related to popular culture and imagination, which makes it difficult to locate studies between disciplines like heritage and film/media tourism studies (Seaton, 2019). However, no matter the discipline, the ethical dimension of dark tourism needs consideration not only the aspect of commodifying and commercializing deaths and their history, but also in regards to the present communities at the sites (Sharp, 2009, p. 8). In this regard, Stone and Morton (2022) recently claimed that dark tourism needs to be reinvestigated from a feminist perspective, as women's deaths tend to be marginalized and their corpses are often turned into morbid beauties or even eroticized for the male gaze. Given these new aspects of the academic discussion, witch-related tourism offers a promising field of investigation.

Methodology

This study employs an ethnographic approach with qualitative semi-structured interviews. During two one-week-long fieldwork stays in the Harz region in September 2022 and March 2023, eight interviews with different stakeholders in the local witch-related tourism were conducted (Table 12.1). The interview questions centered on the witch-related history, tourism, and events on the local level as well as how the community perceives them. Most of the interviewees belong to the two case studied municipalities of Schierke/Wernigerode or Thale, or their activities are connected to these sites. The interviews took around one hour on average and were audio-recorded and transcribed. Additional documents like newspaper articles or governmental papers were consulted for triangulation. Moreover, participant observation was conducted at the tourist sites, such as local museums and witch-related attractions, to gain further insights.

The following findings will focus on three major aspects: (1) global and transcultural context: are there similarities between the witch-themed tourism in Salem and the Harz region? (2) questions of power: how was the witch-themed history handled during the East German era, and how are claims about their dark past of witch trials negotiated? (3) cultural practices: how is the witch-themed tourism embedded into local cultural heritage, and how did the COVID-19 pandemic affect the situation? The whole study was centered on the local communities at the sites and their stakeholders, thus excluding tourists' perceptions. The study, furthermore, doesn't discuss the local witch hunts and trial history in greater detail.

Table 12.1 Data on the interviewees

	Description	Age	Gender	Location
P1	City's tourism official	40s	Female	Thale
P2	Producer of manufactured witch souvenir dolls	50s	Female	Thale
P3	Museum staff	60s	Female	Thale
P4	City's tourism official	50s	Male	Wernigerode
P5	Walpurgis night event participant and witch memorial activist	50s	Male	Dusseldorf (frequent visitor of Schierke/ Wernigerode)
P6	City's guide	70s	Male	Wernigerode
P7	Citizen's center official	40s	Male	Wernigerode
P8	Regional tourism association official	40s	Male	Goslar

Findings

Witch-themed tourism in the global and transcultural context

The situation in the Harz region bares some similarities to that in Salem. However, the two conflicting approaches of Salem seem to be located in two different sites in the Harz region. Wernigerode, to which Schierke has officially belonged since 2009, offers few attractions related to witches besides the annual Walpurgis night party and its side events. The city, as mentioned previously, possesses a castle and a historic old town that serve as tourist attractions. Witches are, nonetheless, quite present in the cityscape as souvenirs, ranging from clothing to locally made liquor or meat products. A witch-themed museum – probably similar to those in Salem – opened there in 2011, but it was not well received by some locals and closed permanently a few years later (Kraus, 2012, P6). The city museum of Wernigerode has a display about tourism to the Brocken, where old train tickets, postcards, and commercial posters are displayed; some of them include references to witches. The two-day-long Walpurgis night festival in Schierke attracts between 6,000 and 8,000 visitors annually. An informant from the tourist office (P4) explains: "Schierke's concept for the events is Middle Ages, including medieval-like stands with traditional handicraftsmen and food. The stage program likewise consists of well-known medieval music acts." Another unique event at Schierke's Walpurgis is a parade of costumed witches and devilish creatures through the village – an event that also creates good photo-ops for the media and attracts journalists from all over the world. However, the approximately 200 participants are not ad-hoc visitors like those at the "haunted happenings" in Salem but organized in local folklore groups or come from such groups in places all over Germany.

The witch-themed tourism of Thale is very different; an informant of the local tourist office (P1) admits: "Witches are the main topic in tourism here. When going to events like tourism fairs, I also wear a witch costume." There are two private museums – one inside the station building and another at the witches' dance floor – that focus on horror and witches. Somewhat comparable to the Witch Museum of Salem, they also employ an entertaining approach with puppets and replicas instead of historical documents. Their similarities to those museums in Salem reflect the idea of modern museums as commodified places under the frame of globalization (Rectanus, 2006); that is, visitors of witch-themed museums might have similar expectations worldwide and more likely want to be entertained than educated. There is also a small witch-themed amusement park for children near the station and a witch house at the witches' dance floor; the latter is a three-floor building turned upside down – as its sign proclaims – due to the witch owner's spell backfiring (Figure 12.2). The Walpurgis night at Thale is primarily a rock festival with gothic and medieval rock acts. It attracts approximately 10,000 visitors and thus more visitors than the attractions in Schierke, although it takes place only approximately 35 kilometers away. The different concepts of the two Walpurgis night events avoid a direct competition between the municipalities, although it is clear that "everybody wants a big piece of the cake," as the informant from the

Figure 12.2 The witch house at Thale's witches' dance floor (author's photograph, 2022).

tourist office (P1) puts it. Since 2010, a second annual event set on Halloween, called Hexoween (from the German word *Hexe,* meaning witch), is also held at the witches' dance floor and attracts approximately 3,000 visitors.

In this sense, the two municipalities of the Harz region demonstrate two different strategies of witch-themed tourism. According to Sharpley's (2009) dark tourism continuum, Schierke/Wernigerode tends to be more at a middle or even "dark" position – historical, or at least with a medievalist "magical gaze" (Lovell, 2019) – in its tourism strategy, in which witches play an important but not the main role. In Thale, the tourism strategy is more on the "light" – entertaining and experience-oriented – side on the continuum, and witches are the essential topic, much like in Salem. An important difference between the Harz region and Salem, however, is the infamous history of witch trials in the American city. The efforts to process this "dark" past in the Harz region will be examined in the next section.

Contested magic and questions of power

The witch-themed tourism and heritage in the Harz region have an unsteady history. The first elephant in the room is the more recent "dark" past: the region is located on the former border between West and East Germany, with most of the cities celebrating the Walpurgis night and Mount Brocken itself on the Eastern side. Thus, the Brocken had been a no-access military area during the period of East Germany between 1961 and late 1989, and nearby places like Schierke could only be entered with special permission. The autocratic SED (Socialist Unity Party) government prohibited witch-themed activities like the Walpurigs night parties. However, there were some small-scale Walpurgis night parties in East Germany at least in the 1960s and 1980s, according to one local informant (P6), and some were continuously held on the Western side of the border as another interviewee stated (P8). The famous events in Schierke and Thale were on hiatus though. With the fall of the Berlin Wall in 1989 and the reunification of Germany one year later, the witch-themed events could be revived in May 1990. The first Walpurgis night parties after the restrictive era of SED policies became major events for the local communities, as newspaper articles from the time testify, with approximately 10,000 participants in Thale (Köhler-Zülch, 1993). As an informant from the Wernigerode tourism office (P4) mentions, the first Walpurgis night party in Schierke also attracted approximately 10,000 visitors back then – more than the ones in recent years – probably because the locals were enthusiastic about reclaiming this part of their cultural heritage and celebrating their freedom after decades of political repression.

The second elephant in the room is common to both the Harz region and Salem: the history of witch trials. The authorities of Thale neglect this topic in their narratives, stating that there is no evidence of local witch hunts or trials. Being a village and later a small city, it is indeed possible that Thale has little to no recorded history on this issue. The same is true for Schierke. But in Wernigerode, the story is more complex, also because the city used to be a larger and more important municipality at that time. In 2011, the leader of a costume group (P5), which regularly participates in the Walpurgis night events of Schierke/Wernigerode, made a request for the local authorities to investigate their history of witch trials. This activist has also sent similar requests to other municipalities, which led to the installation of witch hunt memorials and minute-of-silence ceremonies elsewhere. He states: "One side is the fun of Walpurgis and witch costuming, but another side is the responsibility that should be taken by authorities for [historical] informerism and women's oppression. I want social rehabilitation for the victims of witch trials" (P5).

The City of Wernigerode was reluctant to react at first. But after numerous letters from the activist over the years, continuous media reporting, and finally a regional school class investigating this topic as a project, the city created a volunteer task force with members of the local history club and the city archive to work on this request. Their results were, as the informant of the tourist office (P4) puts it, that "Wernigerode was not a central place of witch hunts and trials." In 2017, a memorial was installed at the city wall stating that in Wernigerode city and county there were 59 witch trials between 1521 and 1708, with 15 executions of innocent people, the last of which was in 1609. The number is, of course, relatively small compared to the estimated 25,000 "witches" executed in total on the territory of today's Germany (Dillinger, 2021). An interviewee of the city office (P7) mentioned the complexity that some of the executed witches were also prosecuted for other "real" crimes, not just for "being witches." Sometimes sources argue that there would have been at least eight more witch executions and deaths due

Figure 12.3 The witch trial memorial in Wernigerode (author's photograph, 2022).

to the trials (e.g., Streckel, 2015), but indeed these cases include individuals also prosecuted for murder, who were excluded from the official count. The memorial in Wernigerode is placed on a walking path to the castle (Figure 12.3), a bit outside of the center, but near the tower where the witches were imprisoned, and thus at a historically relevant location. A local activist (P6), who also helped to push the investigation forward, comments in the interview: "The memorial and its position are fine. I think it has no negative influence on tourism." The memorial's erection gained some attention in the national media, especially because of the city's fame for its festive Walpurgis night events (e.g., Sielaff, 2017).

Historians have emphasized that there is no continuity between historical "witches" as the victims of witch trials and today's popular imaginations of witches in media and society, such as them being progressive and empowered women (Behringer, 2000). Similarly for the field of witch-themed tourism, the Walpurgis night events and witch traditions of the Harz region are probably no older than a few hundred years, and – as far as we know today – neither are directly connected to the witch trials there during the sixteenth and seventeenth centuries, nor to the succession of pagan rituals performed at the site even earlier. Nonetheless, the memorial in Wernigerode serves the purpose of negotiating the tension between this "dark" heritage and "light" witch-themed tourism, and of satisfying the voices who demanded clarification about it, even though it follows the pattern of other dark tourism sites in providing little information on the individual histories of the (mostly) female deaths (Stone & Morton, 2022).

The activist and frequent Walpurgis night tourist (P5), who initiated the discussion for the memorial, described the local residents' ambiguous relationship to the topic of witches by saying: "On the one hand, they love witches and the Walpurgis events, but on the other hand they also hate them."

Historian Dillinger similarly describes the two-sided coin of contemporary witch-themed tourism and culture in peripheral Germany:

> The witch has become a part of regional identity. On the one hand, as a victim of the witch trials, which is commemorated by the public memorial culture. But on the other hand, also as a primarily decorative "piece of homeland." (…) The witch dolls in souvenir stores not only in the Harz region suggest that the witch has also become an emblem of an older, rural Germany.
> (Dillinger, 2008, p. 156, own translation)

Performing and making witches in contemporary times

The two sides of witch-themed tourism can also be negotiated for the visitors by a tour guide. Guided tours play an important role in media tourism by linking visitors' expectations and imaginations with physical places that relate to a certain media content (Schiavone & Reijnders, 2022; Šegota, 2018). Both sites investigated here offer guided tours focused on witches. In Schierke/Wernigerode, the tour only takes place if booked in advance, with a female guide wearing a witch costume. Some other city guides like one of the informants (P6) also mention the memorial and the local witch trials in their tours. In Thale, there is an open tour every Saturday during the summer season with a guide costumed as a witch or a devil. Other tours related to local myths or herbs and traditional medicine can also be booked. This program of guided tours underscores the greater emphasis on entertainment in Thale, also recognizable in monuments like the Walpurgis Hall, stone formations thought to look like mystical creatures, three bronze figures of a witch, a devil, and a homunculus at the witches' dance floor, and other statues based on Germanic mythology. The performative act of guided tours, especially in costume, contributes to the inscription of the Harz region as a witch-themed place, also outside of the Walpurgis night events.

I have mentioned souvenirs numerous times, as they are an important part of the touristic experience, being the "curiosities" that tourists take home, and they function as a link between the visited place and oneself (Adler, 1989). Especially small witch dolls – so-called Brocken witches – are a popular item from the Harz region, but in the globalized age, they are produced abroad, most commonly in China. These Brocken witch souvenirs are usually based on the pop-cultural imagination and feature old women riding on a broomstick with the stereotypical features of a long nose, a headscarf, and grandma-like clothes. Both sites in question here, Wernigerode and Thale, try to find locally produced alternatives to the witch dolls imported from abroad. Some souvenir shops in Wernigerode offer handmade witches produced by a craftswoman from Bavaria, which are certified as being produced in Germany. In Thale, the manufacture of witch dolls dates back to the 1940s and was – according to my informant – a common souvenir in East Germany, but they were eventually supplanted by cheaper alternatives after reunification. In 2017, two local women took over the manufacture and tried to revive this tradition with the support of local authorities, selling their witches and devils (Figure 12.4) in their own shop near the station and in some local tourist places like hotels. The owner (P2) explains: "We don't just want to offer locally made souvenirs for tourists. We also offer classes for children to create their own witches and pass on this local tradition." The locally manufactured witch dolls are, however, two to three times more expensive than the imported ones of the same size. As a result, places like the tourist information office in Thale sell both kinds, while the souvenir shops at the witches' dance floor only offer the imported ones. However, both municipalities, Wernigerode and Thale, want to increase the number of locally produced witch-themed souvenirs as a strategy for more sustainable tourism.

Like the example of the Brocken witches as a traditional souvenir shows, the globalized world and its highly commodified tourism industry challenge manifestations of local culture. In Schierke, the aging population resulted in only three or four residents still engaging in the Walpurgis night events. In the recent years, the COVID-19 pandemic was also a major obstacle to tourism in both municipalities. Schierke scaled back the program in 2021 and 2022, canceled the parade, and limited the number of visitors to 5,000; moreover, in 2021, the location shifted to the more spacious center of Wernigerode. In Thale, all events were canceled in 2021, and in 2022 the number of visitors was downsized to 6,000 with a correspondingly smaller stage program. In 2023, both sites could celebrate their Walpurgis night events without limitations.

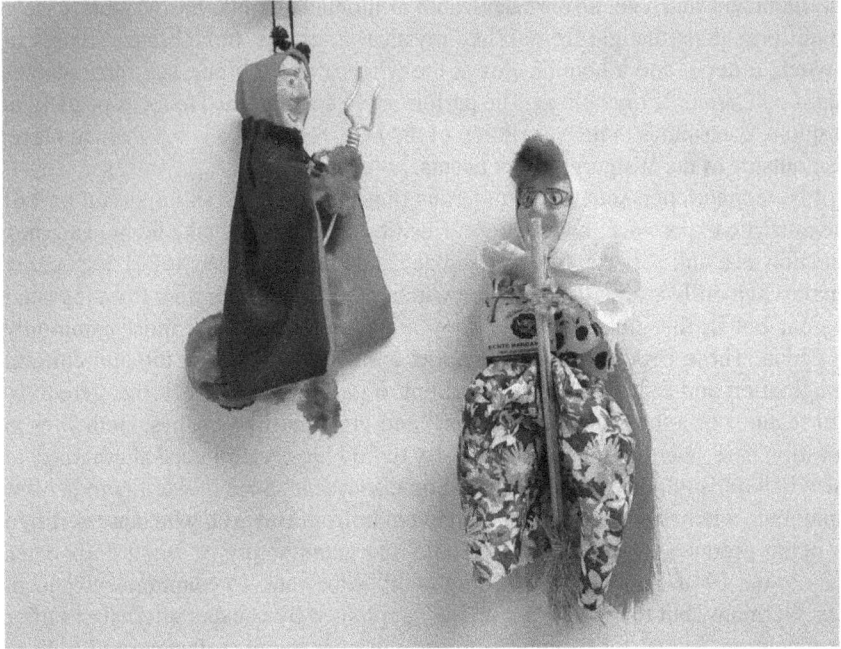

Figure 12.4 Manufactured Brocken devil and witch doll (author's photograph, 2022).

Conclusion

As the studies of witch-themed tourism in Salem (Nugua, 2006; DeRosa, 2009) show, being (in)famous for witch heritage is a double-edged sword. On the one hand, it is a strong tourism magnet, but on the other hand, it can easily overshadow the other qualities of a site. Moreover, the historical facts of "dark" heritage are usually not what tourists want when visiting the site, but rather "lighter" entertainment in the form of museums offering experiences and festive events, nurtured by pop-cultural interpretations. Tourists, especially those interested in fantastic stories of the past like legends of witches, tend to look at sites through a "magical gaze" (Lovell, 2019) that blends expectations and projections with the very present and physical experience of the visited site, which the local communities at the sites also nurture through marketing and performances.

The witch-themed tourism in the Harz region is a contested area, in which two municipalities found ways to avoid a direct competition by setting different focuses on how to implement witches and Walpurgis night events in their tourism strategies. The legends and images of the local witch heritage are predominantly influenced by pop-cultural representations, in other words "light," and thus fit to the visitors' expectations. Although the regional history of witch trials is far less infamous than that of Salem, an investigation of these events in recent years and the subsequent erection of a memorial helped to reconcile critical voices calling for an acknowledgment of the neglected "dark" past.

Recently, an episode of the crime television movie series *Police Call 110*, produced by the German state broadcaster ARD, titled "Witches Burning" was aired on 30 October 2022 (i.e., close to Halloween), at prime time on a Sunday, and was watched by ca. 7.2 million viewers. The movie takes place in the Harz region, mainly in Schierke and Wernigerode. Partly as a parody of American horror and splatter movies, the movie exploits every imaginable cliché of witchcraft from Wiccan herbal magic to Satanic rituals and portrays some of the locals as obsessed with pagan magic and superstition. Here again, the double-edged sword of witch-themed media strikes at the Harz region: advertising the site and fostering its imagination as a place of witches and magic, while reducing the place and its people to the related stereotypes.

Acknowledgments

This research was supported by the Kakenhi Grant No. 22K18100. The author would like to thank the two anonymous reviewers and the editor for their productive input.

References

Adler, J. (1989). Origins of sightseeing. *Annals of Tourism Research, 16*(1), 7–29. doi:10.1016/0160-7383(89)90028-5

Behringer, W. (2000). *Hexen: Glaube, Verfolgung, Vermarktung (2. Auflage)*. CH Beck.

Berger, H. A. (2022). Witchcraft past and present at the Peabody Essex Museum in Salem. *Pomegranate, 23*(1–2), 186–202. https://doi.org/10.1558/pome.22069

Bolan, P., Boy, S., & Bell, J. (2011). "We've seen it in the movies, let's see if it's true": Authenticity and displacement in film–induced tourism. *Worldwide Hospitality and Tourism Themes, 3*(2), 102–116. https://doi.org/10.1108/17554211111122970

DeRosa, R. (2009). *The making of Salem: the witch trials in history, fiction and tourism*. McFarland.

Dillinger, J. (2008). *Hexen und Magie*. Campus Verlag.

Dillinger, J. (2021). Germany – "The mother of witches". In Dillinger, J. (Ed.), *The Routledge history of witchcraft*. Routledge.

Goethe, J. W. (2005). *Faust. Part 1* (B. Taylor, Trans). Retrieved from www.gutenberg.org/ebooks/14591 (Original work published 1808).

Köhler-Zülch, I. (1993). Die Hexenkarriere eines Berges: Brocken alias Blocksberg. Ein Beitrag zur Sagen-, Hexen–und Reiseliteratur. *Narodna umjetnost, 30*(1), 81–81.

Kraus, U. (2012, April 20). Walpurgis: großer Tag für Hexen. *Mitteldeutsch Zeitung.* www.mz.de/mitteldeutschland/walpurgis-grosser-tag-fur-die-hexe-2022459Lee, C. (2012). "Have magic, will travel": Tourism and Harry Potter's United (Magical) Kingdom. *Tourist Studies, 12*(1), 52–69.

Leiste-Bruhn, S. (1995). *Hexentanzplatz und Rosstrappe*. Stadt Thale.

Lovell, J. (2019). Fairytale authenticity: historic city tourism, Harry Potter, medievalism and the magical gaze. *Journal of Heritage Tourism, 14*(5-6), 448–465.

Mencej, M. (2017). *Styrian witches in European perspective: Ethnographic fieldwork.* Springer.

Nugua, A. (2006). Witch city and mnemonic tourism. *Journeys, 7*(2), 55–72.

Rectanus, M. W. (2006). Globalization: Incorporating the museum. In S. MacDonald (ed.), *A companion to museum studies* (pp. 381–397). Blackwell.

Schiavone, R., & Reijnders, S. (2022). Fusing fact and fiction: Placemaking through film tours in Edinburgh. *European Journal of Cultural Studies, 25*(2), 723–739.

Šegota, T. (2018). Creating (extra) ordinary heritage through film-induced tourism: The case of Dubrovnik and Game of Thrones. In: C. Palmer, & J. Tivers (Eds.), *Creating heritage for tourism* (pp. 131–142), Routledge.

Sharpley, R. (2009). shedding light on dark tourism. In R. Sharpley, & P. Stone (Eds.), *The darker side of travel* (pp. 3–22). Channel View.

Sielaff, Y. (2017, January 19). Hexenprozesse: Wernigerode erinnert an Folter und Morde. *Volksstimme.* www.volksstimme.de/lokal/wernigerode/wernigerode-erinnert-an-folter-und-morde-802155

Stone, P. R. (2006). A dark tourism spectrum: Towards a typology of death and macabre related tourist sites, attractions and exhibitions. *Tourism: An International Interdisciplinary Journal, 54*(2), 145–160.

Stone, P. R., & Morton, C. (2022). Portrayal of the female dead in dark tourism. *Annals of Tourism Research, 97*, 1–14.

Streckel, S. (2015, May 21). Verfolgt, gefolgert, verbrannt: Der Hexenfluch vom Schloss Wernigerode. *Bild.* www.bild.de/regional/leipzig/leipzig/das-dunkle-geheimnis-von-schloss-wernigerode-41030544.bild.html

Waysdorf, A. & Reijnders, S. (2016). Immersion, authenticity and the theme park as social space: Experiencing the Wizarding World of Harry Potter. *International Journal of Cultural Studies 21*(2): 173–188.

13 What About the Locals?

Exploring Residents' Interest in and Suggestions for the Development of Film Tourism in Seville, Spain

Deborah Castro

Introduction

All over the world, the increasing number of people interested in visiting locations they have previously seen on films and series has been a cause for celebration among local institutional stakeholders, such as tourism promoters and film commission representatives. In some places, however, it has also provoked resentment among local residents. For example, for Paula, a native-born resident of the city of Seville, the number of people who queue up to visit the Real Alcázar, the city's famous medieval palace that became one of the filming locations for *Game of Thrones* (Ruiz, 2020), has discouraged her from spending time in one of her favorite spots in her hometown. This dissonance between institutional stakeholders' objectives and inhabitants' well-being is the reason why (film) tourism scholars have for decades been stressing the need to involve local residents in tourism development and planning strategies (e.g., Beeton, 2016; Heitmann, 2010; Jaafar, Rasoolimanesh & Ismail, 2017; Malek & Costa, 2015). This chapter aims to broaden the discussion on film tourism, first, by investigating the demographic profile of local film tourism supporters, and second, by exploring local residents' suggestions for (future) film tourism projects in their living place. To this end, it uses descriptive statistics to analyze data from a survey conducted with 407 local residents.

The geographical location chosen for this study is the Andalusian province of Seville in Spain, which has been selected for three main reasons. The first is that Seville is a popular tourist destination (its capital was awarded Lonely Planet's Best in Travel City for 2018 [Lonely Planet, 2017]) where local institutional stakeholders have sought to attract film and television producers as well as film tourists. The province of Seville started to feature on Spanish TV programs in the 1980s and 1990s, produced by the country's sole broadcaster, Televisión Española. After the global financial crisis that hit the country in 2008, the Andalucía Film Commission, the Radio y Televisión de Andalucía, and the regional government's tourism office (Turismo Andaluz) worked proactively to convince the producers of Ridley Scott's *Exodus: Gods and Kings* (2014) to shoot the film in Andalusia, a moment that represented a turning point for the regional audiovisual industry. In October 2014, the production team for the globally renowned HBO series *Game*

DOI: 10.4324/9781003320586-17

of Thrones arrived in Seville to shoot on location at the Real Alcázar. This project has had a remarkable impact on Seville's international visibility and reputation (for more context, see Castro & Cascajosa-Virino, 2022). The Andalucía Film Commission (the first film commission anywhere in Spain) has even mapped out Seville's filming locations for its project "Andalucía Destino de Cine"[1] and released film maps for municipalities such as Carmona, Osuna, and the city of Seville itself.

A second reason for the selection of Seville as a case study is that some recent tourism strategies in the province have incorporated the participation of the local community. For example, the Municipality of Seville created a Local Tourism Council in 2021, made up of neighborhood associations, non-governmental organizations and, notably, the Andalucía Film Commission (Sevilla, 2021), among other entities. This is noteworthy given that residents of southern Spain have traditionally been overlooked in major tourism initiatives (Oviedo-García, Castellanos-Verdugo & Martin-Ruiz, 2008).

The third reason for choosing Seville is that most scholarly work that has considered local community perspectives on film tourism has been conducted at less obvious tourist destinations in English-speaking (e.g., Beeton, 2016; Mordue, 2001) and Asian countries (see Kim & Park, 2023; Yoon, Kim & Kim, 2015). Thus, this study aims to contribute to film tourism scholarship by focusing on a popular tourist destination located in southern Europe, a region as yet largely unexplored in the literature.

Local residents' support for film tourism development

In recent years, anti-tourism protests and movements in different European cities, such as Amsterdam and Barcelona, have expressed the exhaustion and resentment that large tourist numbers have provoked among local communities. Seville has not been immune to this reaction, as evidenced by the existence of organizations such as *Cactus. Colectivo-Asamblea Contra la Turistización de Sevilla.*[2] However, not all types of tourism generate the same type of effects, nor does tourism affect the whole community in the same way (e.g., according to some residents, the *Eat Pray Love* phenomenon in Bali only benefited people directly involved with the film and tourism industries, see Kim, Suri and Park, 2018). This raises the question of whether people living in a popular tourist destination would still be supportive of film tourism initiatives, and, in this context, what the socioeconomic and demographic profile of the film tourism supporter would look like.

Prior research on general tourism suggests that local residents' degree of involvement with the tourism sector constitutes one of the key factors determining their support for tourism development. For example, in a quantitative study conducted in the Sevillian municipality of Santiponce, Oviedo-García et al. (2008) found that local residents who obtained direct advantages from the tourism sector were more supportive of tourism development than those who only obtained indirect benefits. On the other hand, in the specific field of film tourism, Castro, Kim and Assaker (2023) conclude that local residents of the province of Seville are generally supportive of film tourism development regardless of their relationship with the tourism industry. In terms of other socio-demographic factors (e.g., age, gender, education level), scholars have drawn contradictory conclusions about the extent to which these shape local residents' attitudes toward tourism development (e.g., Andriotis & Vaughan, 2003; Bhat & Mishra, 2021; Nunkoo & Gursoy, 2012). Despite these contributions, the profile of the kind of local resident who typically supports film tourism has yet to be clearly defined.

Local residents' involvement in film tourism development

Numerous scholars (Beeton, 2016; Heitmann, 2010; Yoon, et al., 2015) have highlighted the need to involve local communities in film tourism development and planning due to the diverse range of impacts that film tourism has been shown to have on them. These impacts may be environmental (e.g., degradation of natural areas), economic (e.g., new business opportunities, price increases), or sociocultural (e.g., lack of privacy) (Connell, 2012; Croy et al., 2018). Thus, for example, researchers have argued that proper consultation with residents could better prepare the parties affected by film tourism or prevent residents from resenting how institutions use public economic resources (Thelen, Kim & Scherer, 2020).

In practice, however, such consultation is left out of the tourism development process due to the power dynamics and specific interests that characterize both the film and the tourism industries (Beeton, 2016) as well as the lack of both short- and long-term structured planning by local authorities (Kim et al., 2018; Póvoa, 2023). Mordue (2001), who has studied the film tourism phenomenon resulting from the TV series *Heartbeat* in the northern English village of Goathland, explains how the tourism office tried to empower village residents (some of whom felt that tourist flows were threatening their quality of life) by involving them in public meetings with other local stakeholders. However, the initiative failed because of a lack of "consensual agreement" between the parties involved (p. 242). The participation of local community members as consultants in the pre-production phase has also been proposed as a means of involving local residents in film tourism development. For instance, in their analysis of the Japanese drama *Mare*, Thelen et al. (2020) suggest that the involvement of members of the local community as experts in the pre-production phase (e.g., to review the scripts) seemed to improve the authenticity of the media product (e.g., characters' ways of speaking) and local residents' iden-tification with it. In Brazil, Póvoa, Reijnders and Martens (2019) also documented local residents' interest in participating in the creative process of telenovelas shot in their neighborhoods to contribute to more realistic media representations of their living place.

Although the impacts of film tourism are well-documented and for several years scholars have been arguing for the involvement of local communities in film tourism planning and development, few studies have used quantitative data to identify the demographic profile of local residents who would be interested in supporting and contributing to film tourism initiatives. This study aims to fill this gap by offering insights derived from a quantitative survey of 407 local residents of the province of Seville. The method adopted for this research is described in the next section.

Research methods

A survey questionnaire was deemed to be an effective method for collecting data on local residents' interest in film tourism development and developing some initial suggestions for its promotion and development. After a thorough literature review, a first version of the questionnaire was designed and piloted with 30 people living in Seville. This preliminary test helped to refine the instrument and enhance its internal validity. The data collection stage ran from August to October 2020. The structured self-administered questionnaires were distributed online via the panel company, Dynata. The sample size consisted of precisely 407 respondents, all of whom met the following requirements at the time of completing the survey questionnaire: (a) between 18 and 64 years old; (b) Spanish citizens; (c) residents of the province of Seville (in Andalusia, Spain); and (d) viewers of at least one of the popular television series or films shot in Seville listed in the survey (e.g., *Unit 7*, *Spanish Affair*, *Game of Thrones*). The original version of the survey questionnaire was created in English and translated into Spanish by two native Spanish speakers. To double-check the accuracy of the translation, a back-translation was also completed by a professional translator. The frequency table for the demographic profile is presented in Table 13.1. A total of 62% ($n = 261$) of the respondents reported having visited a place after seeing it in a film and/or a series, including locations in other countries but also in their own municipality, province, region, or country; 61% ($n = 249$) of the participants were residents of the city of Seville and the rest were residents of other towns and cities located in the province, such as Dos Hermanas ($n = 22$) and Alcalá de Guadaíra ($n = 14$).

This study forms part of a bigger project for which other aspects, such as local residents' place attachment, were also measured. However, for the specific purposes of this chapter, data resulting from a selection of items will be reported by means of descriptive statistics developed using SPSS Statistics. First, to identify local residents' attitudinal behavior toward the development of film tourism, the following two statements adapted from Rasoolimanesh et al. (2017) were used: "I believe that film tourism should be actively promoted in my town/city" and "I support film tourism and would like it to play an important role in my town/city." Second, to get an approximate idea of the level of familiarity with the notion of film tourism, respondents were also asked whether they had ever visited a place after seeing it in a film or series and, if so, they were asked to indicate the place where they went. Third, the importance given to local community involvement in film tourism development was measured through a multi-item question that included the following statements: "Decisions about the planning and development of film tourism should be made by the community as a whole, independently of the persons' professional background," followed by "Residential communities should have equal opportunities to participate in decision making for film tourism," and finally "It is absolutely necessary for the entire community to be involved in the

Table 13.1 Demographic profile of respondents ($N = 407$)

Characteristics	Frequency	Percentage
Age		
18–34	154	38%
35–49	158	39%
50–64	95	23%
Gender		
Female	215	53%
Male	192	47%
Economic involvement in tourism		
I work in the tourism sector	72	18%
I don't work in the tourism sector, but some of my closer friends and relatives do	135	33%
No, and none of my closer friends and relatives do	200	49%
Education level		
Primary	7	2%
Secondary	128	31%
Vocational training	93	22%
University	188	45%

decision-making process for the success of film tourism." These statements were adapted from Choi and Sirakaya (2005).

All the abovementioned items were answered on a 5-point Likert-type scale, with 1 denoting "strongly disagree" and 5 denoting "strongly agree." It is important to note that in the results outlined below, the word "agree" refers to the combination of respondents who chose "agree" and "strongly agree," while "disagree" refers to both "disagree" and "strongly disagree" responses. Additionally, an open-ended question was included to elicit ideas from respondents about how film tourism could be promoted in their place of residence. This qualitative data has been inductively analyzed using qualitative conventional content analysis (see Hsieh & Shannon, 2005). Finally, it is worth noting that the dataset this chapter analyzes has been partially explored elsewhere through partial least squares-structural equation modeling (PLS-SEM) (see Castro et al., 2023). After citing some of the general insights offered by Castro et al. (e.g., residents support the development of film tourism), this chapter explores the demographic profile of local film tourism supporters and identifies specific suggestions for film tourism development made by survey respondents.

Results

To explore residents' opinions about the development of film tourism in their place of residence, statement 1 ("I believe that film tourism should be actively promoted in my town/city") and statement 2 ("I support film tourism and would like it to play an important role in my town/city") were used. In general, respondents thought that film tourism should be actively promoted in their place of residence (mean score 4.01) (see Castro et al., 2023). In fact, 78% ($n = 317$) of respondents agreed with statement 1. The number of respondents who disagreed with the idea of promoting film tourism represented only 5% ($n = 21$) of the sample. Similarly positive results were obtained for statement 2 (mean score 4.08), with 79% ($n = 320$) of respondents agreeing with the statement.

Most survey participants responded positively to statements 1 and 2 regardless of their relationship with the tourism sector (see Castro et al., 2023). Specifically, 78% of the respondents who neither work in nor have close friends/relatives working in the tourism sector ($n = 156$) agreed with statement 1, a percentage only slightly lower than the 79% of positive responses by respondents who work in the tourism sector ($n = 57$), and slightly higher than the 77% ($n = 104$) by respondents with close relatives/friends working in the sector. Similarly, 80% ($n = 58$) of respondents who work in the tourism sector agreed with statement 2, compared to 79% ($n = 158$) of respondents who neither work in nor have close friends/relatives working in tourism, and 77% ($n = 104$) of respondents whose close relatives/ friends work in the sector.

By gender, the percentage of male respondents who agreed with the active promotion of film tourism in their place of residence was slightly higher (80%, $n = 154$) than it was for female respondents (76%, $n = 163$). A similar majority agreed with the idea of supporting film tourism and with film tourism becoming an important part of their place of residence (80%, $n = 173$ female; 77%, $n = 147$ male). In terms of age, respondents in the 50–64 age group agreed the most with both statements (statement 1: 82%, $n = 78$; statement 2: 85%, $n = 81$, respectively). This was followed by the 35–49 age group (statement 1: 77%, $n = 122$; statement 2: 77%, $n = 121$), and finally, the 18–34 age group (statement 1: 76%, $n = 117$; statement 2: 77%, $n = 118$).

A breakdown by education level shows that respondents with vocational training were the most likely to agree with the active promotion of film tourism in their living place (83%, $n = 78$), followed by university graduates (77%, $n = 140$), high school graduates (76%, $n = 97$), and primary school graduates (71%, $n = 5$). Similar results were observed for statement 2, with 86% ($n = 78$) of respondents with vocational training agreeing, followed by primary school graduates (86%; $n = 6$), university graduates (77%, $n = 140$), and high school graduates (76%, $n = 96$).

To identify local residents' opinions about the need to involve the local community in the development of film tourism initiatives, the statements "Residential communities should have equal opportunities to participate in decision making for film tourism" (statement 3) and "It is absolutely necessary for the entire community to be involved in the decision-making process for the success of film tourism" (statement 4) were used. Both statements received a mean score above 3 (3.68 and 3.66, respectively). In particular, more than half of the respondents (61%, $n = 247$) agreed with the idea that members of the local community should have equal opportunities to participate in this type of decision-making process, and 61% ($n = 246$) also believed that for the success of film tourism, it is absolutely necessary for the entire community to be involved. Respondents working in the tourism sector were the ones most likely to agree with both statement 3 (67%, n = 48) and statement 4 (63%, n =45). This was also the group with the largest proportion of respondents who strongly agreed with both propositions, reflecting the interest that people working in the tourism industry have in other residents' opinions. Among respondents with close friends/relatives working in the tourism sector, 60% (n = 81) and 59% (n =79) agreed with statements 3 and 4, respectively. Finally, 59% (n = 118) and 61% (n = 122) of those who neither work in nor have close friends/relatives working in tourism agreed with statements 3 and 4, respectively.

In terms of gender, male respondents were only slightly more prone to agree with equal opportunities of participation (62%, n = 119 male respondents; 60%, n = 128 female respondents). However, there was a bigger difference by gender in responses to the question about the absolute necessity of involving the entire community in this type of decision-making process, with 65% (n = 125) of male respondents agreeing, in contrast to 56% (n = 121) of female respondents. By age, respondents in the 35–49 age group were the most likely to agree with both statements (statement 3: 64%, n = 101; statement 4: 62%, n = 98), followed by the 18–34 age group (statement 3: 60%, n = 92; statement 4: 60%, n = 93) and the 50–64 age group (statement 4: 57%, n = 54; statement 5: 58%, n = 55).

Finally, the proposition of local communities having equal opportunities to participate in decision making for film tourism development had the largest proportion of agreement among respondents with vocational training (65%, n = 59), followed by university degree holders (61%, n = 111), high school graduates (58%, n = 74), and finally, those with primary education only (43%, n = 3). The idea of involving the entire community in decision-making processes for the success of film tourism received agreement from 61% (n = 78) of high school graduates, 61% (n = 111) of university graduates, 59% (n = 54) of respondents with vocational training, and, finally, by primary school graduates (43%, n = 3).

Nearly half of respondents (49%, n = 200) believed that decisions about planning and development should be made by community members independently of their professional background (statement 5; mean score 3.41). Respondents with less involvement in the tourism sector were more likely to agree with statement 5. Thus, 53% (n = 105) of respondents who neither work in nor have close friends/relatives working in the tourism sector agreed with the proposition, followed by 47% (n = 63) of those with close friends/relatives working in tourism, and finally, 44% (n = 32) of respondents who work in the sector. However, the latter group was the one with the largest proportion of respondents who strongly agreed with the participation of members of the local community regardless of their professional background.

By gender, the percentage of male respondents who agreed with statement 5 was significantly higher than that of female respondents (56%, n = 108 male respondents; 43%, n = 92 female respondents). By age, respondents in the 35- to 49-year-old age group were most likely to agree (54%, n = 85), while less than half of respondents in the other two groups (46% each) agreed with the idea that planning and development should be made by community members regardless of their professional background. In terms of education level, high school graduates were the most likely to agree with this proposition (54%, $n = 68$), in contrast to the other three groups, in which less than half of respondents agreed, with responses ranging from 47% agreement among university graduates ($n = 86$) and respondents with vocational training ($n = 43$) to 43% ($n = 3$) among primary school graduates.

The survey also included an open-ended question to allow respondents to offer ideas for promoting and developing film tourism in their place of residence. A qualitative conventional content analysis of the data identified two basic categories.

The first category covers ideas related to the pre-production phase of the film or series. For instance, respondents emphasized the need to improve the (existing) strategies for attracting international producers, such as better tax rebates, streamlined processes for obtaining filming permits, and also the allocation of film-friendly areas to reduce the impact that film shootings often have on the daily lives of the local community (e.g., interrupted access to some roads and streets). The involvement of the local community in this pre-production phase also included ideas such as hiring more local residents as extras.

Interestingly, a few survey respondents focused their suggestions for promoting film tourism on issues related to media representations of place. These respondents argued that film tourism should be based on the production of films and series that are more realistic, more accurate, less stereotypical, and more sensitive in their representation of a place's unique qualities, or, as one respondent expressed it: "Showing it [the place], as it is." To this end, respondents suggested organizing more casting calls in the place itself, hiring more local actors, and also giving visibility to "less normative profiles," to quote one respondent. Also worth noting is the reference by one of the respondents to the potential impact that having their living place featured in films and series could have on local community members themselves by enhancing their knowledge about their place of residence. Finally, increasing support for junior creators who seek to offer alternative points of view about the towns and cities within the province of Seville was also mentioned.

The second basic category identified by the content analysis relates to initiatives that could be implemented in the post-production phase and after the audiovisual production has been released. In this case, respondents suggested building tourist packages focusing on specific films or series, organizing activities for both tourists and local residents (e.g., film tours with experts), events (e.g., exhibitions), and a permanent visitor center with information on all the films and series shot in the town or city. Also proposed was the erection of street signs to identify where a production was shot and offer some extra information, as well as obtaining more information for the community from local institutional stakeholders about the impact that this type of tourism has on the local economy.

Significantly, the analysis of these responses against the respondents' demographic profiles identified no discernible patterns. For example, suggestions of better tax rebates and the erection of street signs at filming locations were made by both female and male respondents, of all education levels, and varying degrees of involvement in the tourism sector.

However, as the quantitative data has shown, not every respondent was enthusiastic about promoting film tourism in their place of residence, and this opinion is also reflected in some of the responses to the open-ended question. For example, some respondents, particularly those who do not work in the tourism sector, expressed that they find this type of tourism "absurd," while others argued that although they like to see their place of residence featured in films and series, this should not be used to increase tourist numbers since "there are already too many."

Conclusion

The aim of this chapter has been to broaden our understanding of film tourism development by analyzing local residents' opinions about film tourism planning and shedding new light on the demographic profile of local film tourism supporters at a popular tourist destination outside the English-speaking world. To this end, a total of 407 local residents of the Spanish province of Seville participated in a survey questionnaire. Overall, the results of this survey suggest that respondents are supportive of the development of film tourism initiatives. This generally positive attitude toward tourism might be related to residents' positive perceptions of this industry and their interest in diversifying one of the key drivers of their local economy. Despite the heterogeneity of the sample (in terms of gender, age, and education level), the data suggests that the vast majority of respondents support the development of film tourism and the involvement of the local community in its planning strategies. However, in an effort to provide a preliminary idea of the demographic profile of local film tourism supporters in the province of Seville, extra attention has been paid to the differences in the data obtained for statements 1 and 2. In this sense, the predominant profile of Sevillian local residents who believe that film tourism should be actively promoted in their place of residence and would like it to play an important role in it is a male/female in the 50–64 age group with vocational training. Future research could use qualitative methods (e.g., semi-structured in-depth interviews) to better delineate this profile.

Respondents were also supportive of involving the local community in decision-making and planning processes (see, also, Castro et al., 2023). This inclusive attitude is particularly relevant considering that residents of southern Spain have been traditionally neglected in major tourism initiatives (Oviedo-García et at., 2008). The demographic profile of the typical respondent who supports the local community involvement in film tourism planning and development (statements 3 and 4) can be described as a (predominantly) male local resident in the 35–49 age group who works in the tourism sector. According to the survey data, male respondents seem to be more interested in involving the entire community in decision-making processes than female respondents (65% of male respondents agreed, compared to 56% of female respondents), regardless of the professional background of the community members concerned (56% of male respondents agreed, compared to 43% of female respondents) (statement 5). In terms of respondents' relationships with the tourism sector, those working in the sector were the most likely to strongly agree with these inclusive propositions. These results should encourage local institutional stakeholders (e.g., municipalities) to invest efforts in the design and implementation of participatory strategies for the specific development of film tourism initiatives. Future research could thus identify and test effective ways of implementing such participatory mechanisms where different local stakeholders are equally represented.

Survey respondents also exhibited an awareness of the complexities of the film tourism phenomenon. Significantly, 62% of respondents reported taking part in film tourism by visiting local, national, or international locations. Some of the respondents' suggestions to promote film tourism in their place of residence were focused on aspects concerning the pre-production phase, such as attracting film producers by improving the tax incentives. The emphasis that some respondents put on media representations of place is also worth noting (Castro, 2022; Póvoa et al. 2019). In particular, a few respondents argued that film tourism should be promoted through the production of films and series that are more realistic, more accurate, more sensitive to the unique qualities of the filming location, and more focused on involving the local community (e.g., by hiring local actors and using local residents as extras). The suggestions of designing events for both tourists and local residents, together with the visits made by some local residents to nearby filming locations (for example, within their province), point to the potential that film tourism initiatives may also have at a domestic level.

Finally, it is worth noting that the data analyzed in this chapter was collected during the first summer of the COVID-19 pandemic, which may have influenced the enthusiasm with which respondents reacted toward the idea of film tourism development. A comparative study with data collected a few years later might thus also shed some light on the potential impact that crisis situations have on local residents' support for film tourism.

Acknowledgments

I am grateful to Dr. Sean Kim (Edith Cowan University, Australia) and Dr. Guy Assaker (Lebanese American University, Lebanon) for their work on the construction of the survey questionnaire. This investigation was funded by the European Commission through the H2020 Marie Skłodowska-Curie Actions. Grant number 843473.

Notes

1 See   https://andaluciadestinodecine.com/destinos-de-cine/?wpv_view_count=10653& wpv-wpcf-provincias-localizaciones=Sevilla&wpv_filter_submit=Buscar
2 Cactus. Collective-Assembly against the Touristification of Seville.

References

Andriotis, K., & Vaughan, R. D. (2003). Urban residents' attitudes toward tourism development: The case of Crete. *Journal of Travel Research, 42*(2), 172–185.
Beeton, S. (2016). Film-induced tourism. In *Film-Induced Tourism*. Channel view Publications.

Bhat, A. A., & Mishra, R. K. (2021). Demographic characteristics and residents' attitude towards tourism development: A case of Kashmir region. *Journal of Public Affairs, 21*(2), e2179.

Castro, D. (2022). Beyond the Giralda: Residents' interpretations of the Seville portrayed in fictional movies and TV series. *European Journal of Cultural Studies*. doi: 10.1177/13675494221112967

Castro, D., & Cascajosa-Virino, C. (2022). Seville as a filming location in the peak of Spanish TV fiction. In J.F. Gutiérrez Lozano, S. Eichner, B. Hagedoorn, & A. Cuartero (eds.), *New Challenges in European Television. National Experiences in a Transnational Context* (pp. 149–170). Comares Comunicación.

Castro, D., Kim, S., & Assaker, G. (2023). An empirical examination of the antecedents of Residents' support for future of film tourism development. *Tourism Management Perspectives, 45*, 101067.

Choi, H. S. C., & Sirakaya, E. (2005). Measuring residents' attitude toward sustainable tourism: Development of sustainable tourism attitude scale. *Journal of Travel Research, 43*(4), 380–394.

Connell, J. (2012). Film tourism – Evolution, progress and prospects. *Tourism Management, 33*(5), 1007–1029.

Croy, W. G., Kersten M., Melinon, A., & Bowen, D. (2018). Film tourism stakeholders and impacts. In C. Christine & V. Ziakas (eds.), *The Routledge Handbook of Popular Culture and Tourism* (pp. 391–403). Routledge.

Heitmann, S. (2010). Film tourism planning and development – Questioning the role of stakeholders and sustainability. *Tourism and Hospitality Planning & Development, 7*(1), 31–46.

Hsieh, H. F., & Shannon, S. E. (2005). Three approaches to qualitative content analysis. *Qualitative Health Research, 15*(9), 1277–1288.

Jaafar, M., Rasoolimanesh, S. M., & Ismail, S. (2017). Perceived sociocultural impacts of tourism and community participation: A case study of Langkawi Island. *Tourism and Hospitality Research, 17*(2), 123–134.

Kim, S., & Park, E. (2023). An integrated model of social impacts and resident's perceptions: From a film tourism destination. *Journal of Hospitality & Tourism Research, 47*(2), 395–421.

Kim, S., Suri, G., & Park, E. (2018). Changes in local residents' perceptions and attitudes towards the impact of film tourism: The case of *Eat Pray Love* (EPL) film tourism in Ubud, Bali. In S. Kim and S. Reijnders (eds.), *Film Tourism in Asia: Evolution, Transformation, and Trajectory* (pp. 125–138). Springer.

Lonely Planet (2017, October 23). Best in Travel 2018: Top 10 cities. Available at: www.lonel yplanet.com/articles/best-in-travel-2018-top-10-cities (Last accessed on 12 March 2023).

Malek, A., & Costa, C. (2015). Integrating communities into tourism planning through social innovation. *Tourism Planning & Development, 12*(3), 281–299.

Mordue, T. (2001). Performing and directing resident/tourist cultures in Heartbeat country. *Tourist Studies, 1*(3), 233–252.

Nunkoo, R., & Gursoy, D. (2012). Residents' support for tourism: An identity perspective. *Annals of Tourism Research, 39*(1), 243–268.

Oviedo-García, M. A., Castellanos-Verdugo, M., & Martin-Ruiz, D. (2008). Gaining residents' support for tourism and planning. *International Journal of Tourism Research, 10*(2), 95–109.

Póvoa, D. (2023). Film tourism in Brazil: Local perspectives on media, power and place. PhD manuscript, Erasmus University Rotterdam.

Póvoa, D., Reijnders, S., & Martens, E. (2019). The telenovela effect: Challenges of location filming and telenovela tourism in the Brazilian favelas. *Journal of Popular Culture*, *52*(6), 1536.

Rasoolimanesh, S. M., Ringle, C. M., Jaafar, M., & Ramayah, T. (2017). Urban vs. rural destinations: Residents' perceptions, community participation and support for tourism development. *Tourism Management*, *60*, 147–158.

Ruiz, A. S. (2020, January 1). 'Juego de Tronos' dispara las visitas al Alcázar de Sevilla. *La Vanguardia*. Available at: www.lavanguardia.com/local/sevilla/20191230/47260 7806603/juego-tronos-visitas-alcazar-sevilla-dos-millones.html (Last accessed on 17 March 2023).

Sevilla. (2021, 28 September). El Ayuntamiento de Sevilla constituye el primer Consejo Local de Turismo de la ciudad que abre la estrategia turística a una mayor participación para abordar la recuperación desde la sostenibilidad y el beneficio para todos. *Sevilla*. Available at: www.sevilla.org/actualidad/noticias/sevilla-constituye-primer-consejo-local-turismo-estrategia-turistica-participacion-recuperacion-sostenibilidad (Last accessed on 12 March 2023).

Thelen, T., Kim, S., & Scherer, E. (2020). Film tourism impacts: A multi-stakeholder longitudinal approach. *Tourism Recreation Research*, *45*(3), 291–306.

Yoon, Y., Kim, S. S., & Kim, S. S. (2015). Successful and unsuccessful film tourism destinations: From the perspective of Korean local residents' perceptions of film tourism impacts. *Tourism Analysis*, *20*(3), 297–311.

Mentioned Film Tourism Maps

Municipality of Carmona
https://andaluciadestinodecine.com/wp-content/uploads/2022/07/Mapa_Carmona_lo.pdf

Municipality of Osuna
https://andaluciadestinodecine.com/wp-content/uploads/2022/07/Mapa_Osuna_ADdC.pdf

Municipality of Sevilla
https://andaluciadestinodecine.com/wp-content/uploads/2022/07/Mapa_Sevilla_lo.pdf

Part 4

Developing Media Tourism

Policy, Management and Strategies

14 Film Tourism in Brazil

Learning from Local Perspectives

Débora Póvoa

Introduction

From 2016 to 2021, I was part of the European Research Council-funded *Worlds of Imagination* research project, which proposed a comparative analysis of film tourism in various geographies around the world. In my research, I investigated this phenomenon in Brazil, a context where film tourism manifests in spontaneous and unexpected ways, and often in vulnerable locations such as the country's *favelas*, countryside towns and small villages. Intrigued by the incipient and unplanned character of film tourism in Brazil – which contrasts the increasingly professionalized character of other film tourism industries around the globe (Thelen et al., 2020) – I wanted to understand how film tourism developed in these vulnerable locations, and how local communities experienced the sudden interest of filmmakers and tourists in their neighborhoods.

Considering that film tourism research has predominantly focused on success stories (Thelen et al., 2020) mostly occurring in areas of the Global North (Connell, 2012; Beeton, 2016), my goal was to offer novel insights into the challenges that impromptu expressions of film tourism might pose for local communities, particularly in a context in the Global South where film tourism remains underexplored both as an industry and as a research topic (Körössy et al., 2021). In this chapter, I discuss the main results of this project, and what they add to current academic debates on the dynamics between film tourism and local communities.[1]

Film tourism and local communities

In the past couple of decades, film tourism has reportedly become an expanding and highly lucrative tourism niche worldwide (Beeton, 2016; Connell, 2012; Reijnders, 2021). In media, policy and academic circles, this phenomenon has received considerable attention for the potential benefits it could bring to local economies. Positive impacts can arguably already be felt with the arrival of filmmakers to a filming location, since film crews occupy hotel rooms and spend money on local shops just like "usual" tourists (Ward & O'Regan, 2009). Moreover, they also embed themselves in the everyday lives of these communities (Parmett, 2014), having both a tangible effect on the local economy and a symbolic one through interactions with

DOI: 10.4324/9781003320586-19

local people and cultures. If a film generates tourism, then the location experiences another wave of positive effects, from the creation of new tourism-oriented businesses and products, such as film-themed souvenirs and guidebooks (Busby & Klug, 2001), to the alteration of tourism infrastructure and rise in revenues and employment due to new tourist flows (Heitmann, 2010). Particularly in locations in the Global North, success stories of places that highly benefited from film tourism abound, often summarized in celebratory terms such as "Downton effect" (Liu & Pratt, 2019; Mansky, 2019) and "Braveheart effect" (Martin-Jones, 2014).

Even though film tourism can be beneficial to film locations, some authors have questioned the significance of the phenomenon as an economic asset, claiming that very few films actually incite tourism (Croy, 2011) and, even so, film tourism numbers might not be substantial (Croy, 2010). Moreover, the sustainability of film tourism is uncertain since film tourism impacts might not be long-lasting (Beeton, 2016; Connell, 2005). Besides, film tourism might also have detrimental effects on local communities, with the presence of film tourists causing traffic congestions and overcrowding (Mordue, 2001; Tooke & Baker, 1996; Riley et al, 1998), loss of privacy (Bolan et al., 2007; Mordue, 2001), and strains on the local infrastructure (Bolan et al., 2007; Connell, 2005), for example. In the long run, more systemic effects can also be felt, related to the cost of land and housing (Kim et al., 2017), cultural commodification (Heitmann, 2010; Mordue, 2001; Riley et al., 1998), and environmental damage (Koh & Fakfare, 2020; Law et al., 2007).

In order to circumvent these issues, scholars have, time and again, urged for the need to include local communities in film tourism planning and execution (Beeton, 2016; Heitmann, 2010). In practice, however, rarely are residents of a film location involved in the development and management of film tourism (Heitmann, 2010) and their responses to new tourist influxes and opportunities end up being improvised (Thelen et al., 2020). Paradoxically, residents' perspectives on the impacts of film tourism have been neglected in film tourism research (Kim et al., 2017; Thelen et al., 2020). Subscribed to the premise that the success of film tourism initiatives depends on them being constructed collaboratively with the communities involved, this chapter tackles these gaps by presenting the perspectives of multiple local stakeholders involved in three emerging film tourism ecologies in Brazil.

Mapping film tourism in Brazil

The question that guided my research was: How do film tourism initiatives in vulnerable locations in Brazil develop over time, and how do members of the communities involved perceive, experience, and evaluate these initiatives? Following Connell's (2012) broad definition of film tourism, I conceptualized *film tourism initiatives* as any audiovisual media-centered project that generates tourism to a place, regardless of this spillover effect into tourism being planned or not, or occurring before, during or after the completion of the project. I narrowed down these initiatives by focusing on *vulnerable locations* in Brazil, understanding vulnerability as "sensitivity in the well-being of … communities in the face of …

negative changes" that can be "ecological, economic, social and political" (Moser, 1998, p. 3).

Within these parameters, three different cases of film tourism in Brazil were analyzed. The first case study centered on the practice of telenovela tourism in the favelas Complexo do Alemão, in Rio de Janeiro, and Paraisópolis, in São Paulo, after their featuring in the Brazilian telenovelas *Salve Jorge* (2012–2013) and *I Love Paraisópolis* (2015), respectively. Here, I analyzed how local stakeholders involved in film and/or tourism practices, such as tour guides, location scouts, and community leaders, dealt with the use of their neighborhoods as filming locations and tourist attractions. In my second case study, I examined the project Roliúde Nordestina (which translates as "Northeastern Hollywood"), a governmental city branding initiative based on cinema and tourism in the small, economically disadvantaged, and drought-stricken town of Cabaceiras, in Northeast Brazil. My goal was to understand how residents and local policymakers evaluated the attempt of transforming Cabaceiras into a film tourism hub through this initiative. Finally, my third case study focused on the recent Alter do Chão Film Festival (FestAlter), a new film festival in the touristic and environmentally sensitive village of Alter do Chão, in the Brazilian Amazon. Although originally created as a collaborative project, the festival's main organizers – a production company from Rio de Janeiro – ended up being accused of appropriation by sectors of the Alter do Chão community. In this case, I asked how people involved in the production of FestAlter 2019 perceived the changing power dynamics within its organization, and how they saw the implications of these power dynamics for the festival's goals of community participation and empowerment.

In order to advance our understanding of film tourism impacts, Thelen et al. (2020) urged scholars to include "examples that describe the draw-backs and failures of film tourism" (p. 3). In choosing these three case studies, I answer this plea by analyzing instances where film tourism has developed in less-than-optimal conditions. Even though the selected projects should not be considered failures – such a crude assessment would dismiss the positive outcomes of film tourism to these communities as well as the hard work of stakeholders involved – they demonstrate the various challenges in the development of film tourism in vulnerable locations. Importantly, the three case studies prove how these challenges are connected to wider political and institutional frameworks, which points toward the need to look at film tourism within local and national structures of power and inequality. Moreover, the three cases offer a wide range of locations where film tourism occurs in Brazil, local specificities and contingencies that enable or disrupt the development of film tourism projects and different formats that film tourism might take in the country.

Most of the data for this research were collected in visits to these locations from 2017 to 2019, during which I employed participant observation and conducted semi-structured interviews with 61 stakeholders involved in these three film tourism ecologies. Due to the COVID-19 pandemic, interviews related to the Alter do Chão Film Festival were conducted online between December 2020 and

March 2021. In Complexo do Alemão and Paraisópolis, I interviewed ten (former) tour guides or tour company owners, three local entrepreneurs and six community leaders. In Cabaceiras, I interviewed six (former) government members from the Tourism and Culture divisions at local and state level, four tour guides, eight residents who worked as extras in film and TV productions, five residents working in the cultural/hospitality sectors and the creator of the Roliúde Nordestina project. In Alter do Chão I talked to 16 stakeholders involved, in different levels, in the production of FestAlter 2019: three core members of the festival organization, seven stakeholders that participated in the early stages of the festival planning and six festival workers hired either on a paid or voluntary basis during the festival days. Two respondents – one from Complexo do Alemão and one from Cabaceiras – asked to have their participation entirely anonymized.

All interviews were transcribed verbatim (by me or professional transcription services) and analyzed through inductive thematic analysis (Terry et al., 2017). In the remainder of this chapter, I will present the main findings of this project in a chronological structure: as these film tourism initiatives were perceived (when they were being created), experienced (as they were being executed) and evaluated (after they had taken place) by the interviewed stakeholders. It is important to note that the goal of this chapter is not to provide an in-depth discussion of each of the empirical cases; for this, please refer to Póvoa et al. (2019), Póvoa et al. (2021) and Póvoa (2023). Here, instead, I summarize the key insights derived from cross-case comparison and offer overarching arguments that advance our understanding of the dynamics of film tourism in vulnerable locations.

Film tourism planning: Potential for collaboration, development and sustainability

The three film tourism initiatives analyzed in this research project had similar beginnings: all of them were developed because "outsiders" – chiefly film, TV or festival producers coming from the Rio de Janeiro and São Paulo middle and upper classes – took an interest in the location. Telenovela writers saw favelas as a trendy filming location; the success of the film *A Dog's Will* (2000), shot in Cabaceiras, drove more filmmakers to the city throughout the years, and a company from Rio de Janeiro initially visited Alter do Chão for a film project (later turned into a festival project). Once in these locations, these TV, film and festival producers would receive support from community members to navigate these territories, such as community associations and key community figures (Paraisópolis), local government officials (Cabaceiras) and university professors and local politicians (Alter do Chão). Collaboration between locals and "outsiders" took the form of more practical and informal help, for example, research during the pre-production of telenovelas and logistical help during location shootings, but also more meaningful and professional partnerships, for example, teaming up for the conceptualization of a film festival.

The fact that these film tourism initiatives were incited by incoming media professionals attests to the concentration of the audiovisual industry in Southeast

Brazil. Even though some of the researched locations do possess media workers, most of the times they do not have the means, reach or level of professionalization to compete on equal footing with producers from the country's main creative hubs. With this, the favelas, the arid northeastern landscapes and the Amazonian rainforest end up often serving solely as settings for the production of media content. In Paraisópolis and Alter do Chão, though, residents are trying to change this scenario. In the aftermath of the filming of *I Love Paraisópolis*, as well as of my own research period on location, two of the participants of this research started their own location production company in the community. In Alter do Chão, a new film festival was created by members of a local artistic and cultural collective in order to restore the ideals of inclusion and collaboration to which FestAlter failed to live up. These developments highlight the role of independent and local filmmakers and cultural producers in film tourism ecologies, pointing out their agency to foster more community-oriented and sustainable forms of media and event production.

It is important to note that, despite the wish to develop local audiovisual industries, the arrival of external media producers is seen as positive due to their capacity to draw attention to the filmed location and create (at least, temporary) jobs or contribute their expertise to existing cultural projects in the region. The projects that southeastern media professionals bring are seen as opportunities for local empowerment and visibility. This indicates how, for vulnerable populations in Brazil, audiovisual projects might bring hope and new socioeconomic possibilities. In the cases of telenovela tourism and the Roliúde Nordestina project, interviewees also claimed that they were proud that their communities were sought after by media producers and tourists, and that these productions, although not without flaws, increased their self-esteem.

The case studies also demonstrated how, at this initial stage of film tourism projects, two values are of fundamental importance for community members: collaboration and sustainability. To varying degrees, the three sets of interviewees expressed their wish for (more) collaboration with these external crews – what respondents in Alter do Chão called "collective methodology" (Póvoa, 2023, p. 205). Participants also repeatedly mentioned how they would like for these initiatives to occur in a sustained manner through the long-term commitment of filmmakers and festival organizers with their communities.

Existing research on the dynamics of film tourism in local communities tends to focus on what happens in a filming location *after* a film or TV series becomes popular – how visitation numbers increase (Riley et al., 1998; Tooke & Baker, 1996), how existing businesses deal with, or seize on, the sudden attention (Connell, 2005) or how residents perceive these effects (Kim et al., 2017; Mordue, 2001). What these studies often fail to address, however, is how film tourism comes into being and, most importantly, what targeted communities want and expect from film tourism projects in the first place. The findings presented above address these gaps by detailing the origins of film tourism initiatives and unveiling two concrete wishes reported by interviewed community members. As the following section will show, however, film tourism projects often fail to fulfill these expectations.

Film tourism execution: Hierarchies, tensions and neglected goals

Although new film tourism initiatives might initially be seen in an optimistic light, their execution often disappoints and sometimes might even cause conflicts among local stakeholders. In all three case studies of this research project, most interviewees claimed that the execution of these initiatives remained below expectation particularly due to two recurring reasons: the power inequalities created within their production and management, and the failure to accomplish initial goals that these projects set out to pursue.

Film tourism entails various power relationships (Reijnders, 2021), and the three case studies demonstrated how hierarchies tend to materialize in the development and management of film tourism initiatives. The case of telenovela tourism illustrated how only a select group of members of the Paraisópolis community were involved in, and benefited from, the (pre-) production of the telenovela *I Love Paraisópolis*. These residents directly collaborated with the TV crew and had more say in how the telenovela would represent that favela. At the same time, residents not directly involved complained about the lack of community participation, with only a few extras hired and a few scenes filmed on location. This indicates that film and TV crews, in collaborating with community members, might create hierarchies of participation by including and benefiting only specific stakeholders. The case of FestAlter foregrounded similar dynamics. Even though this film festival was supposed to be a community-oriented project, initial partners felt increasingly excluded from the core organization and saw only a few people from the community have an opportunity to voice their opinions. Roliúde Nordestina, in turn, showed how tensions and power inequalities in film tourism also encompass the political realm. While some residents of Cabaceiras blamed the local government for the underwhelming achievements of the project Roliúde Nordestina, this same local government faced various obstacles to further invest in the initiative – including the dependence on higher government levels to obtain support and resources. This indicates the various degrees of power (and powerlessness) involved in film tourism: from residents feeling neglected by local governments, to local governments trying to navigate wider structures of political power within a city, region or country.

Partly because of the power hierarchies and imbalances present in the production of these projects, certain goals that the telenovelas, the project Roliúde Nordestina and FestAlter claimed to strive for were not totally achieved, and this became another reason for interviewees to express discontent with the execution of these initiatives. In the case of the telenovelas, misrepresentation of the favelas in these productions, despite the research conducted by the TV crews in these neighborhoods, was a widespread criticism. As for Roliúde Nordestina and FestAlter, various initiatives within these umbrella projects never fully materialized, such as the promotion of practice-oriented audiovisual workshops for the populations of Cabaceiras and Alter do Chão.

In all locations, though, respondents acknowledged that film tourism initiatives do bring benefits. In line with Ward and O'Regan's (2009) argument, filmmakers and festival producers are themselves tourists, who liven up local economies by staying at hotels, eating at restaurants and hiring locals for different jobs within the production of a film, TV show or event. In doing so, these people symbolically and physically interact with these communities (Parmett, 2014) and engage in *film tourism production*, constructing these locations as film tourism sites for future tourists to visit. For example, the filming and screening of two telenovelas inspired a new sentiment toward otherwise stigmatized neighborhoods in Rio de Janeiro and São Paulo, turning them, at least temporarily, into tourist attractions and a reason for residents to be proud. In Cabaceiras and Alter do Chão, these transformations were also physically tangible, with Roliúde Nordestina creating cinema-related symbols in town and FestAlter transforming the Alter do Chão main square into a festival space. Analyzing the production of film tourism, thus, allows for an understanding of the phenomenon as a longitudinal process whose impacts are felt by communities in various stages – from the arrival of media workers on location to the influx of tourists motivated by the film, TV show of film event in question. Despite their potential to be sustained across time, however, these impacts are often only temporary, as many interviewees regretted.

Film tourism aftermath: The telenovela effect

In my analysis of telenovela tourism in the favelas, I found out that community members considered the benefits brought by the telenovelas, namely an increased local self-esteem and more business opportunities and tourism flows, largely temporary. In light of these findings, I coined the term *telenovela effect* (Póvoa et al., 2019) to express the potentially ephemeral character of the impacts that telenovelas (and other media forms) might have on their filming locations. In this first case study, the *telenovela effect* was conceptualized within a more traditional understanding of film tourism, being restricted to the effects of media productions, for example, films, TV series and music videos, on places. However, the subsequent case studies of this research demonstrated that interviewees in Cabaceiras and Alter do Chão also experienced similar dynamics with the project Roliúde Nordestina and the film festival FestAlter. Therefore, I propose to expand the term *telenovela effect* to also encompass the transiency of the impacts that film tourism initiatives more generally might have on their targeted communities.

With this broader definition, I argue that the *telenovela effect* was perceived in all film tourism initiatives analyzed in this project. In all three cases, the *telenovela effect* proved detrimental to community members, who often felt depleted, neglected or exploited by film tourism initiatives. In this sense, the notion of the *telenovela effect* stands in striking contrast with popular terms such as the "Downton effect" or the "Braveheart effect". While the latter are commonly used by journalists and policymakers to celebrate the benefits that film productions bring to locations, the *telenovela effect* draws attention to the flip side of film tourism – it reveals possibly disappointing and even negative outcomes of film tourism as perceived by members of the affected communities. The *telenovela effect*, thus, not only encourages a more critical approach to the study of film tourism effects on locations but also gives prominence to the voices, opinions and concerns of the ones most affected by this phenomenon since the concept itself is born out of the experiences of local stakeholders with film tourism.

It is important to note that, in the context of this research, the *telenovela effect* was observed even in film tourism initiatives that attempted to generate long-term gains and develop a more productive relationship with the involved communities. This suggests that, although a much-encouraged approach in film tourism research, the goal of including communities in film tourism planning and execution might not be easily accomplished, nor guarantee a project's sustainability. As the case studies illustrated, a series of factors might enable or hamper the development of these initiatives – and it is to a closer inspection of these factors that we now turn.

Lessons from the Brazilian experience

The three case studies analyzed in this research project revealed five factors that might influence the sustainability of film tourism initiatives: (1) the production logic and goals of the project, (2) the infrastructural, geographical and socio-economic characteristics of the location, (3) the political configurations in place to support (or not) the project, (4) the state of the audiovisual industry at country level and (5) the project's internal organization, especially regarding tensions and power dynamics between involved stakeholders. These factors will be further explained below, translated into five lessons that we can learn from the Brazilian experience with film tourism.

1 *Different initiatives (might) enable different outcomes.*
   This research has shown that the production logic of a film tourism initiative, as well as the goals that it sets out to achieve, highly determines the possibilities it offers for local development and empowerment. In the case of telenovela tourism in Complexo do Alemão and Paraisópolis, the cost-effective production of the telenovelas, the timeslots when they aired and their limited screening period restricted possibilities of collaboration with favela residents, the types of favela representations that could be shown and the interest of viewers in the favelas, respectively. Telenovelas are also commercially driven productions whose makers do not necessarily commit to contributing to the filming locations (at least not in the case of the two telenovelas studied in this research). Roliúde Nordestina and FestAlter, however, were very distinct projects compared to telenovelas. Both had clear goals of stimulating regional development through the creation of an audiovisual hub in these locations, which could only be achieved with a consistent engagement with the targeted communities over time. Nevertheless, these initiatives did not escape the same temporariness of the telenovelas. These findings indicate that media workers and policymakers should clearly communicate their plans and goals to the targeted communities in order to establish transparency from the outset of a film tourism initiative. Understanding the characteristics and purposes of film tourism projects is crucial for communities to know what to expect from them, to be able to negotiate

with their makers from a well-informed position and to hold them accountable in case the project fails to live up to its promises.

2 *Film tourism initiatives should take the local context into account.*
The sustainability or unsustainability of film tourism initiatives is also determined by various infrastructural, geographical and socioeconomic characteristics of the locations where they emerge. In the favelas, for example, infrastructural vulnerabilities related to sanitation and street paving, as well as the closure of a main tourist attraction in Complexo do Alemão, were considered to curb the advance of telenovela tourism in these neighborhoods. In Cabaceiras, infrastructural and geographical hindrances, such as the difficult access to some tourist spots, a substandard telecommunication system and the city's distance from the airport, limited the extent to which film tourism could be developed. Moreover, the town's depressed economy did not allow local authorities to offer much more than logistical support to incoming audiovisual productions – let alone to invest in other film tourism initiatives envisioned in the original Roliúde Nordestina project. Lastly, the difficulties in internet access in Alter do Chão made the second edition of FestAlter – held online due to the COVID-19 pandemic – become the last straw in an already deeply strained relationship between organizers and community, with the former seemingly disregarding the local reality when executing this project. These three empirical cases, thus, indicate that producers of film tourism initiatives must take the local context into account in order for these projects to be embraced by local communities, echoing previous research on the topic (e.g., Beeton, 2016; Heitmann, 2010; Kim et al., 2017).

3 *Film tourism initiatives are always political.*
The influence of local political configurations and priorities in the development and sustainability of film tourism initiatives was also found in the three case studies of this research project, albeit to varying degrees. In the favelas, an absolute lack of governmental intervention and investment (in line with the historical governmental neglect that favelas have experienced) was observed. In Complexo do Alemão, telenovela tours were created and conducted by private local entrepreneurs, while in Paraisópolis the local Resident's Union took the forefront in negotiations with telenovela crews and led telenovela tourism initiatives of their own. Later, government security interventions ended up impeding the continuation of these activities, with violent conflicts between policemen and local gangs pushing tourists and investors away and creating an unsafe environment for local tour guides to work, in particular in Complexo do Alemão. In Cabaceiras and Alter do Chão, more significant articulations between private and public partners were observed, with municipalities and local politicians being involved in the conception or, at least, in the execution of film tourism initiatives. In both cases, however, other issues disrupted the continuation of these efforts, such as the need to prioritize other areas for public spending in Cabaceiras and the demobilization of organized collectives in the Amazonian region due to the COVID-19 pandemic. Therefore, these cases

highlight the political dimension of film tourism: how filmmakers, film festival organizers and tour operators are involved in multiple articulations with local political actors and institutions, and how film tourism initiatives often depend on different political levels and mechanisms (e.g. funding allocation) to be sustained. Importantly, these cases showed that even though a more significant articulation with political actors and institutions might imply longevity for these film tourism projects, that is not always the case. Roliúde Nordestina and FestAlter are examples of how various other factors interfere with public–private partnerships fostered in such projects.

4 *Film tourism is part and parcel of local audiovisual networks.*
The longevity of film tourism initiatives is also influenced by the position of the locality in question within the wider organization of the audiovisual industries at the country level. This research revealed how, in a scenario where these industries are geographically concentrated like Brazil, it becomes difficult to develop local audiovisual hubs and decenter audiovisual production. Particularly in locations where financial resources are scarce and other areas of development need attention, like Cabaceiras, building an audiovisual industry almost from scratch is an arduous task for local governments. The case of Alter do Chão indicates similar challenges. The creators of FestAlter had the intention to establish networks of cooperation with film professionals in Rio de Janeiro and São Paulo, which attests to the inequalities in terms of audiovisual investment, resources and infrastructure between the southeast of the country and other regions. In the case of Complexo do Alemão and Paraisópolis, even though these favelas are situated in Rio de Janeiro and São Paulo, they are outside the established audiovisual hubs in these cities. These interregional and intraregional inequalities in the configuration of the audiovisual industry inhibit the long-term stability of film tourism initiatives in vulnerable locations in Brazil. Film tourism stakeholders need to understand the characteristics of existing audiovisual hubs in their cities, regions and countries in order to propose productive initiatives for local communities. Ideally, this understanding will also enable policymakers and citizens to think of better strategies to develop film tourism, such as consistently applying for governmental grants, looking for sponsors in the private sector and pressuring governments to decentralize cultural investments.

5 *Film tourism is, ultimately, done by people.*
Last but not least, the case of FestAlter revealed the power dynamics between the people involved in the management of film tourism initiatives, and how these affect their long-term sustainability. FestAlter illustrated that even projects that have clear goals of community participation and empowerment might still be susceptible to the prevalence of private interests and the concentration of production power. In this particular film festival, such centralization created animosities between organizers, local partners and the Alter do Chão community, and led to rupture and dissent, which prevented the festival to move forward in the same original collaborative spirit. This empirical case, thus, suggests that conflicting visions for a film tourism initiative's mission, goals

and managerial style might pose a threat to its success and sustainability – in the case of FestAlter, even more so than other contextual circumstances. Film tourism stakeholders, thus, should aim to foster transparent and collaborative work cultures and relationships. For researchers, analyzing the organizational structure and dynamics of film tourism is a relevant avenue for future investigation, as it might provide new insights into how different organizational cultures might sustain (or not) film tourism initiatives over time.

Moving forward

The research that I conducted in the context of the *Worlds of Imagination* project moved the academic attention away from locations in the Global North and presented cases of film tourism developed in less-than-optimal circumstances, the results of which differ from film tourism success stories often reported in academic, news and policy circles. Although the findings presented herein are not generalizable to other areas of the Global South, as different localities will present other challenges and opportunities for the development of film tourism, they do create awareness about certain dynamics that might also be encountered in other vulnerable locations around the globe.

The most important of these dynamics is the temporariness of film tourism impacts on locations and communities, encapsulated by the original notion of the *telenovela effect*. This term counters the celebratory narrative around film tourism often summarized in descriptors such as the "Downton effect" or the "Braveheart effect," unveiling the pitfalls of betting on these industries as a means of socio-economic development. Future research on film tourism, thus, should also be aware of the *telenovela effect* in other locations and take more contextual approaches to the study of this phenomenon. As for film tourism stakeholders such as audio-visual and event producers, tourism operators and policymakers, the experiences that the participants of this research had with film tourism initiatives offer evidence that, without an understanding of local cultures, histories and challenges, such initiatives fail to be sustained in the long run. In engaging with local communities, these stakeholders must make an effort to understand their needs, wishes and strengths, instead of imposing strategies that might not work within their particular circumstances, or promising results that will not be delivered. Although involving communities in decision-making processes is still no guarantee of success, it is a first step toward more sustainable forms of film tourism.

Note

1 This chapter is largely based on the introduction and conclusion of my doctoral dissertation, titled "Film Tourism in Brazil: Local Perspectives on Media, Power and Place" (2023).

References

Beeton, S. (2016). *Film-induced tourism* (2nd ed.). Channel View Publications.

Bolan, P., Crossan, M., & O'Connor, N. (2007). Film and television induced tourism in Ireland: a comparative impact study of Ryans' Daughter and Ballykissangel. In *Proceedings of the 5th DeHaan tourism management conference 'Culture, tourism and the media'* (pp. 227–252). Nottingham University Business School.

Busby, G., & Klug, J. (2001). Movie-induced tourism: The challenge of measurement and other issues. *Journal of Vacation Marketing, 7*(4): 316–332. https://doi.org/10.1177/135 676670100700403

Connell, J. (2005). Toddlers, tourism and Tobermory: Destination marketing issues and TV-induced tourism. *Tourism Management, 26*(5), 763–776. https://doi.org/10.1016/j.tour man.2004.04.010

Connell, J. (2012). Film tourism – Evolution, progress and prospects. *Tourism Management, 33*(5), 1007–1029. https://doi.org/10.1016/j.tourman.2012.02.008

Croy, W. G. (2010). Planning for film tourism: Active destination image management. *Tourism and Hospitality Planning and Development, 7*(1), 21–30. https://doi.org/10.1080/ 14790530903522598

Croy, W. G. (2011). Film tourism: Sustained economic contributions to destinations. *Worldwide Hospitality and Tourism Themes, 3*(2), 159–164. https://doi.org/10.1108/ 17554211111123014

Heitmann, S. (2010). Film tourism planning and development – Questioning the role of stakeholders and sustainability. *Tourism and Hospitality Planning & Development, 7*(1), 31–46. https://doi.org/10.1080/14790530903522606

Kim, S., Kim, S., & Oh, M. (2017). Film tourism town and its local community. *International Journal of Hospitality & Tourism Administration, 18*(3), 334–360. https://doi.org/ 10.1080/15256480.2016.1276005

Koh, E., & Fakfare, P. (2020). Overcoming "over-tourism": The closure of Maya Bay. *International Journal of Tourism Cities, 6*(2), 279–296. https://doi.org/10.1108/ IJTC-02-2019-0023

Körössy, N., Paes, R. G. S., & Cordeiro, I. J. D. (2021). Estado da arte sobre turismo e cinema no Brasil: uma revisão integrativa da literatura. *PODIUM Sport, Leisure and Tourism Review, 10*(1), 109–140. https://doi.org/10.5585/podium.v10i1.17212

Law, L., Bunnell, T., & Ong, C. E. (2007). The Beach, the gaze and film tourism. *Tourist Studies, 7*(2), 141–164. https://doi.org/10.1177/1468797607083499

Liu, X., & Pratt, S. (2019). The Downton Abbey effect in film-induced tourism: An empirical examination of TV drama-induced tourism motivation at heritage attractions. *Tourism Analysis, 24*(4), 497–515. https://doi.org/10.3727/108354219X15652651367505

Mansky, J. (2019, 27 September). The "Downton effect" on the English country house tour. *JSTOR Daily.* https://daily.jstor.org/the-downton-effect-on-the-english-coun try-house-tour/

Martin-Jones, D. (2014). Film tourism as heritage tourism: Scotland, diaspora and *The Da Vinci Code* (2006). *New Review of Film and Television Studies, 12*(2), 156–177. https:// doi.org/10.1080/17400309.2014.880301

Mordue, T. (2001). Performing and directing resident/tourist cultures in Heartbeat country. *Tourist Studies, 1*(3), 233–252. https://doi.org/10.1177/146879760100100302

Moser, C. O. (1998). The asset vulnerability framework: Reassessing urban poverty reduction strategies. *World Development, 26*(1), 1–19. https://doi.org/10.1016/ S0305-750X(97)10015-8

Parmett, H. M. (2014). Media as a spatial practice: *Treme* and the production of the media neighbourhood. *Continuum, 28*(3), 286–299. https://doi.org/10.1080/10304 312.2014.900878

Póvoa, D. (2023). "The festival is ours": Power dynamics of community participation in the Alter do Chão Film Festival. *International Journal of Cultural Studies, 26*(2), 200–215. https://doi.org/10.1177/13678779231154288

Póvoa, D., Reijnders, S., & Martens, E. (2019). The telenovela effect: Challenges of location filming and telenovela tourism in the Brazilian favelas. *Journal of Popular Culture, 52*(6), 1536–1556. https://doi.org/10.1111/jpcu.12861

Póvoa, D., Reijnders, S., & Martens, E. (2021). A Brazilian Hollywood in the making? Film, tourism and creative city discourse in the hinterland of Paraíba. *International Journal of Cultural Studies, 24*(5), 691–706. https://doi.org/10.1177/13678779211011635

Reijnders, S. (2021). Imaginative heritage: Towards a holistic perspective on media, tourism, and governance. In N. van Es, S. Reijnders, L. Bolderman & A. Waysdorf (Eds.), *Locating imagination in popular culture* (pp. 19–33). Routledge. https://doi.org/10.4324/9781003045359

Riley, R., Baker, D., & Van Doren, C. S. (1998). Movie induced tourism. *Annals of Tourism Research, 25*(4), 919–935. https://doi.org/10.1016/S0160-7383(98)00045-0

Terry, G., Hayfield, N., Clarke, V., & Braun, V. (2017). Thematic analysis. In C. Willig & W. Stainton-Rogers (Eds.), *The SAGE handbook of qualitative research in psychology* (2nd ed., pp. 17–37). Sage.

Thelen, T., Kim, S., & Scherer, E. (2020). Film tourism impacts: A multi-stakeholder longitudinal approach. *Tourism Recreation Research, 45*(3), 291–306. https://doi.org/10.1080/02508281.2020.1718338

Tooke, N., & Baker, M. (1996). Seeing is believing: The effect of film on visitor numbers to screened locations. *Tourism Management, 17*(2), 87–94. https://doi.org/10.1016/0261-5177(95)00111-5

Ward, S., & O'Regan, T. (2009). The film producer as the long-stay business tourist: Rethinking film and tourism from a Gold Coast perspective. *Tourism Geographies, 11*(2), 214–232. https://doi.org/10.1080/14616680902827175

15 Developing and Managing Film-Related Tourism at Film Studios with the Design of Cultural and Heritage Features

The Case of Hengdian World Studios (China)

Xin Cui, Les Roberts, and Wallis Motta

Introduction

By the early 2000s, the concept of film-related tourism had gained momentum in the research area of tourism with the related knowledge obtained mostly from case studies (Connell, 2012). This study sets the research focus on a Chinese film-related tourism destination – Hengdian World Studios, the world's largest outdoor film studio and film-based theme park – to discuss how film-related tourism is developed and managed at film studios and the role that the Chinese government plays in influencing the Studios' development and management of film-related tourism. The Studios reconstructs the landscapes, streetscapes, and imperial and folk architectures in a number of Chinese past dynasties and early-modern Chinese cities, spanning centuries of Chinese history. In so doing, Hengdian World Studios simulates a number of real heritage sites in different areas of China for screen media productions. Meanwhile, a series of diversified tourism products and activities are also designed and provided for tourists to experience film-related tourism at the Studios.

Based on data collected from ethnographic methods (participant observation and semi-structured interviews) in the research settings between 2018 and 2021 and online interviews with ten tourists who have visited the Studios previously, this chapter will investigate the main characteristics of the Studios' film-related tourism with the design of Chinese cultural and heritage features and tourists' perceptions of their film journeys and experiences of cultural heritage features. The core discussions will be divided into two sections. The first section will show the characteristics of three main types of film-related tourism products and activities, including (a) sightseeing and performative tourism products, (b) 'dynamic' tourism products, and (c) immersive and interactive tourism environments and activities. The aim of this section is to explore how the film culture and Chinese traditional culture are integrally represented at film-related tourism sites. Specific cultural practices and contexts in relation to cultural tourism in China will be examined in this section. The second section will show how tourists perceive their engagement with and consumption of film and cultural heritage when experiencing film-related

DOI: 10.4324/9781003320586-20

tourism. It will take a critical stance to look at the Studios' development and management of film-related tourism and discuss the potential effects of the governmental policies for developing cultural tourism.

Film-related tourism and cultural tourism in China

Film-related tourism can be seen as a form of cultural tourism, which refers to people's journeys to certain cultural attractions with the purpose of satisfying their travel needs and interests in cultural features, activities, and events (Richards, 1996, 24; Jewell and McKinnon, 2008, 153). Over the last three decades, there has been a growing interest in research in the field of film-related tourism (Oviedo-García et al., 2016). People are witnessing the boost of film-related tourism and the popularity of film-related tourist sites around the world. These all support the viewpoint that film-related tourism can be regarded as a growing global cultural phenomenon, catalysed by the growth of the entertainment industry and global travel (Yen and Croy, 2016).

In the way of identifying different types of film-related tourism destinations, this study classifies the research setting and other film-related tourism destinations based on Sue Beeton's classification of on-location and off-location film sites (2005). In Beeton's classification, on-location film sites refer to the existing buildings, built landscapes, and natural landscapes, which are not originally built and designed for filmmaking or film-related tourism purposes. Off-location film sites refer to the constructed set, studio site (separate from the naturally occurring setting of the film or screen production), and the representation of natural landscapes, produced by computer imaging and other techniques, which are deliberately built and designed for filming media works and film-related tourism purposes.

Based on Beeton's classification of film-related tourism destinations (2005), Hengdian World Studios can therefore be classified as an *off-location film-related tourism site*. Hengdian World Studios, the world's largest outdoor filming-making location and film studio theme park, was launched in Hengdian Town, Zhejiang Province in China, in 1996 (Figures 15.1 and 15.2). By 2020, it had around 130 indoor film studios and 14 outdoor film shooting bases and film-themed tourism attractions. These outdoor film studios simulate a number of real (existing or derelict/vanished) heritage sites in different areas of China. The theme of each outdoor film studio is designed based on the architectural style and cultural characteristics of Chinese past dynasties from 221 BC to 1912 AD or a specific Chinese early-modern city in the 1910s to 1940s. Therefore, these film studios can satisfy different media productions and crews to shoot their screen media works with different historical backgrounds and storylines. By the end of 2020, more than 3,200 screen media productions and crews had been completed at the Studios. Meanwhile, in 2010, Hengdian World Studios was classified as the highest-level tourism attraction in China by the China National Tourism Administration (now the Ministry of Culture and Tourism) (Hengdian World Studios, n.d.). From 1996 to 2020, the total number of tourists to the destination during the whole of this period reached an impressive

Figure 15.1 Localization of Hengdian, Zhejiang Province, China.

Source: *China Daily* (2012b).

milestone of 0.2 billion (Hengdian Group, 2020). Simulating real historic sites and cultural heritage settings for filming screen media works, Hengdian World Studios, as a film studio and film-related tourism site, also encourage tourists' consumption of cultural heritage elements. On the one hand, such a theme design at the Studios can enable different media productions and crews to shoot their screen media works with different historical backgrounds and storylines. On the other hand, tourists can also experience off-location film-related tourism involving various historical and cultural heritage elements.

Since the 2000s, China has vigorously developed cultural tourism nation-wide. A series of national guidelines, laws, plans, and guiding opinions have been promulgated by different state departments to develop the cultural and tourism industries. This has been prompted by the Chinese government's realization that the tourism industries can economically and socially impact the country's overall development and play important roles in developing the national economy and representing Chinese traditional culture and socialist culture (China Government, 2007). With the deepening of economic globalization and regional economic integration, Chinese people's demand for tourism has increased significantly, driving the need for new and higher requirements for the development of the tourism industries (op cit.). In addition, since 2007, the cultural industries' contributions to national economic growth in China have increased, and the Chinese government

Figure 15.2 Hengdian town map.

Source: Sogou Pic., translated from Chinese to English by Xin Cui.

has been committed to promoting the cultural industries to become pillar industries of the national economy (China Daily, 2012b). In the '14th Five-Year Plan for Economic and Social Development of the People's Republic of China (2021–2025) and the Long-Range Objectives through the Year 2035' (14th Five-Year Plan and Year 2035 Objectives), the Chinese government introduced a strategic ideology: 'culture as a baseline to develop tourism, tourism as a form to represent culture'. The Chinese government has been committed to expanding the role and significance of tourism in spreading Chinese traditional culture and socialist culture

(China Government, 2007). Tourism in China has become an important tool for promoting and educating people about a destination's local cultures. In this way, culture can be 'consumed' by tourists and bring economic benefits to a tourism destination. Accordingly, Hengdian World Studios is committed to developing its film-related tourism with Chinese cultural and heritage elements, in which both film culture and Chinese traditional culture are represented to tourists. By doing this, the Studios can secure the national government's economic and political support for the management of China's tourism industries.

Film-related tourism products and activities at Hengdian World Studios

Data was collected from ethnographic visits to six outdoor film studios filming sites, and tourism attractions in the Studios from 2018 to 2021. These included the 'Palace of Ming and Qing Dynasties' film studio, the 'Qing Ming Shang He Tu' film studio, 'The Palace of Emperor Qin' film studio, the 'Guangzhou Street · Hong Kong Street' film studio, the 'New Yuanmingyuan' film studio, and the 'Dream Village' tourism site. Participant observation carried out at the Studios by one of the authors (Cui)[1] revealed a variety of film-related tourism products and activities that are provided to tourists (hereafter the subject 'I' refers to 'Cui' and the possessive adjective 'my' is the same as 'Cui's'). These include: (a) sightseeing and performative tourism products, (b) 'dynamic' tourism products, and (c) immersive and interactive tourism environments and activities.

Firstly, film culture and Chinese traditional culture are integrated into the Studios' sightseeing and performative tourism activities and products. As the film locations of more than 3,200 screen media works (Hengdian Group, 2020), the film settings at the Studios (buildings, streetscapes, landscapes, backdrops, and props), which have been used as film locations or represented in famous screen media works, are one of the significant film-related tourism products. Therefore, going sightseeing around the tourist attractions to search for, observe, and take photos of these film settings is a popular film-related tourism activity among tourists. During fieldwork conducted at 'The Palace of Emperor Qin' film studio in 2019, a billboard of the television drama *The Untamed* (2019) was observed, featuring a staged photo of two of the drama's protagonists, with a section of texts 'Here is the shooting location of *The Untamed*' (Figure 15.3). A Chinese female tourist who was taking photos of this billboard explained to her partners: 'It is another way for me to take photos with my favourite actor, as I cannot get an opportunity to do it with the real celebrity, and the photo can prove that I have been here'.

It is worth noting that these film settings at the Studios are essentially copies and simulations of real some Chinese real heritage sites, and thus when tourists experience film-related tourism by going sightseeing in the travel environments, they inevitably experience cultural heritage tourism and consume cultural heritage products at the same time. Take the example of the Forbidden City in Beijing, which was constructed from 1406 to 1420 as the former Chinese imperial palace during Ming Dynasty and Qing Dynasty. As the real heritage site does not allow filming activities inside it and does not allow tourists to visit some areas and spaces

Figure 15.3 A billboard of the television drama *The Untamed* at Hengdian World Studios.
Photography by Xin Cui (2019).

considering heritage protection, Hengdian World Studios reproduced and simulated the Forbidden City at a one-to-one scale and renamed it the 'Palace of Ming and Qing Dynasties' film studio (Hengdian World Studios, n.d.). To some degree, the studios can be seen as physical copies of real heritage and historic sites. During the ethnographic visits to the studios, I also visited both the outside environments and the inside settings and compared them to the Chinese real heritage sites, such as the Forbidden City. No matter the architectural scales, styles, colours, or materials, for general tourists without professional architecture knowledge and educational background, it was difficult to figure out the discernible and significant differences between the simulated and real heritage sites. In this way, film culture and Chinese traditional culture are interwoven in tourists' travel experiences in Hengdian World Studios.

In addition to consuming sightseeing tourism products, based on participant observations at the Studios, tourists are able to watch a number of live stage performance shows in different tourist attractions, which combine Chinese traditional culture and historical stories with various film techniques, such as lighting techniques, sound effects, and 3D holograms. In 2018, when I visited the 'Palace of Ming and Qing Dynasties' film studio, I also watched the live stage show 'Secret Story Happened in Qing Court', which is a 20-minute film-themed live performance that demonstrates the historical stories of emperors in the Qing Dynasty in the

Forbidden City. These live performances are indeed tourism-friendly to a group of tourists whose travel motivations are not largely in relation to consuming film culture. When I talked with some tourists about their travel experiences before and after the performances, a middle-aged female tourist stated that 'the film-related live theatre stage performances with Chinese cultural and historical stories were fantastic'. Moreover, it became one of the most anticipated on-site touristic activities they desired to participate at different film studios, and the female tourist's partners expressed that, 'We are not interested too much in the film and television works made here, but we like watching these performances with cultural heritage elements'. Regarding the representations of Chinese traditional culture in these performances, these tourists stated that 'the performance stories indeed in relation to Chinese traditional culture and history, many historical and cultural stories are heavily modified and falsified, so the function of these performances is far more to entertain than to educate tourists'.

To a certain extent then, Chinese traditional culture and heritage are commercialized as tourism products for entertaining tourists and bringing economic benefits to destinations. The functions of educating people about Chinese history and spreading knowledge of Chinese culture and heritage, one of the core aims for developing cultural tourism in China, are less reflected at the Studios' film-related tourism. Tourism indeed can be used as a form or carrier to represent Chinese culture, but not all forms of tourism, such as film-related tourism, can properly represent Chinese culture and history.

Secondly, a variety of 'dynamic' tourism elements, such as film celebrities at work and film studio tour guides, also contribute to the representation of film culture and Chinese traditional culture. As Hengdian World Studios operates as a working film studio and film theme park, tourists are able to observe the ongoing works of on-site media crews and celebrities. According to Turner (2014, 2), in academic studies, cultural and media researchers have tended to 'focus on celebrity as the product of a number of cultural and economic processes'. This suggests that film celebrities and their ongoing filming activities at the Studios are used as a kind of tourism commodity that tourists can consume during their film journeys through encountering and observing their ongoing works at the Studios. Beeton (2005) specifically classifies journeys that are induced by film celebrities or for encountering film celebrities as 'celebrity film tourism', confirming celebrities' roles as travel attractors and tourism products in film-related tourism.

In addition, as the Studios are simulations of some real heritage sites in China, they are mostly used as film locations of screen media works that depict Chinese heritage stories and show historical figures in a specific Chinese past dynasty. Thus, these film celebrities at work at the Studios also dress in Chinese traditional costumes and ancient accessories to act out a story that took place in one of China's past dynasties. At some film studios, they provide costume rental services, and the studio staff also dressed in different Chinese cultural and traditional costumes in order to highlight its film culture and Chinese heritage culture (Figure 15.4). I observed that a group of tourists made up and dressed in film characters' costumes with distinct Chinese traditional cultural elements for taking photos with the cultural heritage

Figure 15.4 A studio staff dressed in different Chinese cultural and traditional costumes.

Photography by Xin Cui (2021).

settings in one frame (Figure 15.5). Compared with 'fixed' tourism products, such as film settings, these human objects can be seen as 'dynamic' tourism products, as they usually do not appear in a fixed place but move around the Studios.

Based on my participant observations at the Studios, these 'dynamic' tourism objects, especially film celebrities at work, somehow make the commodification of Chinese traditional culture even more serious at the Studios. Namely, the commodification of Chinese traditional culture and heritage is happening during tourists' travel experiences at the Studios. As film celebrities themselves are used as tourism commodities (Beeton, 2005; Turner, 2014), the Chinese traditional cultural elements attached to these celebrities also become a part of tourism commodities. Chinese traditional culture therefore can be 'sold' and 'consumed' in the tourism industries and is represented at tourism destinations with more consideration of commercial purposes. Cultural commodification is happening, which is used to denote a process where culture and the relevant cultural activities and artefacts are packaged and availed for tourists to purchase (Cohen, 1988; Mbaiwa, 2011). At the studios, the meaning and value of Chinese intangible cultural heritage are represented and conveyed through tangible tourism objects, whereby the cultural heritage is endowed on film-related tourism activities and products

Figure 15.5 A tourist dressed in Chinese traditional costumes and took photos with the cultural heritage settings in one frame.

Photography by Xin Cui (2020).

as a part of tourism commodities. Cultural heritage is featured and promoted in film-related tourism that attracts tourists to spend money at tourist sites to generate more tourism income, such as tourism souvenirs or costumes with cultural characteristics. One of the significant negative effects of cultural commodification for tourism purposes is that local culture becomes meaningless and the meanings of cultural products are destroyed (MacCannell, 1973; Mbaiwa, 2011). At the studios, cultural heritage is exploited as a profitable tourism product. Showcasing cultural heritage features and values is regarded as a way for promoting tourism. The emergence of this phenomenon essentially defeats the intentions and objectives of the Chinese government in developing cultural tourism, that is, applying tourism to promote Chinese culture and heritage. It also reflects the challenges that Chinese tourism destinations need to face in developing cultural tourism in future, namely in using tourism as a carrier or form to promote Chinese traditional culture and heritage but avoiding the process of cultural commodification.

In recent years, the Studios has started to upgrade its tourism attractions and operation modes and provided more 'immersive' film-related tourism products. In October 2020, I re-visited the 'Guangzhou Street · Hong Kong Street' film studio for the 'immersive film-related touristic experience'. The first time I visited this film studio was in April 2018, whereas the 'immersive' film-related touristic activities and services have been provided to tourists since October 2018. These were new film-related tourism elements resulting from the upgrade of the tourism site. After the upgrade, the closing time of the film studio was extended from 17:00 to 20:00 in order to provide tourists enough time to join the night tour to watch the light show involving neon lamps on the film settings and buildings and night outdoor live performances to experience Hong Kong's nightlife in the 1910s (Hu and Du, 2018). During the night tour, I also discovered the ways that the film studio had provided 'immersive' tourism activities and products. The film studio reconstructed its film settings and backdrops as well as the locations of some famous film and television works made at the Studios, which depict the stories that took place in Hong Kong and Guangzhou City in the 1910s. Through visiting the film settings and locations, I was brought into a fictional world in which the settings and buildings were designed after fictional scenes from film and television works and with the characteristics of Guangzhou and Hong Kong's city culture and Chinese traditional culture in the 1910s. In addition, in order to emphasize the 'reality' of the fictional environment, the studio also provides film-character and traditional costume rental services and make-up services. In this way, tourists were encouraged to play as film characters or Chinese historical figures to experience both film culture and Chinese heritage culture.

Designing and providing 'immersive' film-related tourism products with the design of Chinese cultural and heritage features conform to the aims and principles of developing cultural tourism in China. Chinese tourism destinations are suggested by the national government to encourage the integration of Chinese culture into modern tourism consumption models in order to satisfy modern tourists' needs and requirements (Huang, 2019, 20). In the case of Hengdian World Studios, the type of tourism site also has upgraded from the 'scenic spot' to the 'cultural space', involving both film-related cultural elements and Chinese traditional cultural elements. Tourism products are no longer sightseeing activities but 'culture-experiencing' activities. By doing this, tourism products at Hengdian World Studios have been expanded to 'interactive and participative' activities. However, if so, not only cultural commodification but also cultural marketization may take place at the Studios' film-related tourism. According to Navarro (2016, 229), the phenomenon that 'governments made cultural industries central to their political schemes' can be understood as the marketization of (patriotic) culture. Tourism is not only a carrier to represent culture but also a promotion tool to propagandize patriotism through leading tourists to immerse in the tourism environments with the design of Chinese cultural and heritage elements. A potential effect is that the development of cultural tourism is highly dependent on or influenced by the ways to promote patriotic culture through tourism. Tourist attractions at the Studios are not only entertainment places but also educational bases for tourists to learn about Chinese history and cultural heritage as well as the Chinese government's state policy in developing cultural and tourism industries.

Tourists' perceptions of film-related tourism with the design of cultural and heritage elements

In order to understand tourists' perceptions of film-related tourism at the Studios, individual interviews were carried out with ten tourists, who visited the Studios previously, by one of the authors (Cui)[2] via two Chinese social media, including WeChat and Weibo, from 12 to 15 April 2020 (hereafter the subject 'I' refers to 'Cui' and the possessive adjective 'my' is the same as 'Cui's'). Informants of the online interviews are the online users of these two Chinese online platforms, who replied to my request post – 'share your touristic experience at Hengdian World Studios' on WeChat, or who posted certain travel notes about their journeys with some keywords or tags on Weibo, such as 'Hengdian' and 'Hengdian World Studios'. Questions in the interview were divided into two sections respectively focusing on informants' film-related tourism experiences and cultural heritage tourism experiences in Hengdian World Studios. This chapter will only demonstrate a part of the informants' responses which are closely related to tourists' understanding of the integration of film culture and Chinese traditional culture at the Studios' film-related tourism. The participants were given pseudonyms as 'Informants 1, 2, 3…10' in this research in order to comply with the policy of anonymity. As all interviews were conducted in Chinese language, a translation was completed by converting textual contents into English as transcripts. The basic background information is displayed in Table 15.1, including the age range, gender, interview platform, and travel year.

Table 15.1 Background information of interview informants

Informant	Age Range (years)	Gender	Interview Platform	Travel Year
1	18–25	Female	WeChat	2016
2	18–25	Female	WeChat	2015
3	26–30	Female	Weibo	2018
4	18–25	Male	Weibo	2017
5	26–30	Female	WeChat	2014, 2015 & 2018
6	18–25	Female	Weibo	2016
7	18–25	Female	Weibo	2018
8	26–30	Male	Weibo	2019
9	18–25	Female	Weibo	2017
10	30–35	Female	Weibo	2019

In terms of the topic in relation to the educational functions of the cultural heritage elements in film-related tourism at the studios, the responses of these ten informants show tourists' different attitudes. Several representative and worth-noting answers were extracted from these informants regarding this topic as below.

Informant 3 is one of the participants that believed that people can learn Chinese traditional culture and history during their film journeys at the Studios. She explained in the interview:

> The historic settings and architectures were built and designed on a one-to-one scale with the real ones, so visually they are reliable for me to learn Chinese heritage and history. Also, compared with some real heritage sites, in which some areas and settings are not accessible, I think Hengdian World Studios provides a better travel experience because more areas are accessible and more touristic activities can be participated in.

As discussed in the former section, in addition to film culture, the studios' tourism activities and products also integrate Chinese heritage culture. Tourists who engage in film-related tourism also inevitably experience cultural heritage tourism and thus can be educated by the representation of Chinese history and heritage culture at the film studios. Compared with Informant 3, Informant 1's attitude to the design of Chinese cultural and heritage elements in film-related tourism at film studios is more neutral, and this could be because of the informant's educational background in architecture studies:

> The settings and architectures at Hengdian World Studios have both strengths and limitations in introducing and representing Chinese culture and history. Regarding the strengths, tourists can be physically close to the toured objects, and this is what we usually cannot do at the real heritage site areas and historic districts. However, based on my knowledge of architecture studies, I have realized a number of incorrect designs and use of some building materials in its settings and architectures that are contrary to historical and real heritage facts.

By contrast, Informant 5 did not agree that the Studios has the educational functions of Chinese culture and history, and the informant explained:

> In terms of representing culture and history, Hengdian World Studios is more like a theme park and an aggregation of several historic sites from different dynasties. Sometimes I cannot exactly know the history and cultural heritage through visiting the attractions unless combining the introduction of tour guides and my imagination of the real sites.

As discussed previously, considering one of the aims of the Chinese government in developing cultural tourism is to represent Chinese culture, I also talked about the topic in relation to the capability of the film studios' cultural heritage tourism in representing and spreading the value of Chinese culture and history with the participants.

Most informants (1, 2, 4, 5, 7, 8, 9, and 10) expressed that in the way of delivering and conveying the meaning and value of Chinese traditional culture and history, the simulations and reproductions at the Studios *cannot* be seen as 'authentic' as the real cultural heritage and historic sites. Some representative responses from these participants are shown below.

Informant 2 in the interview explained: 'Everything at the Studios is the simulation and copy without a strong historical sense, which cannot show the grandeur and status of an imperial heritage site'. For Informant 7, 'it could be better to visit the real heritage and historic sites, because the Studios is more suitable to experience film-related tourism and to have a very general view of history and culture'. Informant 10 also suggested 'Hengdian World Studios is a place for tourists to experience film culture rather than to show the value and significance of Chinese cultural heritage'. Similarly, Informant 8 observed that 'compared with real heritage sites, which create a historic atmosphere, the Studios tends to create an entertaining atmosphere for tourists'. Meng and Tung (2016) suggest that familiarity with the cultural elements in Hengdian World Studios works as an interface that stimulates domestic tourists' interest in exploring, understanding, and appreciating local tourist attractions. However, it could be also the familiarity with Chinese history and heritage culture that makes tourists' perception of the value of film studios to the representation of staged cultural heritage strict and critical. That is why most informants do not believe the staged cultural heritage at the Studios can exactly show the value of Chinese history and culture.

Although the informants in the case of Hengdian World Studios do not form a representative sample of the population of all tourists, the above responses can also demonstrate tourists' various perceptions of developing and managing film-related tourism with the design of cultural and heritage elements for representing and educating Chinese traditional culture and history. These responses can help understand the Studios' film-related tourism with the consideration of national cultural practices and contexts to develop cultural tourism in China from the perspective of tourists. Based on the above interview contents, it can be seen that tourism can, to some extent, be used as an important platform or medium to represent a destination's local culture through the design and organization of different types of tourism products and activities. The representation of the 'non-authentic' Chinese cultural heritage to some degree is beneficial for the public to understand Chinese history and culture, as it makes them more accessible and provides tourists with new ways to feel connected to it. The cultural heritage represented at the film studios is a kind of staged narrative of Chinese cultural heritage, carefully and elaborately designed by the destination. To some extent, this is not difficult for a tourism destination, because a touristic space itself can be called 'a stage set, a tourist setting, or simply a set depending on how purposefully worked up for tourists the display is' (MacCannell, 1973, 597). Even so, tourists may have different attitudes towards the significance and value of the representations of heritage culture, probably because the tourists are a heterogeneous group of people with various demands, motivations, and on-site activities (Cohen, 1988; Pearce, 1995). A group of tourists do not think that they see real heritage at the Studios, yet they still are able to enjoy the history-themed environment and immersion as a bonus to both popular culture and heritage culture they experience there. However, another group of tourists do not think the stage cultural heritage has the capability to educate people about the value of Chinese culture in the same way as real cultural heritage. Thus, while the Chinese government aims to employ tourism to promote Chinese culture, it is crucial to acknowledge that promoting Chinese cultural and heritage features may require distinct strategies depending on the specific tourism destination. The case of Hengdian World Studios suggests that not all forms of tourism are effective in promoting and educating Chinese traditional culture, and not all tourists are receptive to learning about traditional Chinese culture and history during a tour, such as a film tour, where entertainment is predominant. The past in China, including history, traditional culture, and heritage, in the case of Hengdian World Studios, is turned into a tourism commodity and at the same time used as a kind of resource for state propaganda. The case of Hengdian World Studios suggests that tourism destinations employ different ways to represent local culture under the impact of governmental policies, whereas tourists may have diverse attitudes toward their representations. Tourists come to the Studios to experience film-related tourism with the design of Chinese cultural heritage features. This situation poses a potential challenge that Hengdian World Studios needs to deal with in terms of integrating film-related tourism and cultural heritage tourism more effectively, promoting both film culture and Chinese traditional culture, and emphasizing both the entertainment and educational functions of tourism at the same time.

Conclusion

In addition to film elements, film-related tourism can also encourage tourists' engagement with the consumption of cultural and heritage elements at the destination. Based on the ethnographic data at the research setting, this chapter showed the ways in which Hengdian World Studios, the largest outdoor film studio and film theme park, simulates a number of Chinese real heritage sites for filmmaking and tourism purposes, and develops and manages film-related tourism with the design of Chinese cultural and heritage elements. The destination represents the local film culture and traditional heritage culture through providing sightseeing and performative, 'dynamic', and immersive and interactive film-related tourism products and activities. By doing this, tourists can experience film-related tourism and cultural heritage tourism through consuming and participating in these tourism products and activities. Chinese traditional culture and history are also represented to tourists, conforming to the Chinese government's aims and principles to develop cultural tourism at the national level. While film-related tourism that promotes Chinese traditional culture and heritage can have its benefits, there are also potential drawbacks. These could involve adapting the culture to cater to the tourism industries and commercializing Chinese traditional culture and heritage into commodities for tourists' consumption, that is, cultural commodification. Furthermore, representing traditional culture and heritage in this way may also lead to cultural marketization, in which culture is used by governments to propagandize their political scheme when developing tourism. In this regard, tourism is more like a tool to gain economic benefits and promote national policies. A potential effect is that the development of cultural tourism is highly influenced by and dependent on how much economic benefit it brings to a destination and how efficiently it promotes local patriotic culture, while local people and tourists' interests may be overlooked. These all suggest that such a tourist destination has the functions of entertainment, education, and propaganda, and tourism development and management are highly dependent on its economic gain and socio-cultural impacts on the nation.

Based on the data collected from online interviews with ten tourists who have visited Hengdian World Studios previously, this chapter also demonstrated tourists' perceptions of the design of cultural and heritage elements at the Studios' film-related tourism. Different tourists have diverse attitudes to such a design. Some of them believed that people could learn Chinese traditional culture and history during their film journeys, some tourists held relatively neutral attitudes, while other tourists felt that the representations of these cultural elements in film-related tourism had more of an entertainment function than an educational function. Moreover, most of these ten interviewees did not agree that the simulations and reproductions at the Studios can spread the value of Chinese culture in the same way as real cultural heritage and historic sites. Namely, the tourism destination makes claims to the educational functions and value-communication function of its representations of cultural elements in film-related tourism, whereas tourists have diverse attitudes to these representations.

Notes

1 Ethnography was carried out by Xin Cui during the PhD study at the University of Liverpool, supervised by Dr. Les Roberts and Dr. Wallis Motta.
2 Interviews were carried out by Xin Cui during the PhD study at the University of Liverpool, supervised by Dr. Les Roberts and Dr. Wallis Motta.

Bibliography

Beeton, S. (2005) *Film-induced Tourism*, Bristol: Channel View Publications.
China Daily. (2012a) 'The Sixth Plenary Session of the 17th Central Committee of the Communist Party of China (CPC) closes today to further promote the reform of the cultural industries' [online], Available at www.chinadaily.com.cn/dfpd/shizheng/2011-10/18/content_13923036.htm [Accessed on 16 February 2023].
China Daily. (2012b) 'When there is safety in numbers' [Online], Available at http://europe.chinadaily.com.cn/culture/2012-07/09/content_15561233.htm [Accessed on 16 February 2023].
China Government. (2007) 'Opinions on the Further Promotion of Tourism' [online], Available at www.gov.cn/zwgk/2007-09/12/content_746762.htm [Accessed 16 February 2023].

Cohen, E. (1988) 'Authenticity and commoditization in tourism', *Annals of Tourism Research*, 15(3), pp. 371–386.

Connell, J. (2012) 'Film tourism – Evolution, progress and prospects', *Tourism Management*, 33, pp.1007–1029.

Hengdian Group. (2020) 'Corporate Social Responsibility Report 2018–2020' [Online], Available at www.hengdian.com/uploads/documents/%E6%A8%AA%E5%BA%97%E9%9B%86%E5%9B%A22018-2020%E7%A4%BE%E4%BC%9A%E8%B4%A3%E4%BB%BB%E6%8A%A5%E5%91%8A.pdf [Accessed on 16th February 2023].

Hengdian World Studios. (n.d.) 'Theme scenic areas' [Online], Available at www.hengdianworld.com/en/Theme/ [Accessed 16 February 2023].

Hu, D., & Du, Q. (2018) 'The upgrade of the "Guangzhou Street · Hong Kong Street" film studio, bringing tourists into old Hong Kong' [Online], Available at https://zj.zjol.com.cn/news.html?id=1051788 [Accessed 16 February 2023].

Huang, Y. (2019) 'Cultural interpretation and tourism practice of the culture-tourism integrated development', *Renming Luntan·Xueshu Qianyan*, 11(171), pp. 16–23.

Jewell, B., & McKinnon, S. (2008) 'Movie tourism – A new form of cultural landscape?', *Journal of Travel & Tourism Marketing*, 24(2–3), pp. 154–162.

MacCannell, D. (1973) 'Staged authenticity: Arrangements of social space in tourist settings', *American Journal of Sociology*, 79(3), pp. 589–603.

Mbaiwa, J. (2011) 'Cultural commodification and tourism: The Goo-Moremi community, central Botswana', *Tijdschrift voor Economische en Sociale Geografie*, 102(3), pp. 290–301.

Meng, Y., & Tung, V. (2016) 'Travel motivation of domestic film tourists to the Hengdian World Studios: Serendipity, traverse, and mimicry', *Journal of China Tourism Research*, 12(3–4), pp. 434–450.

Navarro, B. (2016) 'Creative industries and Britpop: The marketisation of culture, politics and national identity', *Consumption Markets & Culture*, 19(2), pp. 228–243.

Oviedo-García, M., Castellanos-Verdugo, M., Trujillo-García, M. A., & Mallya, T. (2016) 'Film-induced tourist motivations. The case of Seville (Spain)', *Current Issues in Tourism*, 19(7), pp. 713–733.

Pearce, D. G. (1995) *Tourism Today: A Geographical Analysis*. Harlow: Longman Scientific & Technical.

Richards, G. (1996) *Cultural Tourism in Europe*, Wallingford: CAB International.

Sogou. Pic. (n.d.) 'Hengdian Town Map' [online], Available at https://pic.sogou.com/pic/searchList.jsp?statref=searchlist_hintword_down&spver=0&keyword=%E6%A8%AA%E5%BA%97%E9%95%87%E5%9C%B0%E5%9B%BE [Accessed on 16th February 2023].

Turner, G. (2014) *Understanding Celebrity*, London: SAGE Publications Ltd.

Yen, C., & Croy, W. G. (2016) 'Film tourism: celebrity involvement, celebrity worship and destination image', *Current Issues in Tourism*, 19(10), pp. 1027–1044.

16 Promoting Popular Culture (and) Tourism as National Policies

Comparing "Cool Japan" and "Korean Wave"

Kyungjae Jang and Sean Kim

Abbreviations

ACA	Agency for Cultural Affairs (Japan)
JETRO	Japan External Trade Organisation
JNTO	Japan National Tourism Agency
JTA	Japan Tourism Agency
MLITT	Ministry of Land, Infrastructure, Transport and Tourism (Japan)
METI	Ministry of Economy, Trade and Industry (Japan)
MIC	Ministry of Internal Affairs and Communications (Japan)
MOF	Ministry of Finance (Japan)
KOCCA	Korea Creative Content Agency
KOFICE	Korea Foundation for International Cultural Exchange
KTO	Korea Tourism Organisation
MCST	Ministry of Culture, Sports, and Tourism (Korea)

Introduction

The incurable bitter, hostile relationship between South Korea (hereafter Korea) and Japan is not new to us. The brutal past experiences of Korea during Japanese occupation (1910–1945) have long been the cause of mutual animosities (Kim, 2000). The postcolonial ban on the import of Japanese (popular) cultural products until late 1998 was a counteract (Park, 2005). There may have been a further consideration for the development of policies to promote indigenous popular culture and popular media in Korea (Shim, 2006). Of many schools including 'international cultural flow' and 'importing countries', one argues that the unprecedent recent success of Korean popular culture in the global market, so-called *Hallyu* or *Korean Wave*, is largely due to the fact that Korean government and relevant industries made significant efforts to develop cultural and creative industries (Kim & Nam, 2016; Kwon & Kim, 2014). This central government-led direction is based on an awareness of the importance of culture and its industrial development as well as Korea's resistance to a serious Western challenge and the Japanese occupation history to political sovereignty and cultural integrity.

DOI: 10.4324/9781003320586-21

In comparison, during the last three decades, Japan has traditionally been a leading exporter of popular culture in Asia and beyond and boasted its influence in shaping the region's cultural markets, mediascapes and creating new images of Japan (Otmazgin, 2008). Despite the continuity in the old animosities in many Asian countries including Korea, the Japanese popular culture products had a constrictive circulation and receptive audiences in East Asia in particular (Iwabuchi, 2002). Since the collapse of so-called bubble economy in the early 1990s, it was largely perceived that Japan lost its international presence in many business sectors, but the nation's cutting-edge pop culture rather enjoyed an unprecedented boom during that same time (Matsui, 2014). It ignited the national policy competition by central ministries to promote Japanese popular culture in the name of 'soft power' focusing on *Cool Japan* policies (Nye, 1990).

The two closed but strained neighbouring countries are now developing an integrated national strategy. That combines tourism promotion and marketing and national branding associated with one's popular culture into one axis. Japanese government has implemented the *Cool Japan* campaign as an integrated strategy since the 2000s to disseminate popular culture information, support the export of popular culture abroad, and promote popular culture-induced inbound tourism. Similarly, responding to the *Hallyu* phenomenon, the Korean government has also initiated a national policy of supporting and nurturing them. Through such policies, both countries have attempted to increase the socio-cultural influence of their pop culture by extending it to tourism industry.

In this chapter, the characteristics of popular culture and tourism in the two countries from a policy perspective are systemically compared and analysed using the secondary data, such as policy documents and white papers as well as newspapers headlines, which determines the commonalities and/or differences and how they affect each other. These queries are examined using comparative and cross-case approaches. The chapter reveals the dynamic characteristics of policies that result in a combination of popular culture and tourism.

'Cool Japan' and Tourism Policy

Integrating Industrial, Diplomatic and Tourism Policies through Popular Culture

Cool Japan refers to the national and international "promotion of products, services, and human resources that embody Japanese (popular) cultural appeal overseas, including Japanese anime, manga, games, fashion, and food" (Mihara, 2014, p. 3). The *Cool Japan* policy began as a government strategy that recognised the continuity in global popularity and market penetration of Japanese popular culture products. Drawing on the British Cool Britannia policy of the 1990s, Douglas McGray published "Japan's Gross National Cool in Foreign Policy" in 2002, which emphasised the global influence of Japanese culture including popular music, anime, manga, food, and consumer electronics (McGray, 2002). The article was translated into Japanese and became a catalyst for Japan to recognise the potential of popular culture and pursue *Cool Japan* campaign as an integrated national policy (Seaton et al., 2017). The Japanese government identified cultural exports and the enhancement of Japan's soft power through the popular culture and diplomacy (Nye, 1990) as vital goals of the *Cool Japan* policy (Valaskivi, 2013). The *Cool Japan* policy consists of a three-stage strategy: (1) raising awareness of Japanese culture abroad and creating a boom, (2) consuming Japanese culture locally abroad, and (3) coming to Japan to consume.

Hallyu is also one of the factors that prompted Japan to initiate its *Cool Japan* policy at the government level. Japan perceives its neighbour, Korea, as an immediate rival whose pop culture promotion policies are one step ahead of Japan's and promotes *Cool Japan* as a benchmarking strategy. This is reflected in Japanese government reports on Korea's cultural policies and strategies (JETRO, 2011; Cabinet Secretariat, 2013; Cabinet Office, 2020; Cabinet Secretariat, 2013). The Japanese government analysed the Korea's pop culture promotion policies (JETRO, 2011), noted that related government budgets and markets are declining in Japan but growing in Korea (Cabinet Secretariat, 2013), and used this as a basis for formulating the *Cool Japan* strategy (2019). However, even after the Cabinet Secretariat's report, Japan's *Cool Japan* budget has declined each year, as shown in Figure 16.1. The surge in 2020 from the declining budget was attributed to the 2020 Tokyo Olympics (which was postponed by a year due to the pandemic to 2021), which saw an overall increase in public relations budgeting.

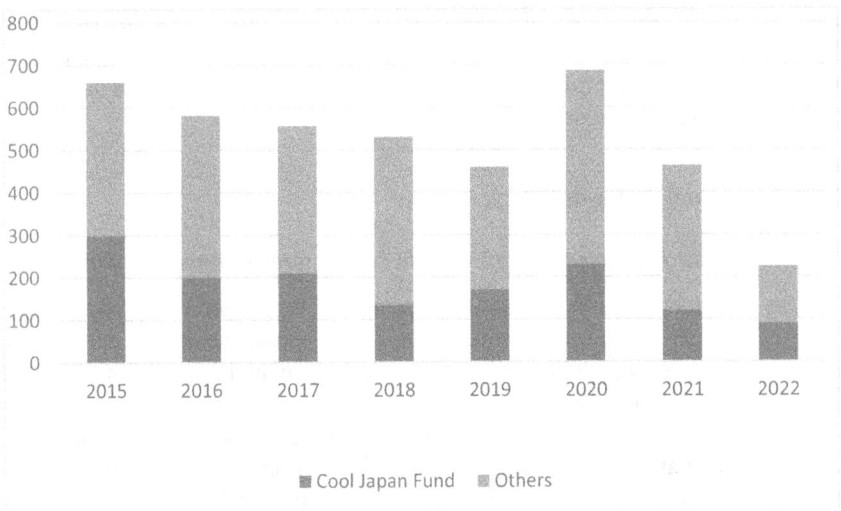

Figure 16.1 Japan's budget for Cool Japan (Cabinet Office, Government of Japan, n.d., USD million).

Source: www.cao.go.jp/cool_japan/platform/budget/budget.html

The largest part of the single budget for the *Cool Japan* policy is its contribution to the Overseas Demand Development Support Organisation, the so-called Cool Japan Fund, which provides insights into the core of this policy. The Cool Japan Fund is a special company established under the Overseas Demand Development Support Organisation Act of 2013 by the Ministry of Economy, Trade and Industry (METI), with the participation of the private sector, including major advertising companies Dentsu and Hakuhodo. As of 2022, the cumulative investment was approximately $10.66 billion from the government and $1.07 billion from 24 private companies. It is an intermediary institution created in a system where it is difficult for the government to support private companies with its budget directly. It aims to invest in media contents, food culture, fashion, and inbound tourism to commercialise Japan's charm, or *Cool Japan*, and link it to overseas demand. The upper limit for each investment proposal is set at 10 billion Japanese yen, and the duration is set at five or ten years, depending on the proposal. Inbound tourism support includes two investments in marketing companies and one each in venture funds, DMOs, accommodation, and online booking sites.

Cool Japan Policy and Tourism Promotion

While the early *Cool Japan* policy had no tourism promotion as one of the major objectives, it gradually merged with the government's tourism revitalisation policy, which began at the same time. In 2003, the Japanese government launched its government-led tourism revitalisation policy called 'tourism-oriented nation' as a new strategy to create economic growth and regional promotion in an era of declining birth rate and a significant rise in an ageing population (JTA, n.d.). The focus is on wealth distribution through consumer economy stimulation by promoting domestic travel and foreign exchange earnings through increased foreign travellers. Both *Cool Japan* and tourist arrivals share that common goal of economic revitalisation, which has facilitated their combination. In particular, the policy for a tourism-oriented nation proposes the promotion of new tourism to develop new areas of tourism outside of existing metropolitan centres and major tourism destinations and attractions. Exemplified are eco-tourism and green tourism as well as elements of *Cool Japan* such as fashion, food, film, and anime (MLITT, 2012).

In 2004, Japan, a tourism-oriented nation, enacted a law on the creation, protection, and use of content for the internal and external promotion of Japanese popular culture. The MLITT, the METI, and the ACA (2005) jointly published a report titled "Investigation into Methods of Regional Promotion by Creating and Utilising Videos and Other Contents". One indication of Japan's policy of promoting popular culture and tourism together was the boom in *Korean Wave* tourism in Japan. In 2003, the popularity of the Korean broadcast television drama, *Winter Sonata* (2002), led to a sudden boom in travel to Korea among middle-aged women, the main target group (Kim et al., 2009). In the seven months from April to July 2004, the boom exponentially increased the number of Japanese inbound tourists to Korea to 187,000 and the country's tourism revenue by approximately $300 million (Kadokura, 2005).

In addition, there was an estimated economic impact of 36.4 billion Japanese yen in increased domestic consumption, including sales of related products and activities of the main actors of *Winter Sonata* in Japan. The Japanese government had considered this phenomenon and highlighted the case of winter holidays in the aforementioned report. Since then, with the implementation of the Basic Law for the Promotion of Tourism-oriented Nation in 2007 and the establishment of the Japan Tourism Agency in 2008, the Japanese government laid its first groundwork for tourism promotion policies in the tourism sector in conjunction with its creative industries as mentioned earlier.

With the organisational structure in place, the combination of *Cool Japan* and tourism promotion began. In 2013, the Japan Tourism Agency (JTA), Japan National Tourism Agency (JNTO), the METI, and Japan External Trade Organisation (JETRO) announced a joint action plan to increase the number of foreign visitors to Japan. This is the first time that the Japanese government linked *Cool Japan* and *Visit Japan*. The aim is to strengthen coordination between the sectors (i.e., tourism and creative industries) to realise the three-stage strategy of *Cool Japan* promotion described above, which was mentioned earlier (JTA, JNTO, METI, JETRO, 2013). Specific initiatives of *Cool Japan* and *Visit Japan* include inviting overseas press and buyers, conducting joint overseas promotions, linking previously separate businesses, and promoting Japanese brands with *Cool Japan* in mind. At the time of the announcement of this joint plan, the focus was on joint participation in events by companies using the Japan Contents Localisation and Promotion Support Grant (J-LOP) provided by the METI to private companies. With the establishment of the Cool Japan Fund at the end of 2013, the cooperation was extended to projects supported by the Cool Japan Fund. In 2014, the JTA and Cool Japan Fund signed a memorandum of understanding for promotion of inbound tourism using video content.

Table 16.1 summarises the Japanese government's diplomatic and economic promotion policies using its popular culture in the chronological order.

Achievements and Challenges of the Cool Japan *Tourism Policy*

Evaluations of the overall *Cool Japan* policy can be divided into internal and external assessments. In 2018, the MIC published its policy evaluation and recommendations for the promotion of *Cool Japan* (MIC, 2018). Focusing on the tourism sector, the evaluation was as follows. The Japan Weekend, a tourism promotion event jointly organised by JNTO and JETRO in Thailand and Malaysia from 2014 to 2016, increased the number of visitors to Japan from these countries, supported the creation of content to promote tourism in local areas, and promoted the Visit Japan Local Linkage Project. However, the government's self-assessment tended to defend the *Cool Japan* policy. For example, the report states that the effect of the Japan Weekend was an increase in inbound tourism to Japan from Thailand and Malaysia, where the events were conducted. However, the data on

Table 16.1 Japanese government's diplomatic and economic promotion policies using popular culture

Year	Author(s) or organisation	Title
2002	Douglas McGray	Japan's Gross National Cool
2003	Japanese Government	Declaration of Tourism-oriented Nation
2003		Popularity of the Korean Drama Winter Sonata (NHK BS2) and the increasing number of Japanese tourists to Korea
2004	Government (Law)	Act on the Promotion of the Creation, Protection, and Exploitation of Content
2005	Ministry of Land, Infrastructure, Transport, and Tourism, Ministry of Economy, Trade and Industry, Cultural Affairs Agency	Report on "Investigation into Methods of Regional Promotion by Creating and Utilising Videos and Other Contents
2006	Ministry of Foreign Affairs	Report on the Use of Popular Culture in Cultural Diplomacy
2007	Government (Law)	Implementation of the Law for the Tourism-oriented Nation
2007	Ministry of Economy, Trade, and Industry	Research Report on Regional Revitalisation by Expanding International Tourism Exchanges Using Japanese Anime
2008		Ministry of Land, Infrastructure, and Transport establishment of Japan Tourism Agency
2010	Japan Tourism Agency	Japan Anime Tourism Guide
2010	Ministry of Economy, Trade, and Industry	Setting up the Cool Japan Office (later Creative Industries Division)
2011	JNTO	Japan Anime Map
2012	Ministry of Economy, Trade, and Industry	Cool Japan strategy with Japanese charm (White Paper)
2013	Government (Law)	Enacted the Overseas Demand Development Support Organisation Act and established the Cool Japan Fund Inc.
2013	Japan Tourism Agency, JNTO, Ministry of Economy, Trade, and Industry, JETRO	Joint Action Plan to Increase Foreign Visitors to Japan
2014	JNTO and Cool Japan Fund	Cool Japan's collaboration with Visit Japan
2017	Cabinet Office	Local revitalisation using Cool Japan's promotion/contents

Note: Authors' additions based on Yamamura (2014).

which this is based is not event-induced, but statistics for the country as a whole, which is not an accurate *Cool Japan* effect, given that the number of overseas tourists tends to increase overall.

However, the position of the Ministry of Finance (MOF) was different. At a meeting of the Finance and Investment Subcommittee of the Fiscal System Review Committee on 20 June 2022, the accumulated deficit of the Cool Japan Fund, which is central to the *Cool Japan* policy, was raised as a major issue (MOF, 2022). In response, the data presented by the METI and the Cool Japan Fund at the next meeting, revealed that the Cool Japan Fund had a cumulative deficit of 30.9 billion Japanese yen as of fiscal 2021. While the fund announced that it will expand its investment targets to include the materials industry, the Ministry of Finance is considering merging the organisations.

In contrast, external evaluations, especially by the public media and independent researchers, are generally critical and sceptical of the success of *Cool Japan* policy. Although the expansion of the Cool Japan Fund's investment targets to include the materials industry can be read as an example, the main problem is that too many targets and segmentations are included in one policy. Consequently, the direction of policy and the setting of priorities became broader and more ambiguous than the original scope and scale of *Cool Japan* proposed by McGray in 2002.

Similarly, in the case of convergence with tourism, support, and cooperation in unrelated sectors such as textile and furniture company have led to difficulties in defining what actually constitutes *Cool Japan*. Commentator Tsunehira Furuya went so far as to call this "the doomed Cool Japan Fund" (Furuya, 2022). Furuya pointed to both the tax-driven mode of operation and the lack of knowledge of those areas in charge of the Cool Japan Fund. Furuya further pointed out that although the fund was structured as a corporation, it was funded by taxes. Thus, there is no accountability, and with that comes arrogance and a lack of basic knowledge of the popular culture that decision-makers use as a basis for identifying investment targets.

Korean Wave and Tourism Policy

From the Private Sector to Government Policy

The *Korean Wave*, the so-called *Hallyu* phenomenon, is defined as the purchase of goods and services related to Korean popular culture, fine arts, and tourism (International Cultural Industry Exchange Foundation, 2008). It is generally accepted that it began in the mid-1990s when Korean dramas and songs became popular in China (Kim et al., 2009; Ministry of Culture, Sports, and Tourism, 2013). The term *Hallyu* is a neologism coined by the *Beijing Youth Daily* in November 1999 to describe the popularity of Korean culture in China, which has spread to Korea and beyond (MCST, 2013).

The Korean government evaluates and categorises the *Korean Wave* as 1.0 to 3.0, depending on the era and core genre: Hallyu 1.0 (the late 1990s to the mid-2000s), centred on the popularity of TV dramas in Asia; Hallyu 2.0 (mid-2000s to the early 2010s), centred on the popularity of K-pop idols in parts of Asia, Europe, and the Americas, and Hallyu 3.0 (since the early 2010s), characterised by a global rise in popular culture, arts, and traditional culture (MCST, 2013). However, Hallyu 3.0 is an ideal goal presented by the government, and thus, it is necessary to examine whether Korean culture is being consumed worldwide. The Japanese Trade Organisation (2011) analysed the goals of Korea's cultural promotion policy, including the spread of the *Korean Wave*, the export of cultural industries, the protection of infant industries, and the enhancement of cultural diplomacy capabilities through soft power.

The Korean government's policy of promoting cultural industries began in the 1990s. In 1994, President Kim Young-sam established the Cultural Industries Bureau under the MCST to promote cultural industries in earnest. This marked a shift in cultural policy from a regulatory to an economic emphasis. Later, in 1998, the Kim Dae-jung government emphasised the twenty-first century as a policy goal and seriously promoting the development of cultural industries. The *Korean Wave* coincided with this period and began with the popularity of Korean TV dramas and K-pop idols in China and Southeast Asia. The *Korean Wave* then spread to Japan and the rest of Asia, as well as to South America and the Middle East, and developed in line with the government's cultural industry policy (Kim et al., 2009; Kim & Nam, 2016).

Specifically, the proportion of the MCST's budget allocated to the content industry (i.e., cultural and creative industries) increased from 2.2% in 1998 to 21.6% in 2021, reflecting the government's commitment to promoting the cultural industry (Figure 16.2). As of 2021, the breakdown of the 750 billion Korean won budget for the cultural industry is as follows: 331.2 billion Korean won for the promotion of the video content, music, and game industry and 275.9 billion won for the creation of a foundation for international cooperation and export of cultural content and the activation of investment (MCST, 2021). Rather than being used to

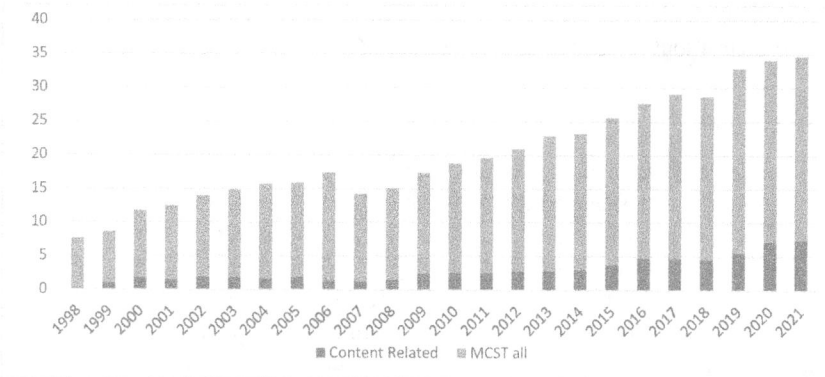

Figure 16.2 South Korea's content-related budget (MCST, 2021, in billions of USD).

directly promote *Hallyu*, the government budget was spent to support *Hallyu* and promote Hallyu-related industries.

The Korean government's approach to promoting *Hallyu* emphasises that it is essentially a private sector-led phenomenon. This is due to the government's adherence to the principle of "support but not interfere" as the core operating principle of the cultural industry support system and its focus on infrastructure support as a burden on the use of public funds for private events (MCST, 2007). Nevertheless, the government has indirectly supported *Hallyu* promotion policies in the form of quasi-governmental organisations and dedicated private foundations established by law and funded by the government. These include the Korea Foundation for International Cultural Exchange (KOFICE) (a foundation for international cultural exchange under the Ministry of Culture, Sports, and Tourism), the Korea Creative Content Agency (KOCCA) (a statutory corporation under the Ministry of Culture, Sports, and Tourism), the Korea Foundation (under the Ministry of Foreign Affairs), the Sejong Hakdang Foundation (a public institution for Korean language dissemination), and the Korea Arts Council (a public institution).

However, since the mid-2010s, there has been a growing trend to implement the dissemination and promotion of the *Korean Wave* as part of the central government policy. This is evident in the changes in the organisational form of the aforementioned agency, the KOFICE (see Table 16.2). With the implementation of the International Cultural Exchange Promotion Act in 2017, the organisation was renamed the Korea International Cultural Exchange Promotion Agency and became a buffer organisation to implement the policy of spreading the *Korean Wave* in the form of a private organisation. Furthermore, according to the Second Comprehensive Plan for the Promotion of International Cultural Exchange, which will be implemented in the second half of 2023, consideration is being given to transforming the organisation from a private foundation into a full-fledged government agency such as the KOCCA (MCST, 2022). Meanwhile, in 2022, the MCST established the Hallyu Support and Cooperation Division to promote *Hallyu* more directly and effectively, which is expected to increase the role of government organisations dedicated to *Hallyu*. Table 16.2 presents the chronological changes in the organisations in charge of *Hallyu* since 2002.

Promoting Hallyu Tourism

Hallyu tourism is defined as outbound travel to Korea in which a preference for Korean culture has influenced the decision to travel. This is divided into consultative *Hallyu* tourism, driven by a preference for popular culture, and broader *Hallyu* tourism, influenced by traditional culture (KTO, 2019a). Alternatively, activities such as shopping and fashion and *Korean Wave*-inspired tourism are sometimes referred to as the "new Korean Wave" to distinguish them from the direct consumption of Korean popular cultural contents that is the subject of the *Korean*

Table 16.2 Changes in organisations in charge of Hallyu

Year	Author(s) or organisation	Role and responsibility
2002	Asian Cultural Exchange Council	Organisation formed around popular music
2003	Asian Cultural Exchange Council -> Asia Culture and Industry Exchange Foundation (name change)	Promoted the Hallyu business in various ways. Participated in broadcasting, film, game, and animation-related organisations.
2006	Asian Cultural and Industrial Exchange Foundation – International Cultural and Industrial Exchange Foundation (name change)	Expanded its business scope from Asia to the world and expanded its content to include pure culture.
2007	Ministry of Culture, Sports and Tourism	Study on establishing a system to promote international cultural exchanges
2009	National Assembly	Passed the Basic Act for the Promotion of Cultural Industries
2009	International Cultural and Industrial Exchange Foundation – Korea Culture & Industry Exchange Foundation (name change)	
2009	Korea Creative Content Agency	Statutory corporation
2012	National Assembly	Proposed legislation on the promotion of international cultural exchange
2013	National Tasks	Promoting cultural diversity and expanding cultural exchange cooperation
2016	National Assembly	Bill withdrawn due to the end of the session and reintroduced
2017	National Assembly	Enforced the International Cultural Exchange Promotion Act and established the legal basis for establishing a dedicated agency
2018–2022	1st Comprehensive Plan for the Promotion of International Cultural Exchange	
2018	Korea Culture and Industry Exchange Foundation – Korea International Cultural Exchange Agency (name change)	A dedicated organisation based on the above law, in the form of a private foundation. Supported by the government budget. Budget for 2022: KRW 2.632 billion
2023–2027	Second Comprehensive Plan for the Promotion of International Cultural Exchange	Promote the conversion of the Korea Council for International Cultural Exchange into a statutory corporation

Wave (KTO, 2003). Since its inception, the *Korean Wave* has been associated with tourist behaviour. Examples include the popularity of the dramas *Winter Sonata* (2002) and *Daejanggeum* (2003), which led to visits to locations and filming sites and K-pop concerts attended by overseas fans (Kim et al., 2009).

Although the Korean government recognises tourism as one of the economic spill overs of the *Korean Wave*, it has not developed a comprehensive and systematic strategic plan for tourism policy (Lee & Kim 2012). Tourism policy-related materials include research reports published by the Korea Tourism Organisation (KTO, 2001; KTO, 2003; KTO, 2005; KTO, 2012; KTO, 2014; KTO, 2019b; KTO, 2022), providing insight into the policy, current analysis, and outlook for *Hallyu* tourism in the previous period, the *Hallyu* in Tourism section of the Hallyu White Paper, which KOFICE has published annually since 2013, taking over from the Hallyu White Paper published by the MCST in 2012, and the Economic Impact of Hallyu study, which KOFICE has published annually since 2014.

The Korea Tourism Organisation (KTO) is a commissioned research body that analyses the current situation of *Hallyu* tourism based on TV drama filming locations, and suggests ways to revitalise it (KTO, 2005), marketing and promotional measures to expand *Hallyu* tourism linked to K-pop (KTO, 2012), the *Hallyu* tourism market and marketing strategies (KTO, 2014; KTO, 2019b), and the need for industry, regional, and administrative cooperation to promote *Hallyu* tourism (KTO, 2022). *Hallyu* tourism policy is rarely mentioned, and as noted above, it is suggested that there is no systematic system or strategy for *Hallyu* tourism policy (Kim & Nam, 2016).

Annually, KOFICE's Hallyu White Paper summarises the most important issues related to *Hallyu* tourism. From 2013 to 2016, the white paper was written by KOFICE without an author; however, since 2017, it has been written by designated experts from the Korea Visitors Bureau or the KTO in the form of columns, and the content varies in the same format each year since then, which is common for white papers. While inbound tourist numbers are sometimes cited (KOFICE, 2014), they are mainly an introduction to issues or new business items related to *Hallyu* tourism, which has been popular since 2014, as well as to factors affecting the number of travellers, such as China's ban on travel to Korea (KOFICE, 2017). In particular, the need to secure future tourists through online platforms during the COVID-19 pandemic in 2020 and beyond is raised. As with the KTO reports, there is no mention of tourism policies.

Furthermore, *Research on the Economic Impact of the Korean Wave*, published by KOFICE since 2014 (titled Research on the Impact of the Korean Wave, depending on the year), is the only annual report that is comprehensive and thus makes policy recommendations related to the *Korean Wave*, including tourism. The reports define the economic impact of the *Korean Wave* as exports of cultural content products, consumer goods, and tourism exports (KOFICE, 2019, p. 59), whereas tourism exports are defined as Korean domestic income from foreign tourists (KOFICE, 2019, p. 42). After 2020, it is assumed that tourism will not be addressed due to the impact of the COVID-19 pandemic. This report highlights the importance of improving infrastructure. It points out that current *Hallyu* tourism policies focus on demand development, such as marketing; however, the actual flow of visits to Korea is in the form of content consumption, that is, liking and intention to visit resulting in actual visit, suggesting that a demand-fulfilment approach, such as improving infrastructure and providing information, is needed rather than demand development (KOFICE, 2017, p. 91). The 2018 and 2019 reports point to an increase in outbound tourism through the *Korean Wave* and the higher value-added and employment-induced coefficients of tourism products than those of other consumer goods, implying that policy support is required to stimulate product development (KOFICE, 2018, 2019).

Achievements and Challenges of Hallyu Tourism Policy

The global consumption of the *Korean Wave* influences tourism decisions and plays a critical role in generating new tourism demand (KTO, 2019a), as documented (Kim et al., 2009; Kin & Nam, 2016). The question is, to what extent do tourism policies influence these decisions and contribute to repeat visits? It can be argued that there has been no sound system to support *Hallyu* tourism other than ad-hoc marketing and one-off events (Kim & Nam, 2016). This is because, as in other sectors, the government's stance on politicising the *Korean Wave* has been passive and indirect, as described above. This passivity can be attributed to the influence of the cultural industry development principle created in the 1990s and the negative view of it becoming a national policy. This is supported by the negative views and voices of concern that emerged as soon as the *Korean Wave* became a policy after 2020 (Kim, 2022). Nevertheless, the government has begun to promote *Hallyu* policy by implementing the Comprehensive Plan for the Promotion of International Cultural Exchange and establishing a department dedicated to the promotion of *Hallyu*. In terms of tourism policy, as mentioned above, the mid- to long-term strategy for *Hallyu* tourism marketing is scheduled to be implemented in 2023, and it is expected that the marketing scope and scale of *Hallyu* tourism, organisation of existing projects, and coordination of related organisations and marketing strategies will be newly created.

Conclusion: Comparison of *Cool Japan* and *Hallyu* Tourism Policies

This chapter examined the *Cool Japan* and *Hallyu* tourism policies implemented by Japan and Korea using a comparative and cross-case approach. These policies as integrated national strategies, which were implemented around the same time, have similarities as both aimed to promote the global diffusion of national popular culture and the expansion of the cultural and creative industries through it. In terms of mutual influences, the design of Japan's *Cool Japan* policy and promotion of pop culture tourism, locally known as contents tourism (Seaton et al., 2017), were stimulated by the Korean boom driven by the popularity of Korean culture in Japan, and the same phenomenon has led to the active promotion of *Hallyu* tourism in Korea. However, while Japan's budget support for *Cool Japan* has stagnated, Korea's budget support for *Hallyu* has increased, which demonstrates two dissimilar pathways over the years since its inception.

This is the first study that undertook a systematic, direct comparison of the tourism policies of *Cool Japan* and *Hallyu*, responding to the lacuna of the literature on the government or policy from a critical comparative cultural policy study approach (Kim & Reijnders, 2018). It revealed some differences in the scope and scale of the pop culture targeted policy and the methods of support, leading to different outcomes. In the case of *Cool Japan*, it is evident that there is a tendency to expand its scopes and scales beyond the popular culture and creative industries such as agricultural products, small and medium-sized enterprises, and even the materials industry, resulting in the blurring and confusion of the policy's directions regarding *Cool Japan*. Even when discussing *Cool Japan* tourism, it became difficult to understand what the relevant policy was for, as it was not aimed at not only focusing on popular culture and media but also rural tourism. In addition, the indirect method of support is common; however, this can point to the problem and/or issue of how to support private companies by creating a fund. This is because despite being capitalist companies that have broken away from the bureaucratic system, it has become a system in which the state budget is involved, leading to the lax management and the lack of accountability. As noted above, the fact that none of the Cool Japan Fund's inbound tourism investments was related to popular culture, reflects both the ambiguity of its direction and continued problems with the fund's format and structure.

In the case of *Hallyu* however, the basic scope and direction of the fund are pop culture-focused. It is yet to mention that the government's ultimate goal is to broadly introduce the Korean lifestyle to the rest of the world, similar to Japan, and this is something that Korea should consider by referring to the case of *Cool Japan* as a good lesson. The *Korean Wave* policy is primarily focused on supporting the cultural industry, and tourism is regarded as a by-product; therefore, there is not much of a policy approach other than tourism promotion and marketing. Above all, *Hallyu* has not emerged as a national policy per se, for it has received indirect support with minimal government intervention. However, after 2020, the Korean government has been shifting forward more aggressively implementing the *Hallyu* as a government policy and national strategy in earnest; therefore, it is expected that future *Hallyu* tourism policies will be more proactive and integrating in terms of marketing, infrastructure construction, and cross-sector linkages.

In summary, this chapter identified the directions and outcomes of popular culture and tourism promotion policies in Korea and Japan, two leading countries of Asian popular culture. The chapter provides implications for the extent to which government intervention in cultural policy is effective (*Hallyu*), how the scope and degree of such policy should be established (*Cool Japan*), and how cultural policy can be linked to tourism as an integrated national strategy. Thus, it contributes to our better understanding of the connection and interplay between culture and tourism policies for researchers in the field, as well as for policymakers and those involved in tourism and cultural and/or creative industry businesses.

References

Cabinet Office. (2020). *First meeting of the Create Japan WG Secretariat briefing material.* Cabinet Office.

Cabinet Secretariat. (2013). *Discussion paper on intellectual property strategy (related to strengthening content) (draft).* Cabinet Secretariat.

Furuya, T. (2022, 11 December). Reflections on the Failure of the Cool Japan Organisation. The arrogance of "officials" who know nothing about Japanese anime and manga, *Newsweek Japanese Edition.* www.newsweekjapan.jp/furuya/2022/12/post-32_1.php.

International Cultural Industry Exchange Foundation. (2008). *A comprehensive research study for the sustainable development of the Korean Wave.* International Cultural Industry Exchange Foundation.

Iwabuchi, K. (ed.) (2002). *Recentering globalization: popular culture and Japanese transnationalism.* Duke University Press.

JETRO. (2011). *Analysis of Korea's content promotion policies and their direct and indirect effects in foreign markets.* JETRO.

JTA. (n.d.). *About Japan Tourism Agency.* JTA. www.mlit.go.jp/kankocho/kankorikkoku/index.html

JTA, JNTO, METI, and JETRO. (2013). *Joint Action Plan for increasing the number of foreign visitors to the country.* JTA, JNTO, METI, and JETRO.

Kadokura, T. (2005). The impact of Japan's "Winter Sonata" boom on the economy. *Daiichi Institute of Life Economy Report* 8(11), 12–14.

Kim, E. (2000). Korea loosens ban on Japanese pop culture. *Billboard,* 112(40), 68–69.

Kim, Y. (2022, May 7). Korean Wave, a policy product or an "unplanned success"? *Hangyore Sinmun.* www.hani.co.kr/arti/economy/economy_general/1041887.html.

Kim, S., Long, P., & Robinson, M. (2009). Small screen, big tourism: The role of popular Korean television dramas in South Korean tourism. *Tourism Geographies,* 11(3), 308–333.

Kim, S., & Nam, C. (2016). *Hallyu* revisited: Challenges and opportunities for the South Korean tourism. *Asia Pacific Journal of Tourism Research,* 21(5), 524–540,

Kim, S., & Reijnders, S. (Eds.). (2018). *Film tourism in Asia: Evolution, transformation and trajectory.* Springer.

KOFICE. (2014). *Korean Wave White Paper.* KOFICE.

KOFICE. (2017). *Korean Wave White Paper.* KOFICE.

KOFICE. (2018). *Study on the Economic Impact of the Korean Wave 2018.* KOFICE.

KOFICE. (2019). *Study on the Economic Impact of the Korean Wave 2019.* KOFICE.

KTO. (2001). *Hallyu Tourism Marketing Strategy Report.* KTO.

KTO. (2003). *Marketing the New Wave of Tourism.* KTO.

KTO. (2005). *How to promote Daejanggeum tourism products based on comparative analysis of drama winter season and Daejanggeum.* KTO.

KTO. (2012). *A Study on Preventing the Prolongation of Hallyu Tourism.* KTO.

KTO. (2014). *Research on the Hallyu Tourism Market.* KTO.

KTO. (2019a). *Hallyu Tourism Market Research Study.* KTO.

KTO. (2019b). *Research on the Hallyu Tourism Market.* KTO.

KTO. (2022). *Study on how to activate Hallyu and Hallyu Tourism Cooperation.* KTO.

Kwon, S. H., & Kim, J. (2014). The cultural industry policies of the Korean government and the Korean Wave. *International Journal of Cultural Policy,* 20(4), 422–439.

Lee, W., & Kim, S. J. (2012). Promotion of inbound tourism utilizing the K-pop Neo Korean wave. *The Journal of Tourism Sciences,* 36(2), 31–56.

Matsui, T. (2014). Nation branding through stigmatized popular culture: The "Cool Japan" craze among central ministries in Japan. *Hitotsubashi Journal of Commerce and Management*, 48, 81–97.

Mcgray, D. (2002). Japan's gross national cool, *Foreign Policy, 40,* 44–54.

MCST. (2007). *Researching ways to establish a system to promote international cultural exchange*. MCST.

MCST. (2013). *Korean Wave white paper*. MCST.

MCST. (2021). *Content Industry white paper 2021*. MCST.MCST. (2022). *Action Plan for the Promotion of International Cultural Exchange 2022*. MCST.

MIC. (2018). *Policy Assessment Paper on the Promotion of Cool Japan*. MIC.

Mihara, R. (2014). *Why is Cool Japan so hated?* Chuo-Koron-Shinsha.

MLITT. (2012). *Basic Plan for Tourism-oriented Nation Promotion*. MLITT. www.mlit.go.jp/common/001299664.pdf.

MLITT, the METI, & the ACA. (2005). *Methods of regional promotion by creating and utilising videos and other contents*. MLITT, the METI, and the ACA.

MOF. (2022). *Proceedings of the Fiscal Investment and Loan Subcommittee of the Council on Fiscal Institutions*. MOF.

Nye, J., (1990). Soft power, *Foreign Policy*, 80, 153–171.

Otmazgin, N. K. (2008). Contesting soft power: Japanese popular culture in East and Southeast Asia. *International Relations of the Asia-Pacific*, 8, 73–101.

Park, S. (2005). The impact of media use and cultural exposure on the mutual perception of Koreans and Japanese. *Asian Journal of Communication*, 15(2), 173–187.

Seaton, P., Takayoshi, Y., Sugawa-Shimada, A., & Jang, K. (2017). *Contents tourism in Japan*. Cambria.

Shim, D. (2006). Hybridity and the rise of Korean popular culture in Asia. *Media, Culture & Society*, 28(1), 25–44.

Valaskivi, K. (2013). A brand new future? Cool Japan and the social imaginary of the branded nation. *Japan Forum*, 25(4), 485–504.

17 Destination Nollywood

The Connections between Film and Tourism in Nigeria, 2012–2022

Emiel Martens and Edmund Onwuliri

Introduction: Setting the Scene

In November 2022, the first Global Conference on 'Linking Tourism, Culture and Creative Industries' was held in Lagos, Nigeria. The conference was organized by the United Nations World Tourism Organization (UNWTO) with the aim to explore the connections between tourism and culture, and to promote partnerships between the tourism and creative industries in the country and across the African continent. As Lai Mohammed, Nigeria's Minister of Information and Culture, put it in his address, "Today, more than ever, tourism and the creative industry, due to their economic viability, are in the global spotlight and have their place at the forefront of the national and international development agenda" (*The Guardian Nigeria*, 19 November 2022). Inspired by the conference, the Nigerian government announced the start of a new collaboration with the World Tourism Organization to "strengthen Nigeria's tourism sector" (*Business Insider Africa*, 11 February 2023). According to Mohammed, the UNWTO had decided to team up with the Nigerian government due to the significant strides the country had made in the creative industries in recent years. Apart from plans to build a tourism academy, the minister stated to have "plans to work with Nollywood," Nigeria's prolific film industry, to "positively change Africa's image" (*BIA*, 11 February 2023). This reflected the remarks made on the conference by UNWTO's Secretary General, Zurab Pololikashvili, who stated that the partnership between Nollywood and UNWTO would serve "the promotion of African culture" and "attract tourists to the continent" (*AVI News*, 17 November 2022).

Over the past three decades, Nigeria's film industry has become a "global powerhouse" releasing hundreds of films each month (*The Circular*, 15 March 2022). Although Nigerian cinema dates back to the 1960s, when the first post-independence era films were made by pioneering Nigerian filmmakers such as Hubert Ogunde, Eddie Ugbomah and Ola Balogun, Nollywood, a portmanteau of Nigeria and Hollywood as well as Bollywood that was coined by *New York Times* journalist Norimitsu Onishi in 2002, came to mark "the plenitude of film-making happening in Lagos" from the early 1990s onwards (*TC*, 15 March 2022). Following the success of the 1992 straight-to-VHS film *Living in Bondage*, the Nigerian film industry experienced a popular "home video boom" (Atilola &

DOI: 10.4324/9781003320586-22

Olayiwola, 2012) that gained a huge following in Nigeria, across Africa and among the global African diaspora. In fact, with "an estimated worth of $6.4 billion as of 2021" (*TC*, 15 March 2022), Nollywood is nowadays considered the world's second largest film industry, and even the largest in terms of output and popularity. Producing about 2,500 films annually and reaching an estimated audience of over 200 million people worldwide, Nollywood has seemingly not only surpassed Hollywood but also Bollywood regarding production and consumption size.

Due to its commercial success and cultural impact, from the early 2010 Nollywood has often been heralded as a "rising star in Africa's tourism industry" (Abiola, 2016, p. 209). For example, in the *Global Trends Report 2012*, the World Travel Market predicted a gross domestic product (GDP) growth in Africa as a result of "domestic and regional tourism to Nollywood locations" (Abiola, 2016, p. 209). More generally, it was anticipated that the popularity of Nollywood would become "a major growth driver in the leisure sector," attracting both "film fans and business travelers fascinated by the growing … importance of the Nigerian film industry" (Abiola, 2016, p. 210). In 2014, Nigerian writer Ekenyerengozi Michael Chima similarly stated that Nollywood had the potential to turn Nigerian tourism – which is currently a relatively small industry facing infrastructure and security issues[1] – into "a billion dollar industry" (*ModernGhana.com*, 2 August 2014).

However, despite much talk about the *potential* impact of Nollywood on Nigerian and African tourism in recent years, relatively little is known about Nollywood's *actual* impact on film tourism and the existing connections between Nigeria's film and tourism industries more generally. Already over a decade ago, Heitmann (2010) observed that it "is yet to be researched … to what extent film tourism in connection to Nollywood in Africa has created" (p. 31) – and to date this is still largely the case. This chapter, then, seeks to discuss the reported connections between Nollywood and tourism and to review the efforts and opportunities related to Nollywood film tourism as suggested in Nigeria's public domain. Based on a content analysis of over 50 newspaper articles and other online available materials, such as websites, blogs and forums, this chapter explores the practices and discourses surrounding Nollywood connections with tourism in Nigeria and across Africa over the past ten years.

Nollywood Studies: Themes and Trends in Studying Nigeria's Film Industry

Due to its unique development and popularity, Nollywood has garnered significant scholarly interest since its inception in the 1990s.[2] From the early 2010s, the phrase 'Nollywood Studies' started to be used to refer to the growing "body of work and academic field" examining the Nigerian film industry (Haynes, 2012, p. 3). As "a dedicated transdisciplinary research niche" (Oguamanam, 2020, p. 518), Nollywood Studies consists of a "conglomerate [of] research practices" (Makhubu, 2014, pp. 416–417) looking at the Nigerian film industry. While covering a wide variety of issues and perspectives, research in the field of Nollywood Studies has largely focused on the three traditional branches of the film industry: production, distribution and consumption. In addition, over the years, quite a lot of work has been done on the history and content of Nollywood films.

While most scholars in the field locate the beginnings of Nollywood in the 1990s, and hence evaluate "the evolutionary interface between technology and entrepreneurship" (Oguamanam, 2020, p. 518) from that decade onwards, others have traced the history of Nollywood back to "the older national cinema which came into existence upon Nigeria's independence in 1960" (Akande, 2021, p. 34) and even to cultural practices during the colonial era in particular "theatrical practices that have long existed among Nigerian peoples" (Akande, 2020, p. 456). Specifically, many have claimed that Nollywood grew out of "the Yorùbá Travelling Theatre tradition, which was at its most prominent between the 1950s and early 1980s" (Akande, 2020, p. 456). In fact, Yorùbá performers, such as Herbert Ogunde, Ade Afoloyan (Ade Love) and Moses Olaiya (Baba Sala), were largely responsible for the rise of Nigerian cinema in the years after independence (Azeez, 2019). Turning to celluloid film production in the 1970s, they produced and inspired a string of films featuring Nigerian actors, settings and themes, including *Kongi's Harvest (1970)*, *Thing Fall Apart* (1971), *Amadi* (1975), *Bisi-Daughter of the River* (1977), *Kanta of Kebbi* (1978) and *The Mask* (1979). The growth of the Nigerian film industry more or less came to a halt in the mid-1980s, when the country plunged into a severe recession. The resulting crisis not only "put an end to celluloid film production" but also "closed the cinema houses" in Nigeria (Haynes, 2014, p. 53). However, the crisis did mark the beginning of the production of home video films that came to characterize early Nollywood. Although they never achieved the success of *Living in Bondage* and the home videos that followed in its wake, films like *Orun Mooru* (1982) and *Aare Agbaiye* (1984) did pave the way for the home video model that revolutionized Nigeria's film industry in the 1990s.

From the onset, Nollywood films have been examined in terms of their content, particularly regarding cultural representation and identity formation. Over the years, textual analyses of Nigerian video films have scrutinized their stories and styles; portrayals of religion and spirituality; representations of gender and women; narratives of ethnicity and nationalism; discourses of youth, class and disability; images of Africa and the African diaspora; and discourses of tradition and modernity. While the stereotypical representation of Nigerian and African women has traditionally received quite some attention in content studies of Nollywood films, the depiction of modern (metropolitan and transnational) life, either in Nigeria or abroad, seems to be on the rise in light of the increasingly diasporic and global audience of Nollywood movies. In addition, the (potential) connection between the projection of Nigeria in Nollywood films and the country's nation image has at times been discussed, with that bringing Nollywood film content into the realm of cultural heritage, nation branding and cultural diplomacy. Still, although historical and textual analysis has drawn considerable interest among film scholars over the years, the "budding research field" (Ugochukwu, 2013, p. 4) of Nollywood Studies has focused even more so on the political economy of Nigerian filmmaking due to its distinct production, distribution and consumption methods. Being understood as "unabashedly commercial and apolitical, informal, low-budget, and wildly heterogeneous and syncretic" (Garritano, 2016, p. 280), the operational arrangements of the Nollywood industry have been the subject of ongoing scrutiny since its emergence in the 1990s. Following Nigeria's "home video revolution" (Evuleocha, 2008, p. 407), Nollywood researchers have focused on the technological determinants, funding structures, production practices and distribution networks of the country's film industry. Particularly, the industry's "distinctive informal system of production and distribution" (McCall, 2012, p. 9), which developed without any state support, has received considerable academic attention in Nollywood's first 20 years, known as "the video boom era" (Ezepue, 2020, p. 2).

From the mid-2000s, Nollywood entered a new era of filmmaking. According to Haynes (2014), around 2010 "the phrase 'New Nollywood' began buzzing in Lagos and other places where people talk about Nollywood", describing the attempt by Nigerian producers and directors "to 'take Nollywood to the next level' by making better films with bigger budgets" that are aimed for projection in cinemas "rather than being released ... for home viewing" (p. 53). Apart from the return of movie theatres in Nigeria (various multiplex cinemas were built in the country since 2004), this strategy largely depended on "transnational distribution circuits including both African diasporic audiences and international film festivals" (Haynes, 2014, p. 53). While the transition from "an informal industry" to "the New Nollywood phenomenon" (Ernest-Samuel & Uduma, 2019, p. 45) has been studied in various publications, particularly Nollywood's "Afrocentric transnational and diasporic mediascape" (Jedlowski, 2013, p. 171) has attracted considerable scholarly interest. Here, the development of Nollywood as a "global cinema" (Agba, 2014), enabled by digital technologies and global networks, both alternative and dominant, and nowadays including Netflix, has been scrutinized, as well as, and increasingly, the "transnational consumerism" (Garritano, 2014, p. 48) of Nollywood video films. Being watched "in many African countries, in African Diasporas and in some Caribbean climes" (Edong, 2018, p. 77), the consumption and reception of Nigerian films by various populations of Nollywood's "vast global audience" (Ebelebe, 2019, p. 466) has been the subject of a number of academic publications. At the same time, the "audience perception of Nollywood films in Nigeria" (Agba & Ineji, 2011, p. 259), where homegrown films have remained hugely popular to this day, has also been studied, including how watching Nollywood films (potentially) influences behaviour change and national security.

In recent years, the connections between Nollywood and tourism have been discussed, or at least mentioned, in several academic publications. However, while there is said to be "evidence of Nollywood's impact on local tourism" (Abiola, 2016, p. 209), most of these publications highlight the *potential* impact of Nollywood on Nigerian and African tourism, that is, the opportunities of Nigerian films to motivate travel to the country and the continent, "thus *possibly* stimulating indirectly indigenous and foreign tourism" in the future (Chowdhury et al., 2008, p. 24, emphasis added). In other words, these publications tend to bring forward predictions and forecasts of Nollywood's impact on tourism and provide insights and advice on how to develop what could eventually become "Nollywood tourism" (Kim & Reijnders, 2018, p. 3). For example, Iwuh (2015, p. 28) discusses the tourism potential of the Nigerian film industry, arguing that Nollywood "contributes to cultural exports and arrivals in various ways". According to Iwuh (2015, pp. 28–29), "Nollywood films have traversed the world" and are, in turn, inspiring people from these parts of the world to visit Nigeria and particularly "the sites where these products are created". Similarly, Oguntoyoyinbo and Adesemoye (2019) argue that Nollywood films "should be utilised as a tool for popular movement in promoting tourism" (p. 156), thereby recommending "the need for a synergy between the Nigerian film industry and the tourism industry" (p. 164). However, like most others, they mainly point to the "yawning opportunities" (Iwuh, 2015, p. 37) and "gross prospects" (Oguntoyoyinbo & Adesemoye, 2019, p. 164) for Nollywood tourism, while the already existing or attempted connections between Nigeria's film and tourism industries, either spontaneous and incidental or intentional and structural in nature, remain largely out of the picture. This chapter, then, offers an exploration of these connections as they were made in the past decade of New Nollywood and shows how practices and discourses of Nollywood film (and) tourism have been increasingly presented in Nigerian (and international) newspapers and magazines throughout this period.

The Emerging Discourse on Nollywood Film (and) Tourism in the Early 2010s

The idea that Nollywood could promote Nigerian tourism seems to enter popular discourse in the early 2010s. As early as 2012, two industry reports resulting from the World Travel Market appear to have particularly put forth the potential of the Nigerian film industry in positioning Nigeria as "an emerging destination in Africa for tourism" (*Daily Trust*, 24 November 2012) on the country's public agenda. According to the first report, the Global Trends Report (2012), Nollywood was poised to boost intra-regional tourism in Africa, which was subsequently reprinted in various newspapers in Nigeria and across the African continent. According to the report, arrivals to Nigeria were expected to increase by 3% in the coming years, and this was in part attributed to the success of Nigerian films abroad, particularly in African countries: "The popularity of Nollywood will be a major growth driver with the leisure sector, attracting [both] film fans and business travel" (*How We Made It in Africa*, 4 December 2012). In fact, the report already signalled "a boom in hotel openings in Lagos as well as hotel expansion plans in the country" in the year 2012 due to "increasing tourism flows", which were linked to the many film productions shot in Lagos, "the heart of Nollywood", as well as to the several film villages that were being developed across the country (HWMIIA, 4 December 2012). According to the Global Trends Report, emerging film villages such as the Abuja Film Village, Plateau Film City and Lagos Film City were not only aimed to stimulate Nigeria's film industry by "providing modern film, TV and audio production facilities", but also to "encourage tourists to visit" the country (HWMIIA, 4 December 2012).

In their second report (2012), the World Travel Market similarly stated that Nollywood was to make Nigeria "the highlight of the African tourist industry with Africans ... visit[ing] the country made famous in the movies" (*Daily Trust*, 24 November 2012). Although most Nollywood films did not feature tourism settings and themes, but rather aspects of daily and religious life, the report stated that Nollywood could contribute to building the country's "market for pilgrimage and religious tourism" (*DT*, 24 November 2012). Two Nollywood films from the late 2000s that reportedly did feature Nigerian tourism sites, were *White Waters* (2007) and *The Figurine: Araromire* (2009), both displaying waterfalls (*Trip & Tours*, 25 May 2013). While it was stressed that Nollywood filmmakers usually did not capitalize on this aspect, and instead mainly used "people's homes" and "eat, clothes and accessories stores", it was argued that these two films successfully captured "the beauty of waterfalls" and as such created "the much desired tourism effect" (*T&T*, 25 May 2013). The article ended with the call for Nollywood producers to "take advantage of various tourism spots to shoot their movies", to partner with government officials to "promote tourism sites within their jurisdiction", and to unlock "the untapped power of film tourism" (*T&T*, 25 May 2013). Interestingly, not much later, it was reported that the Association of Nollywood Core Producers had pledged to use more tourism sites as film locations. According to Nollywood actor Emeka Ike, the association wanted to "create awareness about the enormous tourism potentials in the country" (*BizWatch Nigeria*, 27 September 2013). Ike continued that showcasing Nigeria's "rich tourism resources" in Nollywood films would arouse "the interest of both local and foreign tourists" (*BWN*, 27 September 2013). This way, Nollywood would be able to confirm its assumed status as the "leading promotor of Nigeria[n] tourism" (*Vanguard News*, 16 November 2012).

A prominent case in point used at that time was *Streets of Calabar* (2012), an early New Nollywood film that was "produced on a significantly higher budget and took months to shoot" (*Nigeria A-Z.com*, 18 September 2012). This acclaimed comedy thriller featured various "tourist magnets" in Nigeria as well as images of a safe nation due to "its portrayal of strong local law enforcement" (*Chanters Lodge Livingstone Blog*, 23 February 2013). According to Paz Casal, at the time travel and tourism analyst at Euromonitor, *Streets of Calabar* portrayed Nigeria as "a good tourism destination ... with adequate security" (*CLLB*, 23 February 2013). The film, which was supported by the Cross Rivers State government, was considered by some as a "signature for change" in terms of the obvious promotion of Nigerian tourism and the explicit collaboration between the country's film and tourism industries (*CLLB*, 23 February 2013). Others, however, argued that Nollywood films like *Streets of Calabar* would not increase tourism as long as Nigeria's public safety would remain a critical issue. According to Peter Tarlow, a specialist in tourism risk management, for tourism to "really grow, you need political security, political security and physical security" – "just filming a film is not going to bring people to Nigeria," first "get your safety under control" (*CLLB*, 23 February 2013).

In 2014, Ekenyerengo Michael Chima, editor and publisher of the Nollywood Mirror Series, reinforced the call for Nigerian filmmakers to use "exotic tourist attractions in Nigeria for their film locations" (*ModernGhana.com*, 2 August 2014). According to Chima, Nollywood should follow Hollywood and Bollywood in using film as "a vehicle for the promotion and appreciation of Nigerian tourism" (*MG*, 2 August 2014). While there "may not be any big film studio" and "no awesome movie sets" in the country, he argued that "the entire Nigeria," with all its tourist attractions, should be considered "the location of Nollywood" (*MG*, 2 August 2014). More concretely, Chima advised the Nigerian Film Corporation and Nigerian Tourism Development Corporation to synergize with Nollywood filmmakers to "produce movies that will boost the Nigerian film industry and tourism" (*MG*, 2 August 2014).[3] In the days after the publication of Chima's article, the question of Nollywood as a tourism promotion tool was widely discussed on *Nairaland Forum*. Various participants in the discussion challenged the idea that Nollywood would be able to advance Nigerian tourism, thereby stressing the film industry's quality and piracy issues as well as the lack of tourism infrastructure, organized cities and safety and security provisions in the country. A few years later, PricewaterhouseCoopers, in their Global Entertainment & Media Outlook 2017–2021 (2017), more or less identified the same challenges on the side of the film industry, notably the level of professional and creative capacity, the lack of formal distribution channels, copyright infringements and piracy.

Within the Nollywood film industry, the capacity, distribution and piracy issues were increasingly being acknowledged and addressed, this time with the help of the Nigerian government. In the early 2010s, when Nigeria allegedly "became Africa's largest economy", several state-supported efforts were undertaken "to combat piracy and to improve the legitimacy of the film industry" (*The Worldfolio*, 19 March 2015). In 2011, then President Goodluck Jonathan "pledged a $200 million government loan for the film industry" to help "distributors to establish new legitimate distribution channels" (*TW*, 19 March 2015). In addition, the government invested millions in training to enhance "the quality of scripts, acting and other technical aspects of Nigerian film", foster "a more business-oriented approach among filmmakers" and "improve IP awareness within the film industry" (*TW*, 19 March 2015).[4] At the same time, the early 2010s witnessed the rise of various "legal online streaming sites" (*TW*, 19 March 2015) specializing in African content, such as iROKOtv, iBAKAtv and Pana TV. These and other streaming services were reportedly supported by the "proliferation of the internet in Africa" and "growing popularity of Nollywood movies" across the continent and the rest of the world, and stimulated the making of higher budget, higher quality and higher profit Nollywood productions (*TW*, 19 March 2015). However, since the majority of Nollywood movies in the 2010s still went "straight to DVD", the "illegal distribution of DVDs" remained rampant, with pirates "selling boot leg copies for a fraction of retail price[s]," often "within hours of a film's release" (*TW*, 19 March 2015).[5] It was believed that, until a serious "crackdown" by means of regulation and IP laws was initiated by the government, pirates would continue "to thrive and to take the lion's share of [the] profits" within the Nollywood film industry (*TW*, 19 March 2015).

The Call to Government Intervention in Nollywood Film and Tourism in the Late 2010s

With the growing production of high-concept and star-led blockbusters for international theatrical or streaming release, in the late 2010s, the idea of viewing Nollywood films as tourism promotion tools became increasingly widespread among government, both inside and outside Nigeria. In 2017, allegedly for the first time, the Nigerian government entered into a partnership to "promote tourism in the country through the Nollywood industry" (*The Defender*, 4 May 2017). Apart from the government, the "tripartite partnership" involved the UNWTO and CNN and was intended to exploit the country's "competitive advantage in film production, through Nollywood, to promote tourism in Nigeria," as Alhaji Mohammed, Nigeria's Minister of Information and Culture, announced (*TD*, 4 May 2017). Although the Nigerian government had supported Nollywood before, boosting the film industry for tourism promotion purposes was, according to Mohammed, "a path we have not trodden before" (*TD*, 4 May 2017). Two years later, Otunba Segun Runsewe, Director-General of Nigeria's National Council for Arts and Culture, similarly stressed the need to use Nollywood as a vehicle to promote Nigeria's cultural heritage to attract tourists. He called on Nollywood filmmakers "to use their works in showcasing Nigeria's rich cultural tourism" (*Independent*, 11 September 2019).

Around the same time, other governments also took notice of Nollywood's tourism potential as well. In 2017, the Nollywood romantic action comedy *10 Days in Sun City* (2017) not only became the "highest-earning film in Nigeria" that year so far, but also marked "the official partnership between South African Tourism and the Nigerian film industry" (*Eyewitness News*, 25 August 2017). Shot on location in and around Lagos and Johannesburg, and specifically at Sun City, a luxury holiday resort in South Africa, the country's state tourism agency celebrated the Nigerian production as "a shining example of how entertainment and tourism can converge" (*The African Courier*, 21 August 2017). The CEO of South African Tourism, Sisa Ntshona, was "thrilled" that so many of the country's tourism attractions had "a prime starring role" in the film (*EN*, 25 August 2017). Apart from the fact that *10 Days in Sun City* was being watched all over Nigeria, "one of South Africa's most important tourist source markets in Africa," the film was released in "20 cinemas across South Africa" (*EN*, 25 August 2017) and was "slated for distribution" in various countries across Africa, Europe and North America (*TAC*, 21 August 2017). Such worldwide distribution could, it was believed, potentially stimulate South Africa's domestic, intra-regional and international tourism. According to Ntshona, the partnership between the Nigerian blockbuster and South Africa Tourism spoke to "the core functions and priorities set by the two countries in future collaborations" in film (*EN*, 25 August 2017).

In a similar vein, Dubai's Department of Tourism and Commerce Marketing (DTCM) entered into a partnership with Nollywood production company EbonyLife to support their new comedy dramas *Wedding Party* (2016) and *Chief*

Daddy (2018). Apart from being in line with DTCM's "mission to help boost the Nigerian film industry", the agency's sponsorship was intended to show-case "the beauty of Dubai to Nigerians as one of the leading visitors to the mag-nificent city in the United Arab Emirates" (*This Day Live*, 23 December 2018). While *Wedding Party* and *Chief Daddy* were both set in Nigeria, their sequels, *Wedding Party 2: Destination Dubai* (2017) and *Chief Daddy 2: Going for Broke* (2018), which were also supported by Dubai Tourism, were both partly shot at "glamourous locations" in UAE's biggest city (*The Nation* Online, 5 December 2018). The promotion of Dubai tourism through Nollywood films continued with *Omo Ghetto: The Saga* (2020), one of Nigeria's highest grossing movies of 2020, which became the first Nollywood film to receive a theatrical release in the United Arab Emirates. While being largely set in Lagos, the film contained "a few scenes shot in Dubai" and was therefore deemed to be of "high quality and interest to UAE film distributor Gulf Film" (*The National News*, 21 January 2021).

Not only Nollywood locations, but also Nollywood stars increasingly came to be seen as tourist motivators in the late 2010s. Already in 2013, Penny Biram, a spokeswoman of Nigerian hotel company, Starwoord Hotels and Resorts, predicted that Nollywood movie stars would become "another attraction for tourism in the region" (*Chanters Lodge Livingstone Blog*, 23 February 2013). One year later, then President Jonathan even described Nollywood's actors and actresses as "the greatest ambassadors of Nigeria" (*All Africa*, 24 December 2014). While these statements were made with the idea that tourists from other African countries and beyond would come to visit Nigeria because of their appreciation of Nollywood stars, Nigerian actors and actresses also started to tour abroad to promote their films and having film fans travel to these locations. For example, in 2019, Nollywood film studio, ROK Studio, took "a grand tour across Southern Africa" where "superfans in key locations across South Africa and Zambia" could connect with "their favourite Nollywood stars," before "proceeding to West African cities" (*Glitz Africa*, 21 May 2019). The so-called ROK ME Tour was joined by various "top stars" from the Nollywood television series *Festac Town* (2014–present), *Husbands in Lagos* (2015–2017) and *Single Ladies* (2017–present), who visited several cities in southern and western Africa to connect with "Nollywood lovers via exclusive Meet and Greet events and VIP dinners" (*GA*, 21 May 2019). While the fans who came to see them could be seen as film tourists, the Nollywood stars also became tourists in their own right, experiencing "some of the best tourist destinations in each city" (*GA*, 21 May 2019). Nollywood had, at least to a certain extent, become a tourist attraction within Nigeria and across Africa, with Nollywood films and stars being increasingly recognized and called upon by industry and government for their potential to encourage "intra-African tourism" (*GA*, 21 May 2019).

In 2020, like in most countries around the world, the film and tourism industries in Nigeria were brought to an abrupt halt by the COVID-19 pandemic. Arguably in combination with the rise of "online streaming giant" Netflix, Nollywood entered a "deep-set demise" as a result of the measures introduced to contain the spread of the virus (*Curiosity Shots*, 11 November 2020). Still, despite the nationwide lockdown, behind the scenes the collaboration between Nollywood and tourism remained sought after by the Nigerian government. In June 2020, for example, the Lagos State Ministry of Tourism, Arts and Culture sought to engage "some key Nollywood personalities" to see how the film industry could work together with the state government to portray "Lagos and Nigeria in a positive light" (*The Culture Newspaper*, 29 June 2020). Uzamat Akinbile-Yusuf, at the time Nigeria's commissioner for Tourism, Arts and Culture, once again urged Nollywood practitioners to be "deliberate in filmmaking" by having "the strength, values and the resilience of Nigerians" reflected in their films (*TCN*, 29 June 2020). In addition, several directions were proposed to institutionalize Nollywood with the aim to advance the film industry in the post-COVID-19 era. First, Akinbile-Yusuf revealed advanced plans of the Lagos State Government to "establish a film village to assuage the challenges faced by the filmmakers during film production" (*TCN*, 29 June 2020). In fact, the commissioner disclosed that "100 hectares of land" were already "earmarked and secured for the village" in Lagos State (*TCN*, 29 June 2020). Second, Bolajo Amusan, President of the Theatre Arts and Motion Pictures Practitioners Association of Nigeria, urged state governments to "revive community cinemas" as a way of "rekindling the interest of residents" and generating "revenue for the government" (*TCN*, 29 June 2020).

One month later, various Nigerian tourism stakeholders called on Nollywood to "showcase Nigeria's tourists' sites in movies to attract visitors and grow the [tourism] industry" (*Independent Newspaper Nigeria*, 19 July 2020). According to them, Nollywood had "all it takes to project Nigeria's cultural heritage and tourism potential beyond the shores of the nation" (*ING*, 19 July 2020). Apart from projecting Nigeria's tourism sites in Nollywood films, Ngozi Ngoka, Founder of the Tourism Consultant Forum, advised tourism operators to engage renowned Nollywood actors as "influencers to visit tourism sites within the country" (*ING*, 19 July 2020). At the same time, the view was expressed that the Nigerian government needed to "render some level of support" by upgrading the tourism infrastructure, such as "the road network to each site" (*ING*, 19 July 2020). Not much later, Goge Africa, a Nigerian-based, pan-African tourism and cultural TV show "with over a 40 million viewers" (*The Nation Online*, 13 June 2022), brought together tourism stakeholders, Nollywood filmmakers and government representatives to discuss film as "a vantage point for promoting tourism" (*Goge Africa*, 12 August 2020). Fidelis Ducker, former President of the Directors Guild of Nigeria, argued that Nollywood had indirectly been promoting tourism through featuring "key monuments in films," but that government intervention was needed to "effectively boost tourism traffic" through Nollywood (*GA*, 12 August 2020). According to Ducker, the Nigerian government could, for example, provide subsidies for films featuring tourism destinations, infrastructures to get to these destinations, and legislation making it easier to film on locations that are considered a security risk. These collaborative efforts, according to Rollas, would steer "tourism through films" to "a new phase" (*GA*, 12 August 2020).

Concluding Remarks: Nigeria's Unstable Nexus of Film and Tourism

By the early 2020s, most of the recommended government interventions to establish "the nexus between film and tourism" in Nigeria (*Goge Africa*, 12 August 2020) have not (yet) got off the ground. For example, to date, none of the planned film villages across the country, which were to be public–private partnerships, reached full completion.[6] However, the Lagos State Government (LASG) did recently sign an agreement with Del-York International Group and its US partner Storyland Studios for an "ultra-modern film city in the heart of Lagos" (*News Agency of Nigeria*, 26 October 2022), with the construction of the "$100m African Film City" being slated for October 2023 (*Vanguard*, October 14, 2023). The film village, which has been named Kebulania, is designed as a 100 hectare "fully-resourced entertainment industry campus and film lot" in the Epe area of Lagos where filmmakers and other industry personnel "can live and work" (*Blooloop*, 20 October 2022). Upon completion, which will not be before the late 2020s, Kebulania would become the "first-of-its-kind studio city on the African continent", set to "revolutionise the Nigerian film industry" and further establish the country as "a leading destination for filmmaking and creativity throughout all of Africa" (*Blooloop*, 20 October 2022). Notably, apart from being an "all-in-one facility" with "multiple sound studios, backlot sets, water tanks, workshops, and production offices," the film village will also have a tourist district in a "theme park setting" featuring "rides, tourist attractions, movie museums, and a walk of fame" (*Punch*, 26 December 2021). Aiming for "5,000 to 10,000 daily visitors", Kebulania is predicted to generate significant tourism revenue for Nigeria and, according to Matt Ferguson from Storyland Studios, to become "one of the most popular tourist destinations in Africa" (*Blooloop*, 20 October 2022). While the realization of the film village remains uncertain, with its construction having already faced multiple delays, it does seem to indicate a growing political will of the Nigerian government, or at least the Lagos State Government, to support the film and tourism industries.

Still, most of the sporadic film tourism initiatives in Nigeria seem to emerge without state support. For example, Nollywood filmmaker Kunle Afolayan has nearly completed the Kunle Afolayan Production (KAP) Film Village Studio and Resort. The objective of this private film village in South-West Nigeria is to provide professional support to the Nigerian film industry as well as encouraging tourism, since the facility doubles as a hotel resort. In early 2023, Afolayan stated that KAP Film Village and Resort had attained the two-thirds completion mark – and his latest film, *Anikulapo* (2022), was already entirely shot at the film village. Then, in 2022, Nollywood star and talkshow host Monalisa Chinda-Coker announced "a deal to boost tourism in West Africa" (*The Nation* Online, 13 June 2022). According to Chinda-Coker, her popular TV show *You and I with Monalisa* had partnered with Goge Africa and others to visit tourist attractions in Nigeria and across West Africa "to shine the light on destinations" and "revamp and revitalise tourism" there (*TNO*, 13 June 2022). Finally, in 2023, TripAdvisor, for the first time, made mention of a Nollywood film tour in Lagos. The two-day tour, called 'Explore Nigeria's Nollywood Industry', reportedly consists of visits to Nollywood film locations, theatres and villages as well as the Idumota Film Market, traditionally one of the major distribution hubs for Nollywood home videos. Although the tour has not received any reviews on TripAdvisor yet, according to the description it affords participants "the opportunity to have a one-on-one and direct assessment of the Nollywood industry", including a meet-and-greet session with Nollywood filmmakers.

In the past decade and into the 2020s, the idea of using Nollywood as a tool to boost tourism has gained traction in Nigeria's public domain. From the early 2010s onwards, the potential role of Nollywood in positioning and promoting Nigeria and Africa as a tourist destination has become increasingly commonplace in the country's news and political discourse. Since most Nollywood films did not focus on tourism settings, Nollywood filmmakers were called upon to utilize tourism spots as film locations. In addition, there have been calls for Nollywood celebrities to become tourism ambassadors and endorse Nigeria's natural and cultural tourist attractions in their productions and promotions. To-be-built film villages across the country were expected to become tourist attractions in themselves. However, despite these high expectations, which form part of a wider celebratory discourse of the creative industries worldwide (Martens 2023), film tourism related to Nollywood seems to have remained limited. Despite growing government attention and support to Nollywood and its potential for tourism, most state initiatives in this direction did not (yet) materialize. In fact, it seems that other countries, such as South Africa and Dubai, have seized upon more opportunities for collaboration with Nollywood films and stars to attract Nigerian and other African visitors to their destinations.

Within Nigeria, both the film and tourism industries continue to face various challenges that hinder the development of Nollywood tourism. While the Nigerian film industry still suffers from quality and piracy issues, the country's tourism industry has to deal with so many challenges – notably a lack of tourism infrastructure and security issues – that the promotion of film locations and the creation of film villages alone do not seem to be able to establish Nigeria as a major tourist destination. The current plan of the Kebulania film village sounds promising for the country's film and tourism industries, but it has to be seen to what extent 'Destination Nollywood' is really able to come to fruition as long as the structural issues underlying the development of Nigerian film (and) tourism are not sufficiently addressed.

Notes

1 In 2019, before the global COVID-19 pandemic emerged, "the travel and tourism industry of Nigeria represented 4% of total GDP" and "employment supported by tourism totaled 3.3 million jobs" (Tourism Economics 2021, p. 12). Although figures vary per source, it seems that Nigeria, with over 220 million people by far the most populous African country, attracted around 4 million domestic and 2 million international tourists that year. In the years following the pandemic, tourism flows have

remained below pre-pandemic levels, in part due to the increased deterioration of the security situation throughout the country since 2022 (resulting in many negative travel advisories).

2 Following Akande (2021, 34) and others, we acknowledge that "the term 'Nollywood' is problematic in its use as a cinema category." Most researchers in the field see Nollywood as "arising from the confluence of economic and technical conditions in Nigeria in the 1990s, and ... therefore, [as] an outgrowth of the older national cinema which came into existence upon Nigeria's independence in 1960" (Akande 2021, pp. 34–35). However, Nollywood at times has come to refer to the (Igbo and Yoruba) cinemas of Southern Nigeria or the English-language Nollywood films only, excluding the cinemas of Northern Nigeria and the majority of the Nigerian-language industries made and consumed in the country. For the propose of this chapter, we adhere to the most common use of the term, that is, "the various regional video industries that together make up the Nigerian video economy" (Lobato 2010, p. 337), while at the same time being "careful to register its distinctive usage" (Akande 2021, p. 35).

3 Two years later, Adeniyi Ogunfowoke gave similar advice on how Nollywood could promote Nigerian tourism. Apart from the familiar recommendation to have Nollywood films shot at tourist sites, Ogunfowoke shared three other ideas on "how Nollywood can promote the hospitality industry", that is, write scripts that portray the "culture, unity, and diversity of Nigerians", have Nollywood stars become "tourism ambassadors", and showcase "Nigeria's food and dress culture" (*Business Post*, 8 October 2016).

4 In addition, in 2012, the Lagos State Government launched the so-called Nollywood Upgrade Project, which was aimed to "improve the production quality and formal distribution of movies" (*The Worldfolio*, 19 March 2015).

5 Although investors "spotted an opportunity in producing big-budget blockbusters" following the "growing popularity of Nigerian movies across Africa and throughout the world", by the mid-2010s, most of the Nollywood films were still being "made in as little as 10 days on shoestring budgets of between \$25,000 to \$70,000" (*The Worldfolio*, 19 March 2015).

6 The Abuja Film Village, a public–private partnership with Abuja's Federal Capital Territory Administration, has not witnessed significant development since its announcement in 2012. The film village was envisioned as a 50,000-hectare one-stop shop for film production featuring, among other things, "a studio for film and television" and "university for film, television and radio" (*Oxford Business Group*, 13 June 2012). However, at present the village is still at the planning stages with the owners seeking investors to take up construction. Similarly, the Plateau Film City, a 100-hectare film village planned in the Miango district of Plateau State, featuring "film production facilities, upmarket hotels, residential estates and leisure facilities", would be facilitated by the Nigeria Film Corporation and the Plateau State Government, but seemingly never moved beyond the "under development" stage (*OBG*, 13 June 2012). In addition, the Lagos Film City, a 100-hectare project located in Lagos' Epe area at Ejirin City, did not materialize in the remainder of the 2010s and first years of the 2020s.

References

Abiola, A. O. (2016). "Reviewing Nollywood." *Innovation Africa: Emerging Hubs of Excellence*, edited by Olugbenga Adesida, Geci Karuri-Sebina, and João Resende-Santos. Bingley: Emerald Group Publishing, pp. 207–238.

Agba, J. (2014). Creating the Locale for Nigerian Film Industry: Situating Nollywood in the Class of Global Cinema. *Academic Journal of Interdisciplinary Studies, 3*(6), pp. 155–162.

Agba, J., & Patrick I. (2011). Audience Perception of Nollywood Films. *LWATI: A Journal of Contemporary Research, 8*(1), pp. 259–271.

Akande, L. (2020). Nollywood Cinema's Character of Recurrence. *Journal of African Cultural Studies, 33*(4), pp. 456–470.

Akande, O. (2021). *Nollywood Film Industry: Informal Film Practices and their Cultural Formations.* Unpublished Diss. (Ph.D. Thesis). York University, Toronto.

Atilola, O., & Olayiwola, F. (2012). The Nigerian Home Video Boom: Should Nigerian Psychiatrists Be Worried? *International Journal of Social Psychiatry, 58*(5), pp. 470–476.

Azeez, A.L. (2019). "History and Evolution of Nollywood: A Look at Early and Late Influences", edited by Bala A. Musa. *Nollywood in Glocal Perspective.* Basingstoke: Palgrave Macmillan, pp. 3–24.

Chowdhury, M., Landesz, T., Santini, M., Tejada, L., & Visconti, G. (2008). Nollywood: The Nigerian Film Industry. *Harvard School of Business.* pp. 3–35.

Ebelebe, U.B. (2019). Reinventing Nollywood: The Impact of Online Funding and Distribution on Nigerian Cinema. *Convergence: The International Journal of Research into New Media Technologies, 25*(3), pp. 466–478.

Endong, F.P.C. (2018). Cinema Globalization and Nation Branding: An Exploration of the Impact of Nollywood on the Nigerian Image Crisis. *Journal of Globalization Studies, 9*(1), pp. 77–90.

Ernest-Samuel, G., & Ngozi, E.U. (2019). From Informality to 'New Nollywood': Implications for the Audience, edited by Bala A. Musa. *Nollywood in Glocal Perspective.* Basingstoke: Palgrave MacMillan, pp. 45–66.

Evuleocha, S. (2008). Nollywood and the Home Video Revolution: Implications for Marketing Videofilm in Africa. *International Journal of Emerging Markets, 3*(4), pp. 407–417.

Ezepue, E. (2020). Political Economy of Nollywood: A Literature Review. *Journal of African Cinemas, 12,* Special Issue the Filmic and the Photographic, pp. 247–262.

Garritano, C. (2014). Introduction: Nollywood - An Archive of African Worldliness. *Black Camera, 5*(2), pp. 44–52.

Garritano, C. (2016). Virtual Encounters in Postcolonial Spaces: Nollywood Movies about Mobile Telephony. *The Postcolonial World,* edited by Jyotsna Singh and David Kim. London: Routledge, pp. 280–293.

Haynes, J. (2012). Reflections on Nollywood. *Journal of African Cinemas, 4*(1), pp. 3–7.

Haynes, J. (2014). Close-Up: Nollywood – A Worldly Creative Practice. *Black Camera, 5*(2), pp. 53–73.

Heitmann, S. (2010). Film Tourism Planning and Development: Questioning the Role of Stakeholders and Sustainability. *Tourism and Hospitality Planning & Development, 7*(1), pp. 31–46.

Iwuh, J. (2015). Nollywood, Cinema Culture and Tourism Potential of the Movie Business in Nigeria. *Dynamics of Culture and Tourism in Africa: Perspectives on Africa's Development in the 21st Century,* edited by Kenneth Nwoko and Omon Osiki. Illishan-Remo: Illishan–Babcock University Press, pp. 13–41.

Jedlowski, A. (2013). Exporting Nollywood: Nigerian Video Filmmaking in Europe. *Behind the Screen: Inside European Production Cultures,* edited by Petr Szczepanik and Patrick Vonderau. Basingstoke; Palgrave Macmillan, pp. 171–185.

Kim, S., & Reijnders, S. (2018). Asia on My Mind: Understanding Film Tourism in Asia. *Film Tourism in Asia, Perspectives on Asian Tourism*, edited by Sangkyun Kim and Stijn Reijnders. New York: Springer Nature, pp. 1–18.

Lobato, R. (2010). Creative Industries and Informal Economies: Lessons from Nollywood. *International Journal of Cultural Studies, 13*(4), pp. 337–354.

Makhubu, N. (2014). Review of *Nollywood. Safundi: The Journal of South African and American Studies, 15*(2–3), pp. 416–419.

Martens, Emiel. (2023). The Failing Promise of the Audio-visual Industries for National Development: The History of Seventy Years of Film Policy in Jamaica, 1948–2018. *Creative Industries Journal,* 1–28.

McCall, J. (2012). The Capital Gap: Nollywood and the Limits of Informal Trade. *Journal of African Cinema, 4*(2), pp. 9–23.

Oguamanam, C. (2020). The Nollywood Phenomenon: Innovation, Openness, and Technological Opportunism in the Modeling of Successful African Entrepreneurship. *The Journal of World Intellectual Property, 23*(3–4), pp. 518–545.

Oguntoyoyinbo, A.S., & Steven, A.A. (2019). Film as an Expressive Tool for Popular Movement in Tourism and Development: The Nigerian Paradigm. *The Creative Artist, 13*(1), pp.157–165.

PricewaterhouseCoopers. (July 2017). Spotlight: The Nigerian Film Industry. *PWC's Global Entertainment & Media Outlook 2017–2021.*

Tourism Economics. (2021). *Data and Digital Platforms: Driving the Tourism Recovery in Nigeria 2021.* Oxford: Tourism Economics.

Ugochukwu, F. (2013). *Nollywood on the Move. Nigeria on Display.* Literaturen und Kulturen Afrikas.

World Travel Market. (2012). *Industry Report 2012.* London: WTM.

Index

For Product Safety Concerns and Information please contact our EU
representative GPSR@taylorandfrancis.com
Taylor & Francis Verlag GmbH, Kaufingerstraße 24, 80331 München, Germany